BIRD NAVIGATION

the solution of a mystery?

BIRD NAVIGATION:
the solution of a mystery?

R. Robin Baker,
Reader in Zoology, University of Manchester

HODDER AND STOUGHTON
LONDON SYDNEY AUCKLAND TORONTO

To my parents

British Library Cataloguing in Publication Data

Baker, R. Robin
   Bird navigation.—(Biological science texts)
   1. Bird navigation
   I. Title   II. Series
   598.2'18   QL698.8

   ISBN 0 340 33416 9

First published 1984

Copyright © 1984 R. Robin Baker

All Rights Reserved. No part of this publication may be reproduced or transmitted in any form or by any means, electronic or mechanical, including photocopy, recording, or any information storage and retrieval system, without permission in writing from the publisher.

Typeset in 11/12 pt Bembo (Monophoto) by Macmillan India Ltd., Bangalore
Printed in Great Britain for Hodder and Stoughton Educational a division of Hodder and Stoughton Ltd
Mill Road, Dunton Green, Sevenoaks, Kent,
by Richard Clay (The Chaucer Press) Ltd, Bungay, Suffolk

Preface and acknowledgements

The theory that each year some birds migrate hundreds or thousands of kilometres from their nesting site finally gained acceptance toward the end of the eighteenth century. At the same time came the discovery that despite the distances involved some birds returned year after year, often to the same nest. The question of how birds perform such navigational feats has ever since provided a major puzzle to tease the inquiring minds of biologists and laymen alike. As a result, the phenomenon has provoked perhaps the most imaginative theorizing and dedicated research in the whole field of biological investigation.

Now, for the first time in over two centuries, there are signs that at last the mystery has yielded. At the end of a powerful lecture at the last international meeting to discuss bird navigation, at Pisa, Italy, in September 1981, Professor Floriano Papi felt confident enough to suggest that a memorized map of familiar smells, in combination with sun and magnetic compasses, is sufficient to explain the navigational performances of homing pigeons (Papi 1982). I am inclined to agree with him, except that I would argue in favour of a more general 'landscape' map (which includes smells). If we then add, for birds that also fly at night, compasses based on stars and the Moon, I suggest we can describe the way in which all birds find their way from place to place. The puzzle seems as good as solved.

Undoubtedly, many people will disagree, thus illustrating what has been one of the most endearing features of the study of bird navigation. It is probably true to say that not a single major suggestion as to how birds might navigate has ever received widespread support when first proposed. The result has been that each new hypothesis has generated both intense controversy and careful scientific evaluation. In consequence, theories have been tested, contested, moulded and, in many cases, rejected almost as soon as they have been conceived. The study of bird navigation thus provides a number of case histories of the scientific method in action. At the same time, from the example of those theories which have survived, prospective researchers can gain encouragement that ridicule in one decade can often be transformed into world-wide acceptance and respect in the next.

I have written this book with the serious biology student uppermost in my mind. At the same time, I hope that the use of non-technical language and the assumption of no basic knowledge of theories of navigation, of the movements of sun, stars and Moon, or of the form of the Earth's magnetic field will enable the book also to be read without difficulty by both high school students and the interested layman; indeed, by anybody who is interested in birds, navigation or migration.

I expect that the biology student will read this book for one of four reasons: (1) from basic interest; (2) as part of a course in behaviour or ornithology; (3) as a source of references in the preparation of a dissertation or even at the start of a research project; or (4) to obtain case histories to illustrate the development and testing of theories or the need for care in experimental design. The bibliography has been compiled in accordance with these likely types of use. Although priority has naturally been given to the most recent publications on bird navigation, I have tried also to retain references that illustrate the conception and development of each particular idea and line of research.

My aim was a reference list that was both comprehensive and as up to date as the delay between compilation and publication would allow. If I have succeeded in either of these aims it is due in large part to the generosity of those colleagues actively engaged on research into bird navigation who sent me comprehensive reprint sets of their work and who have allowed me to see unpublished manuscripts or to quote work that is still in progress at the time of writing. Among those who sent me reprint sets were the late Bill Keeton, Charles Walcott, Hans Wallraff, Wolfgang and Roswitha Wiltschko, Joe Kirschvink, John Richardson, Jorgen Rabøl and the Italian group headed by Floriano Papi. Ken Able and Joe Kirschvink allowed me to see unpublished manuscripts and Verne Bingman, Willie Beck and Janice Mather allowed me to quote unpublished results. To all of these I owe my sincere thanks. I am particularly grateful to the Wiltschkos and others in their research group at Frankfurt for showing me their current experiments on homing pigeons and migratory birds. My thoughts on bird navigation matured considerably during the many 'heady' conversations and discussions on my recent visit.

Thanks are due once again to Hodder and Stoughton for their interest in this project. In particular I appreciate their financial support in the purchase of a word processor, which removed so much of the mechanical drudgery involved in preparation of the manuscript. Line drawings of the birds and other animals in the main text are by Josephine Martin and Mandy Stiller. The illustrations at the opening of each chapter are by the late Charles Frederick Tunnicliffe, RA, and are reproduced by courtesy of Hodder and Stoughton.

Contents

Preface and acknowledgements ... v

Chapter

1 Introduction: the past two thousand years ... 1

2 What is navigation? ... 9
   2.1 Navigation, pilotage and other terms ... 9
   2.2 Homing pigeons and racing pigeons ... 12
   2.3 Life within a familiar area ... 14
   2.4 Exploration and navigation ... 17

3 Major types of experiment ... 19
   3.1 Homing experiments ... 19
   3.2 Homing success ... 20
   3.3 Vanishing points ... 21
   3.4 'Pointing' home ... 22
   3.5 Circular statistics ... 22
   3.6 Separating goal orientation from orientation ... 25
   3.7 Testing for orientation ... 26

4 How could birds navigate? ... 33
   4.1 Mental maps ... 34
   4.2 Following the outward journey ... 37
   4.3 Reading the landscape ... 41
   4.4 Reading a grid ... 44
      4.4.1 Finding 'latitude' ... 45
      4.4.2 Finding 'longitude' ... 48
   4.5 Reading gradients ... 52

5 The avian landscape — 55

- 5.1 Visual landmarks — 56
- 5.2 Smells — 63
- 5.3 Sounds — 69
- 5.4 Fields of magnetism and gravity — 73
- 5.5 Landscape differences between species — 73

6 Bird compasses: by day — 77

- 6.1 The sun compass — 77
- 6.2 Polarization patterns — 83
- 6.3 A role for ultraviolet light — 85

7 Bird compasses: by night — 87

- 7.1 Stars — 87
- 7.2 Moon — 92

8. Bird compasses: magnetism — 95

- 8.1 Evidence for a magnetic compass — 95
- 8.2 The magnetic compass: polarity or inclination? — 97
  - 8.2.1 Alternatives — 98
  - 8.2.2 Evidence — 98
  - 8.2.3 Advantages and disadvantages — 99
- 8.3 Critical factors — 103
  - 8.3.1 Field intensity — 104
  - 8.3.2 The search for other critical factors — 104
  - 8.3.3 Electric fields? — 107
  - 8.3.4 Orientation during rest? — 109
  - 8.3.5 Exposure to strong magnetic fields? — 111
  - 8.3.6 Magnetic storms? — 113
  - 8.3.7 Time of day? — 115
- 8.4 The search for the magnetic receptor — 116
  - 8.4.1 Induction — 117
  - 8.4.2 Optical resonance — 118
  - 8.4.3 Magnetite — 120

9 Bird compasses: relationships — 125

- 9.1 The compass hierarchy and redundancy — 125
- 9.2 Evidence for an inborn compass — 127
- 9.3 Compass development from birth — 134
  - 9.3.1 Day-time compasses — 134
  - 9.3.2 Night-time compasses — 137
- 9.4 The use of compasses by experienced birds — 140
  - 9.4.1 By day — 140
  - 9.4.2 By night — 142
- 9.5 Links — 145
  - 9.5.1 The setting sun — 145
  - 9.5.2 Wind — 149

10 Grid maps: fact or fiction? — 151

- 10.1 'Latitudinal' gradients — 152
  - 10.1.1 Coriolis force — 152
  - 10.1.2 Altitude of the sun — 153
  - 10.1.3 Night sky — 157
  - 10.1.4 Geomagnetic field — 158
- 10.2 'Longitudinal' gradients — 165
  - 10.2.1 Geomagnetic field — 165
  - 10.2.2 Night sky — 166
  - 10.2.3 Local time from the sun's arc — 166
- 10.3 Fact or fiction? — 169

11 Navigation in action: short-distance homing — 171

- 11.1 Map and compass? — 172
- 11.2 Following the outward journey — 175
  - 11.2.1 By magnetism — 176
  - 11.2.2 By smell — 180
  - 11.2.3 Age effects — 182
- 11.3 Regional effects — 184
  - 11.3.1 Preferred compass direction — 185
  - 11.3.2 Release-site bias — 187

		11.3.3 Release-site bias, preferred compass direction and landscapes	190
	11.4	Information at the home	193
	11.5	Navigation and distance	195
	11.6	Correcting mistakes: navigation while homing	197
12	Navigation in action: long-distance migrants		201
	12.1	Inborn programs	202
	12.2	Models of migration	204
	12.3	Exploration and navigation after fledging	210
	12.4	Autumn migration	213
	12.5	Arrival at the winter range	219
	12.6	Spring return	224
	12.7	Migration as an adult	226
	12.8	Are landscapes and compasses enough?	228
13	Epilogue: the next ten years		229

References and author index	231
Subject Index	252

1 Introduction: the past two thousand years

Ever since the human lineage emerged from its African cradle and spread slowly across the temperate and subpolar lands of the Earth, people have probably been aware of the seasonal migrations of birds. Among even the earliest of human writings, such as the *Book of Job* and Homer's *Iliad* nearly three thousand years ago, were references to birds escaping the approaching winter by flying to warmer climes.

Observing such seasonal disappearances was one thing, trying to understand them quite another. The first evidence of such an attempt appeared in the works of Aristotle, who lived about 2200 years ago. In Book Eight of his *Historia Animalium*, Aristotle wrote that some animals move south after the autumn equinox to avoid the coming cold of winter and north again after the spring equinox to avoid the coming heat. He also noted that 'all creatures are fatter in migrating', an early observation that birds lay down fat to provide fuel for their journey.

Had Aristotle confined his thoughts to migration, the study of bird navigation might now be further advanced than it is. Unfortunately, he offered for some birds two further explanations of their seasonal disappearance and these alternative suggestions were to form the basis of an argument destined to last for 2000 years. They were: (1) that some birds avoid the cold of winter by hibernating; and (2) that others disappear for the winter, not because they migrate or hibernate but because they transmutate into other species (e.g. redstarts (*Phoenicurus phoenicurus*) into robins (*Erithacus rubecula*)). Aristotle's alternatives to migration were preferred by the majority of Natural Historians that came after him until nearly the end of the eighteenth century. Indeed, the idea of hibernation received additional impetus in the sixteenth century when Olaus Magnus, Archbishop of Uppsala, wrote with conviction that swallows pass the winter under water and even illustrated his thesis with a drawing of fishermen reeling in a netload of hibernating birds (Fig. 1.1)

Fig. 1.1. In 1555, Olaus Magnus published this illustration of two fishermen standing at the edge of ice in winter and hauling in a net containing a mixed catch of fish and hibernating birds. The idea that swallows hibernated under water persisted long afterwards.

[Redrawn from Baker (1980b), after Magnus]

Many well-known naturalists accepted the immersion theory of hibernation unquestioningly throughout the eighteenth century. Linnaeus, John Reinhold Forster, Baron Cuvier, Geoffroy de St. Hilaire, and the Honourable Daines Barrington all gave the idea their support, many claiming to have seen swallows in hibernation or being taken from a river in winter. Even Gilbert White, although convinced that swallows did fly south in winter, conceded that hibernation also occurred. Even as late as the early nineteenth century, Edward Jenner, the man credited with developing vaccination, still felt it necessary to argue against the hibernation theory. On the whole, however, acceptance that the disappearance of birds in winter was due to migration rather than hibernation was more or less complete by the beginning of the nineteenth century.

These early discussions of bird migration (reviewed by J. Cherfas, in Baker 1980b) involved little contemplation of navigation. Not until the advent of bird ringing, the placing of a small lightweight and uniquely

Fig. 1.2 Golden plover (**Pluvialis apricaria**) wearing individually numbered 'ring' or 'band' (Photo by San Diego Zoo, courtesy Frank W. Lane)

coloured or numbered ring on a bird's leg, was the navigational feat accomplished by a migrating bird realized. One of the first ringing 'experiments' to be recorded was carried out by a nobleman in hiding from the mob during the years following the French Revolution. A copper ring placed on the leg of one of a pair of swallows (*Hirundo rustica*) that nested in his chateau in the Lorraine revealed that the same bird returned on three consecutive years to the same nest. Such information, coming in the late eighteenth century at a time when people were just beginning to accept that bird migration was a reality, brought a new dimension to discussions of migration. Not only did some birds seem to fly relatively long distances from their nesting sites each winter, they were also capable of finding their way back to the very same site, year after year. By the mid-nineteenth century, this ability of birds to 'home' dominated discussions of migration. Nor was the discussion confined to birds, for by this time people had begun collecting anecdotes about homing by a wide range of animals, but particularly by mammals, such as dogs, cats, sheep and cattle. The pages of the scientific journal *Nature* abounded with stories of spectacular homing feats and even Charles Darwin contributed to the discussion (Baker 1981a).

Many people at the time considered that such homing involved the use of a special sense which several decades later, in the early twentieth century, was to become known as a 'sixth' sense. This sixth sense was possessed by migrating birds and those mammals known to perform feats of homing but was denied to all humans except perhaps 'primitives', children and the exceptional westerner. Over a century ago, in 1855, Dr. Von Middendorf proposed that this navigational sense was in fact a sense of magnetism, and this view became the favoured interpretation of the sixth sense, should it exist. Even so, the idea of a magnetic sense was generally ridiculed. Professor Alfred Newton who, in the 1870s, was the first Professor of Zoology at Cambridge University, when accused of believing in the existence of a magnetic sense in animals, replied 'I had no need to declare my disbelief in Dr. Von Middendorf's magnetic hypothesis, for I never met with any man that held it' (Cherfas in Baker 1980b). Although Newton could not offer an alternative explanation, he thought that the solution was likely to be 'simple in the extreme'.

The nature and involvement of the sixth sense in migration and navigation was one hotly disputed area at the turn of the twentieth century. The other was whether migration occurred by instinct or by learning. As the young of some birds, such as the European cuckoo (*Cuculus canorus*), migrate alone with no parental guidance, many people thought that some, if not all, migrant birds must be born with an instinctive knowledge of when and where to go. Yet other people preferred to believe that birds migrated as a result of practice or experience, with some knowledge being handed down from one generation to the next. With the passing of the first half of the twentieth century, these two views tended to polarize. Eventually, all birds that were residents, staying in the same area throughout the year, were thought to find their way around by learning local landmarks and by subsequent pilotage. Long-distance seasonal migrants, however, had the need of a qualitatively different set, or program, of instincts, including a sophisticated repertoire of senses and other aids to navigation. This expectation, that long-distance migrants have senses superior to those of the so-called non-migrants or residents, is evident even in the writings of the 1970s and 80s (see Bellrose 1972 for a review of the evolution of navigation abilities along these lines). Yet the view arose and persists in the face of an enormous paradox.

Although some homing experiments are carried out on 'wild' birds, the most common experimental animal in the study of bird navigation is the homing pigeon, the domesticated descendant of the rock dove (*Columba livia*). Pigeons have been used to carry messages since the days of the ancient Egyptians and were used by the Greeks, by the Roman Empire, and throughout the Middle East. Their most regular use was as a 'pigeon-post' service, but there is an equally long history for the return of messages from military operations. Pigeons were used in the siege of Paris in 1870–1871 when 150 000 official and a million private messages were carried. In the 1939–1945 World War some 200 000 birds were supplied to the British Services by private breeders and 50 000 were reared by the US Army. Nearly 17 000 were parachuted to the Resistance in occupied Europe and 2000 returned safely (Matthews in Landsborough Thomson 1974).

Fig. 1.3 Homing pigeon in harness for parachuting and carrying messages from behind enemy lines
(Photo by US Army, courtesy of Frank W. Lane)

Industrial and technological revolutions changed the civilian relationship with pigeons. The development of telegraphy and radio diminished the use of pigeons as the carriers of messages. On the other hand, the advent and growth of a rail network, with the associated ease and speed of transporting birds over long-distances, led to the development of the sport of pigeon racing. The first pigeon race over 100 miles was held in 1818 in Belgium, and similar races had become regular in England by 1875 (Matthews in Landsborough Thomson 1974). There are perhaps 100 000 pigeon fanciers in Britain alone, with something like two million pigeons in their lofts. In continental North America, pigeons often race over distances longer than 1000 kilometres, though no birds return in less than a day.

Other species of birds have also been used for carrying messages. The Romans used swallows (*Hirundo rustica*) to convey the winning colours of chariot races, but these were wild birds caught at their nests on the day required. Pacific Islanders used tamed frigatebirds (*Fregata* spp) for communication between islands. No other species, however, lent itself quite as readily to domestication and the carrying of messages as the pigeon. It was just about the optimum size for the conflicting demands of housing in numbers yet having legs large enough for the attachment of messages. It was a grain-feeder, could be eaten, could be handled easily, homed readily, and most important and ironic of all, unlike the swallow, it remained resident in the same place all year round. It is not quite true, however, to say that the rock dove, from which the homing pigeon is descended, does not perform seasonal migrations. In the Saharan parts of its range (Fig. 1.2), such migrations do occur, but in most places the species is well-and-truly resident (Goodwin 1967). Such sedentary behaviour aided domestication, led to the long association with man, and made the homing pigeon the obvious choice when it came to selecting a bird for use in experiments on homing. Yet at the same time it ran counter to the expectation that only long-distance migrants would have true navigational ability. Researchers who found it convenient to use homing pigeons argued their way round this conflict on the grounds that over 2000 years man had selected for good homers amongst his stocks and thus artificially had produced a strain of 'resident' birds with 'true' navigational ability. Other ornithologists, with no direct involvement in orientation argued that, interesting though the results obtained with pigeons may be, they were of minimal relevance to the navigational mechanism of long-distance migrants.

Recent developments in orientation studies have removed the paradox of the homing pigeon, but as is always the case in a healthy subject, they have replaced that conflict with another. The 1970s saw growing interest in a comparative approach toward orientation (e.g. Adler 1971; Schmidt-Koenig and Keeton 1978). This led eventually to a categoric proposal (Baker 1978, 1981a, b, 1982) that all vertebrates, along with at least some invertebrates, would be found to use navigational mechanisms that in all essential respects were similar, whether the animal concerned travels short

Introduction: the past two thousand years 7

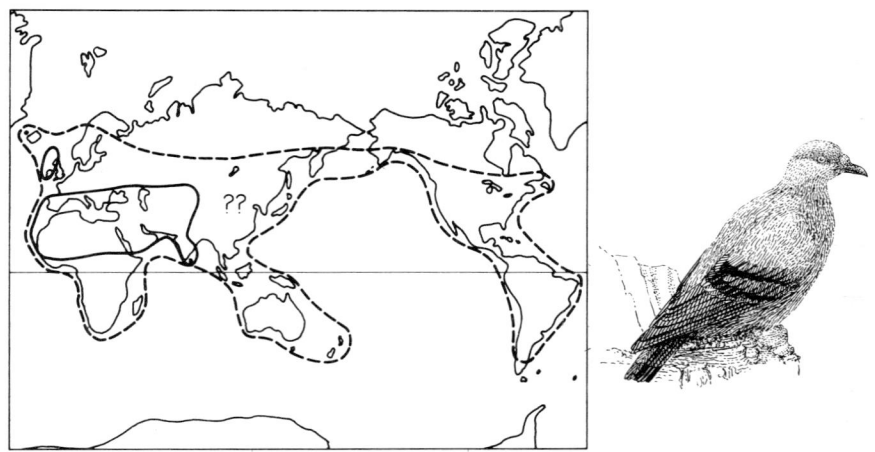

Fig. 1.4. The range of **Columba livia**

The solid line shows the present distribution of the rock dove, the 'wild' form of **C. livia**. The dashed line shows the approximate distribution of the feral form of the species, free-living descendants of birds that have failed to return home or otherwise escaped from captivity.

[Redrawn from Baker (1978), after Goodwin]

distances (e.g. bees, mice, house sparrows), intermediate distances (e.g. humans, homing pigeons), or long-distances (e.g. salmon, tuna, swallows).

Such an extreme of the comparative viewpoint is unlikely to be welcomed by all ornithologists, many of whom feel that bird migration and navigation is unique and as such requires a unique explanation. After all the debate and research on migration that has centred on birds since the time of Aristotle, it would indeed be ironic, and perhaps a little unjust, if the final pointers to an understanding of avian abilities were to come from animals as unspectacular as, say, mice, or even people.

Yet many researchers would argue that such pointers are still needed. Indeed, some workers in the field still anticipate the discovery of new, as yet unsuspected, senses. Only in such discovery can they see final understanding. A few people, however, consider that we now have all the most important facts and that the road to understanding lies in a reconstruction of the way in which known senses and abilities are used in the solution of navigational problems. Yet others (e.g. Gerrard 1981) consider that there is not, and never has been, a mystery to bird navigation. Indeed, Gerrard argues with conviction that the idea of mystery was fabricated by the researchers themselves in a conspiracy to extract money for research from governments and tax-payers.

Such is the contentious nature of the subject. Indeed, the entire history of man's search for an understanding of bird navigation from the time of Aristotle to the present day has been one of heated debate, intense argument and strong feeling. As one conflict has been resolved, another has emerged. The motivation to prove the correctness or otherwise of cherished and rival hypotheses has led researchers to the limits of imagination and ingenuity in the design and execution of experiments. The field of orientation research, more than any other branch of biology, has made major contributions to the discovery and understanding of the sensory world of animals, including man. Perhaps there are still major discoveries to be made, perhaps not. Either way, for many of us an understanding of the nature of bird navigation remains as appealing, exciting and challenging as ever.

This book has two major aims. The first is to present the facts and interpretations on which our modern appraisal of the methods by which birds navigate is based. The second is on the one hand to place these facts, and particularly their interpretation, in the perspective of past theories and on the other hand to identify those aspects of research that seem to be the key areas for future progress. Chapter 3 describes the major experimental techniques and tools that have been used in the study of bird navigation and Chapter 4 paves the way for the wide range of information to follow by discussing the various ways that birds might navigate. Chapter 5 describes experiments that tell us the nature of the landscape within which a bird spends its life. Chapters 6–8 discuss the wide range of compasses by day and by night that birds have been discovered to use and Chapter 9 considers how these different compasses are developed during the bird's lifetime and the way they are integrated and used once the bird has become an experienced traveller. Chapter 10 addresses the question that for many people is the most interesting of all: can birds use some form of absolute grid map, based roughly on latitude and longitude, to work out the direction of home from areas they have never before visited?

After a review of the evidence with regard to the senses available to birds for orientation and navigation, Chapters 11 and 12 then look at the way these senses are used when navigation is in action. Chapter 11 concentrates on homing pigeons and Chapter 12 applies the conclusions reached to long-distance migrants in an attempt to decide whether there really is any difference in the methods used by resident and seasonally migrant birds. Finally, Chapter 13 looks at the future of research into bird navigation over the next ten years.

We begin, however, with much less ambitious questions. Chapter 2 considers just what types of behaviour involve navigation and what range of birds, indeed, what range of animals, can usefully be studied. In other words: what is navigation and who navigates?

2 What is navigation?

2.1. Navigation, pilotage and other terms

To a mariner, navigation means finding his way to a destination by the use of astronomy and geometry. To the designer of a modern jet aeroplane or space ship, navigation means using a gyroscope to monitor and control all the twists and turns of the journey. To a passenger in a car, navigation means reading a map and staying alert for road signs.

Different though each of these uses of the word navigation may seem, they do have a common element. In each case, the destination (or 'goal') is known but the route to the goal is unfamiliar. To the car passenger, the route is unfamiliar because he or she has not travelled the route before and cannot recognize what would otherwise be key landmarks. The mariner and space-ship designer have a different problem. There are no obvious landmarks in the featureless voids of ocean or space. Nearer to shore, when ships take on a person with an intimate knowledge of local routes and landmarks, he is known, not as a navigator but as a pilot.

Navigation, then, is the art of finding one's way to a known destination across *unfamiliar* terrain, by whatever means. Pilotage is the art of finding

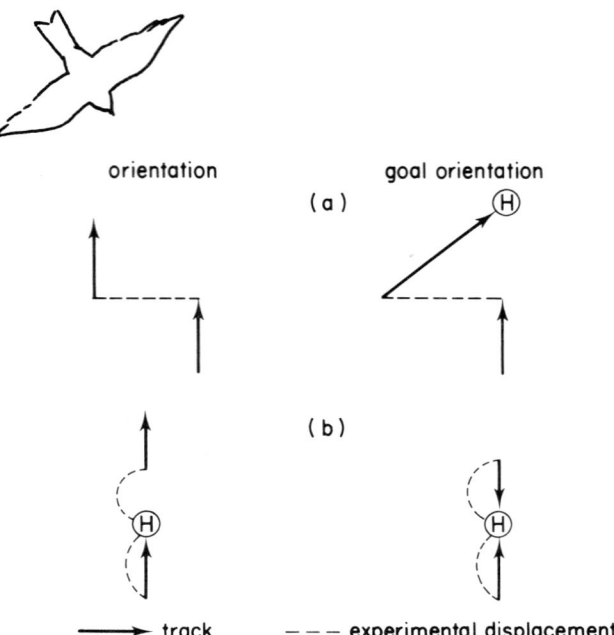

Fig. 2.1 Displacement as a means of distinguishing between orientation and goal orientation

(a) lateral displacement of a moving animal
(b) displacement in opposite directions of an animal with a 'home' (H)

[Modified from Baker (1981a)]

one's way to a known destination across *familiar* terrain through the use of landmarks. Both are forms of goal orientation and contrast with simple orientation (Fig. 2.1), in which the animal is concerned only with travelling in a particular direction. Put another way, goal orientation involves orientation towards a goal or destination by means of either navigation or pilotage. Orientation, on the other hand, involves taking up a particular direction (e.g. south) irrespective of destination.

The test of whether an animal is showing goal orientation or just orientation depends on what the animal is doing just before testing (Fig. 2.1). If it is travelling across country, the test is to displace the animal laterally. If, on the other hand, it is resident at some site ('home'), the test is to take it in opposite directions from that site. In the first type of test, the animal is showing goal orientation if it compensates for the displacement so as to continue to travel towards a particular point in space. If it continues to travel parallel to its original track, it is showing orientation. In the second type of test, the animal is showing goal orientation if from both sites it points toward home, but only orientation if instead it points in the same

What is navigation? 11

direction irrespective of the location of the place where it was caught.

Goal orientation is shown by homing pigeons if, when taken elsewhere, they fly away from their site of release in the general direction of their loft. Orientation is shown in their first autumn by the young birds of seasonally migrant species (Chapter 9). When transported sideways from the migration route traditional for their species or population, they continue to

12 *Bird navigation: the solution of a mystery?*

travel parallel to that route and arrive at wintering grounds to the side of those they would otherwise have occupied (Fig. 2.2). It so happens that orientation is also the mechanism used by racing pigeons upon release.

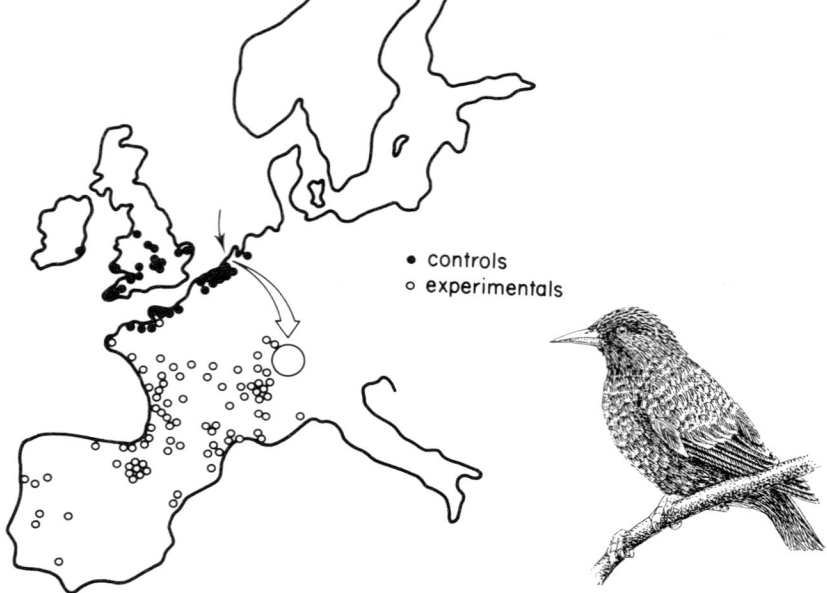

Fig. 2.2 Recoveries of young starlings (**Sturnus vulgaris**) displaced from the Netherlands to Switzerland

Young starlings captured and ringed in the Netherlands (arrow) in October and November were divided into two groups. Controls were released at the site of capture. Experimentals were displaced by aeroplane to Switzerland before release. Recaptures in the first winter after release are shown by dots.

[Redrawn from Perdeck (1958). Photo p. 11 by S. C. Porter, courtesy Bruce Coleman Ltd]

2.2. Homing pigeons and racing pigeons

It often comes as a surprise to people to learn that homing pigeons and racing pigeons in fact use two quite different mechanisms upon release. Ideally, experiments on homing pigeons are designed so as to force the birds to show goal orientation. Racing pigeons, on the other hand, are encouraged through training to use only orientation for as much of their homeward journey as possible. The adoption of a simple compass direction is obviously a great deal easier than having to work out the direction of home from an unfamiliar location by navigation. The seconds or minutes saved by the bird through simple orientation in a particular compass direction instead of going through the complexities of navigation could obviously make all the difference between winning or losing a big race.

The way that pigeon owners encourage their racers to orient rather than navigate is to put them through a period of intensive training before ever entering them for a race. This training involves a series of releases at increasing distances from the home loft but always in the same direction: that from which eventually they will have to return during a race. Trained in this way to fly in a particular direction, a bird only ever takes part in races involving homing from that direction.

Experiments have shown that if pigeons are trained to associate the experience of being displaced and released with the fact that home always lies, say, to the east, but are then taken in the opposite direction, upon release they automatically adopt their training direction. When radio transmitters are fitted to such birds so that their flight paths can be followed by observers on the ground or in a plane, it is found that some birds may travel 80 km or so before apparently they 'realize' their mistake and turn to head in the correct direction of home (Fig. 2.3).

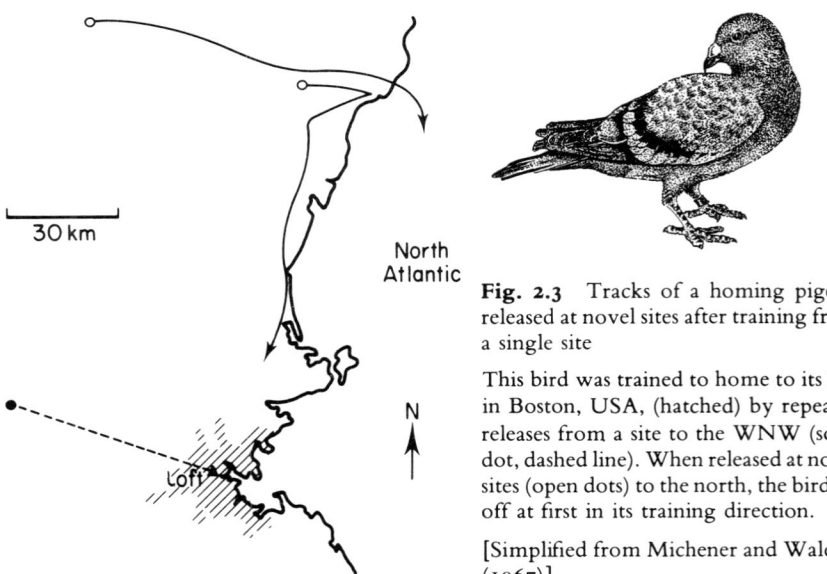

Fig. 2.3 Tracks of a homing pigeon released at novel sites after training from a single site

This bird was trained to home to its loft in Boston, USA, (hatched) by repeated releases from a site to the WNW (solid dot, dashed line). When released at novel sites (open dots) to the north, the bird set off at first in its training direction.

[Simplified from Michener and Walcott (1967)]

Homing pigeons are also subjected to a series of training flights before being used in experiments but usually these flights take a different course. As a bird grows in experience during its first weeks and months, free flights in the vicinity of the loft give way to releases at increasing distances, usually in a variety of directions.

Racing pigeons can tell us little, therefore, about the mechanisms of goal orientation. They do tell us, however, what can be achieved in terms of homing speed. Moreover, accurate and reliable measures are available, largely because of the elaborate precautions that are taken during races to ensure correct timing and to eliminate fraud. Officials mark each bird before release with a temporary rubber leg band with a code number. On the bird's return to its home loft, the owner removes the band and drops it into a sealed clock which marks the hour and minute of arrival. Officials measure the distance from the release point to the home loft to the nearest yard or metre (depending on the country) on a Great Circle route and then express the speed of the bird in terms of the distance flown per minute. In good conditions, racing pigeons commonly achieve 1100 metres per minute. With tail winds, speeds may be in excess of 2000 metres per minute. In addition, it seems that racing pigeons may fly for up to 16 hours per day (Matthews, in Landsborough Thomson 1974).

It should be pointed out, however, that, despite such impressive figures, the domestic pigeon is, in fact, not as infallible at returning home as popular folk-lore might often have us believe. Even racing pigeons, trained and raced always in the same direction, have only perhaps 5 per cent of their number come through all the initial stages of training and final return from long-distance races.

2.3. Life within a familiar area

Both homing and racing pigeons have a home loft where they sleep, eat and breed and to which they show goal orientation when out on exercise or training flights or when released in homing experiments. Their descendants, the feral pigeons of city centres, living testimony to the frequent failure of their ancestors to home, also have a regular sleeping site and nest site. In addition they have favoured feeding sites elsewhere at places such as railway stations and granaries, some birds even flying out to feed in the surrounding countryside. All of these pigeons, whether feral, racing, or homing, spend their lives engaged in sequences of goal orientation: from nest to feeding site, from one feeding site to another, back to the nest or roosting site, and so on. They organize their lives more or less entirely around an area of familiarity, or familiar area: a mosaic of sites familiar as a result of previous visits, with each site offering one resource or another.

Pigeons are no different from other birds in the way they organize their lives and movements around a familiar area. Indeed, they may be no different from most, if not all, other vertebrates and at least many invertebrates. We have to look very hard at animals if we wish to find a species with a different way of life. When we do, it seems that all exceptions are invertebrates (Baker 1978, 1981b, 1982). Every single vertebrate may spend the major part of its life living and moving within a familiar area.

As humans, we have no difficulty in understanding the practicalities of life within a familiar area, for that is how we organize our own movements. We have an intimate knowledge of where and when is best to obtain this resource or that, whether food, drink, information, mate, or sleeping place, and we spend the major part of our time in goal orientation from one such site to the next. This is clearly true if we are a city dweller but it is no less true if we are a long-distance seasonal migrant, such as a Lapplander, Bedouin, or North American cotton picker, moving from one seasonal resource to another. We might suspect, therefore, that the same may also be true for all birds, again no matter whether they are a city dweller, such as the feral pigeon, or a long-distance seasonal migrant, such as a swallow.

Evidence is accumulating steadily (Baker 1978; Wiltschko and Wiltschko 1978) that among birds not only do many seasonal migrants return year after year to the same breeding site, but also many use the same wintering sites each year and, during autumn and spring migration, even the same staging posts. Sea birds also often visit the same area of ocean at the same season, year after year. It seems that when a bird has travelled its year's migration circuit once or a few times, that circuit is as much a familiar area despite the distances involved, as a city and its environs are to a feral pigeon.

Life within a familiar area is a life of continual goal orientation as the individual travels from one familiar site to another. Invariably, however, it is primarily a life of pilotage, for not only are the goals familiar but the routes have also been travelled before. Where, then, in this way of life, might a bird find a need for goal orientation by navigation rather than by pilotage? The answer seems to be, as for humans, that navigation is only necessary during exploration, those occasional forays during which an animal travels to areas that it has never before visited.

All animals that live as adults within a familiar area have to give up a proportion of their time when young, often a large proportion, to exploration. After all, when an animal is born, it has no familiar area. Yet, by the time it is ready to reproduce, it should ideally be living within a familiar area that is both suitable and productive. To achieve this, it has at some time to visit and assess a variety of potentially useful sites. Of these, some presumably will be suitable, others will not. Some will juxtapose in just the right way to allow economy of travel between them, thereby allowing the bird access to different resources at different times of day or different seasons, others will not.

Of all the areas that a bird or other animal visits during the course of its explorations, only a small proportion is likely eventually to be used by the individual in the formation of its home range, that part of its familiar area actually visited and utilized during any given period of time. Obviously a bird's home range during the course of a day will be smaller than its month's home range, which in turn will be smaller than its season's or year's home range, Moreover, the way that home ranges of successive days, months and seasons juxtapose to form the year's home range will differ between individuals, populations and species. Some, such as the feral pigeon, may occupy essentially the same home range throughout the year. Others, such as the barn swallow (*Hirundo rustica*) or wandering albatross

(*Diomedea exulans*), may occupy a succession of seasonal home ranges that fit together to form huge and complex migration circuits (Fig. 2.4). All year's home ranges, however, whatever their shape, size and complexity, will have crystallized from a much larger familiar area that was built up during the individual's earlier life by a process of exploration, habitat assessment . . . and navigation.

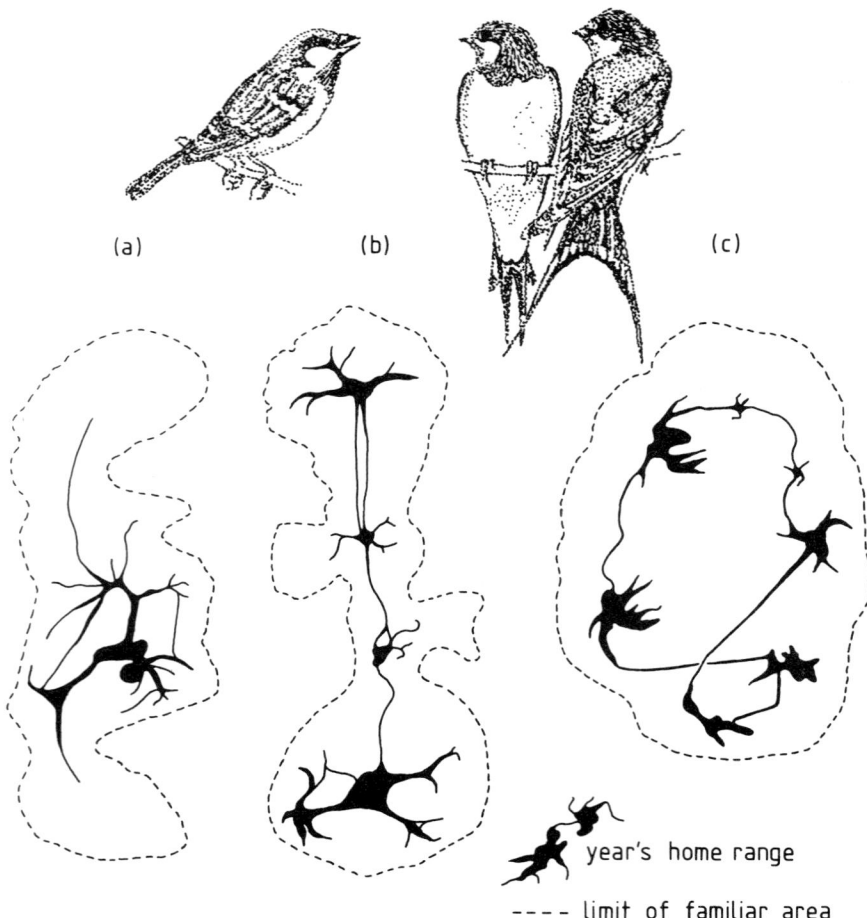

Fig. 2.4 Three types of year's home range: (a) static home range; (b) seasonal to- and-fro return migration; (c) annual migration circuit

Solid black shows the area used by a bird during the course of a year. This is its year's home range. The dashed line shows the limits of the familiar area it built up when younger. No matter what the form of the year's home range, it crystallises out from a much larger familiar area.

[From Baker (1981a)]

2.4. Exploration and navigation

Exploration could, of course, be carried out solely by pilotage, without involving navigation, simply by retracing steps past landmarks learned during the outward journey. To do so, however, would make either the exploration itself or later movements within the familiar area that results, or both, less efficient than they need to be. Figure 2.5 shows a typical exploratory foray beyond the previous limits of the familiar area. The foray shown is curved. Had the outward path been straight, the animal could, of course, return home perfectly well by pilotage, retracing its steps past landmarks that it had memorized during the outward journey. Such straight-line forays, however, either limit the area that can be explored before returning home or else rapidly lead the animal to sites much further than necessary from the already established home range. Some form of curved track is obviously more efficient for thorough exploration of the areas near to home. The outward path could still be retraced by pilotage, but the journey home would be less economical than it need be. Some means of pioneering the most economical, which for birds is most often the most direct, route home following exploration is of clear advantage, particularly if there are to be frequent future journeys between the old home and the newly discovered site. Exploration followed by navigation will often be far more efficient than exploration followed by pilotage.

All birds that live within a familiar area would seem to gain, therefore, from being able to navigate during exploration. So, too, would any animal that lives within a familiar area. If it is correct, then, that all vertebrates and many invertebrates live within a familiar area that they build up by

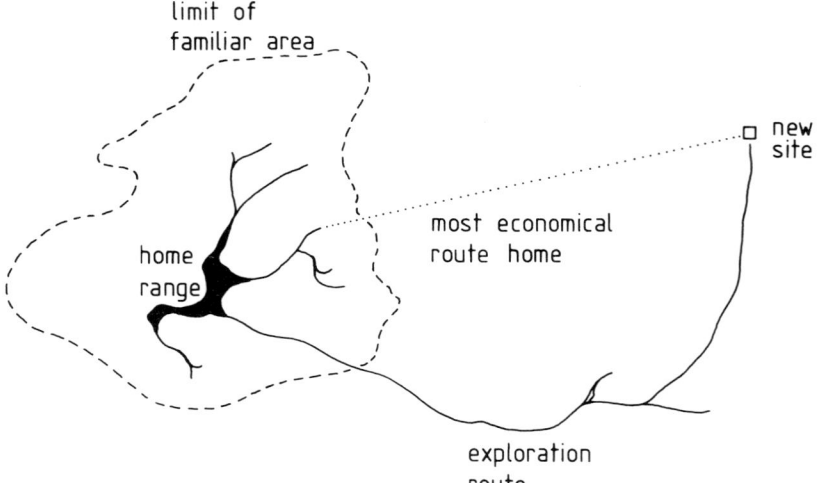

Fig. 2.5   When a bird explores, the direction of the most economical route home is determined by navigation.

[From Baker (1981a)]

exploration, all should, to some degree or other, have evolved the ability to solve navigational problems. It follows that, although in this book we are interested primarily in the way that birds solve their navigational problems, it would be short-sighted in the extreme to ignore any evidence there may be (which is very little) concerning the way that other animals solve their navigational problems. This comparative approach becomes particularly important when there are experiments we should like to carry out on birds but cannot, because birds are birds, but can carry out easily on some other animal.

This chapter has stressed the distinction between goal orientation and orientation and pointed out the division of goal orientation into navigation and pilotage. It helps to have these categories clear in our minds at this stage, for the distinction between them is often crucial in the design of experiments and interpretation of results. Yet, as we shall see in later chapters, it is by no means always clear just which form of orientation is being shown by the birds under particular test conditions. In the field of animal orientation, if the results are to be interpretable in a meaningful way, experimental design is paramount. Before we can come to grips with the results of experiments on bird navigation, therefore, we have first to understand why the experiments are designed and analyzed in the way they are. This is the subject of the next chapter.

3 Major types of experiment

3.1. Homing experiments

If navigation is an integral part of exploration, it follows that experiments on navigation should be an attempt by the experimenter to mimic the exploration process. At the same time, by controlling the amount and types of information available to the animal at different stages during 'exploration', the experimenter can aim to find out when the different types of information are most important for successful navigation.

Effectively, this is what displacement–release or 'homing' experiments try to do (Fig. 3.1). Animals are taken from their home range, displaced beyond the limits of their familiar area, and released. Usually, at some stage during the experiment, they are deprived of particular information (e.g. by preventing them from seeing the route over which they are being displaced). At the same time, 'control' individuals are treated in exactly the same way except that they are not deprived of the information being evaluated. Some measure of navigational ability is then made and the performances of control and experimental animals compared to see if the deprivation has had any influence.

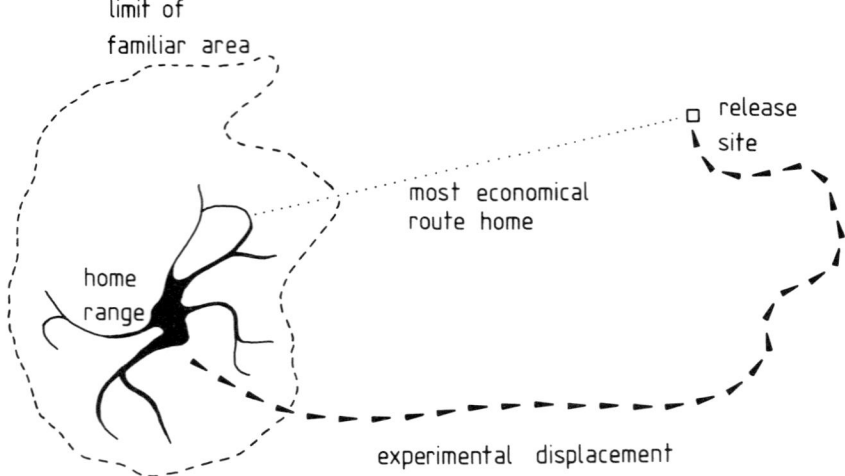

Fig. 3.1 Displacement–release experiments mimic natural exploration
[From Baker (1981a)]

3.2. Homing success

In the early days of experiments on navigation, the measure of performance most widely used was homing success. The proportion of released birds that returned home and the speed with which they did so was used as the measure of navigational efficiency. Homing success was abandoned as the primary measure, however, when it was realized that most often it failed to provide critical evidence for even the existence of navigation.

When a group of animals is released away from home, a lucky few might return successfully, not because they could work out the direction of home from the release site but because by chance they happened to set off in the correct direction. One of the characteristics of homing experiments is that the proportion of birds that returns home decreases as the distance of release increases (Fig. 3.2). This effect could be due to any number of factors, few of which relate to navigational ability. For example, the further an individual has to travel, the less motivated it may be to attempt to home. Similarly, the further it has to travel over unfamiliar terrain, with no knowledge of where to seek shelter or hide, the more prone it may be to predation. More importantly, however, the further from home the release site, the less likely is an animal to arrive back in its familiar area purely by chance.

Wilkinson (1952), for birds, and Saila and Shappy (1963), for salmon, developed mathematical models in an attempt to work out what proportion of individuals would return home and at what range of speeds if they had no navigational ability at all but instead employed various

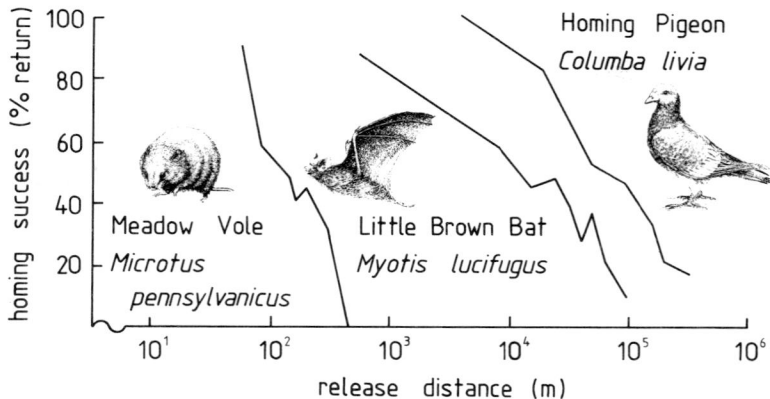

Fig. 3.2 For birds, as for other animals, homing success declines with increase in release distance

[From Baker (1981a), after Bovet]

patterns of random search. The close fit between observed homing performances and the predictions of their models led to a revolution in experimental design. No longer was it enough simply to release a group of animals and to wait for them to return. Some measure had to be found of the ability of birds to recognize home direction at the release site. After all, if it were found that more individuals set off in the home direction than in any other, random search could not be involved no matter what the characteristics of homing performances. For example, even if no birds returned home, an ability to recognize home direction from the release site could be inferred if animals set off in the correct direction. If we were to displace snails and release them 100 km from home, they may be perfectly able to work out the direction of home and set off in that direction, but we may doubt that any would actually return.

3.3. Vanishing points

In homing experiments on pigeons, the measure that is now universally adopted in measuring their ability to solve navigational problems is the compass bearing of their vanishing point. Pigeons are released singly, from the hand, and then watched through binoculars until they vanish from sight. The compass bearing of the last sighting before the bird vanishes is recorded. The next bird is not released until the previous bird has vanished and successive birds are faced in different directions to be released from the hand. The use of vanishing points requires that a release site has equally good visibility and open aspect in all directions. Pigeons are a convenient size for recording vanishing points but the same measure has been taken on much smaller birds, such as swallows.

3.4. 'Pointing' home

There are a number of disadvantages in having to release the experimental animal and watch it disappear in order to measure its 'answer' to the navigational problem set by the experimental conditions. One of these disadvantages is that unless the individual homes successfully, it cannot be used again. Another is that during the homeward flight it gains unknown experience that may influence its performance on any subsequent flights. The major disadvantage, however, is that it is usually difficult and often impossible to manipulate experimentally the information available to the bird at the release site. There would be a tremendous advantage in being able to restrain the animal at the release site and have it 'point' towards home rather than having to watch it set off in that direction.

Of all experimental animals, humans are the most convenient in this respect. They can be asked literally to point towards home. Even more conveniently, they can also be asked to state or write down their estimate of the compass direction of home to see if this differs from the direction in which they pointed. It is very difficult to separate these two different estimates of home direction using other animals, particularly birds. Yet, given both of these measures, a great deal could be inferred about the navigational process.

Animals other than humans have been restrained at the release site and, in effect, asked to point toward home. Most often, this technique involves the use of a so-called orientation cage to restrain the animal. Such orientation cages have been used with particular success for mice (Fig. 3.3) but have also been used for birds (e.g. bank swallows (*Riparia riparia*) – Sargent 1962). Unfortunately, homing pigeons have so far proved to be totally uncooperative in orientation cages. The latest attempt to design a cage in which a pigeon will indicate home direction is being made by the Frankfurt group.

Measures taken in an orientation cage are either the frequency of movements or the lengths of time spent in different directions. These measures are then processed in some way to calculate the mean direction of the animal's movements. This mean direction is then taken to be the animal's estimate of home direction.

3.5. Circular statistics

It is not enough, of course, simply to collect a number of estimates of home direction, whether from vanishing point bearings or from a set of mean directions in an orientation cage, and to judge them by eye for evidence of an ability to recognize home direction. Some statistical technique is required to calculate the probability of a given set of estimates arising by chance. Unfortunately, as estimates of home direction are angles and thus restricted to values on a circle of 360 degrees, the more familiar statistical techniques are not applicable. Instead, we have to use a group of statistical

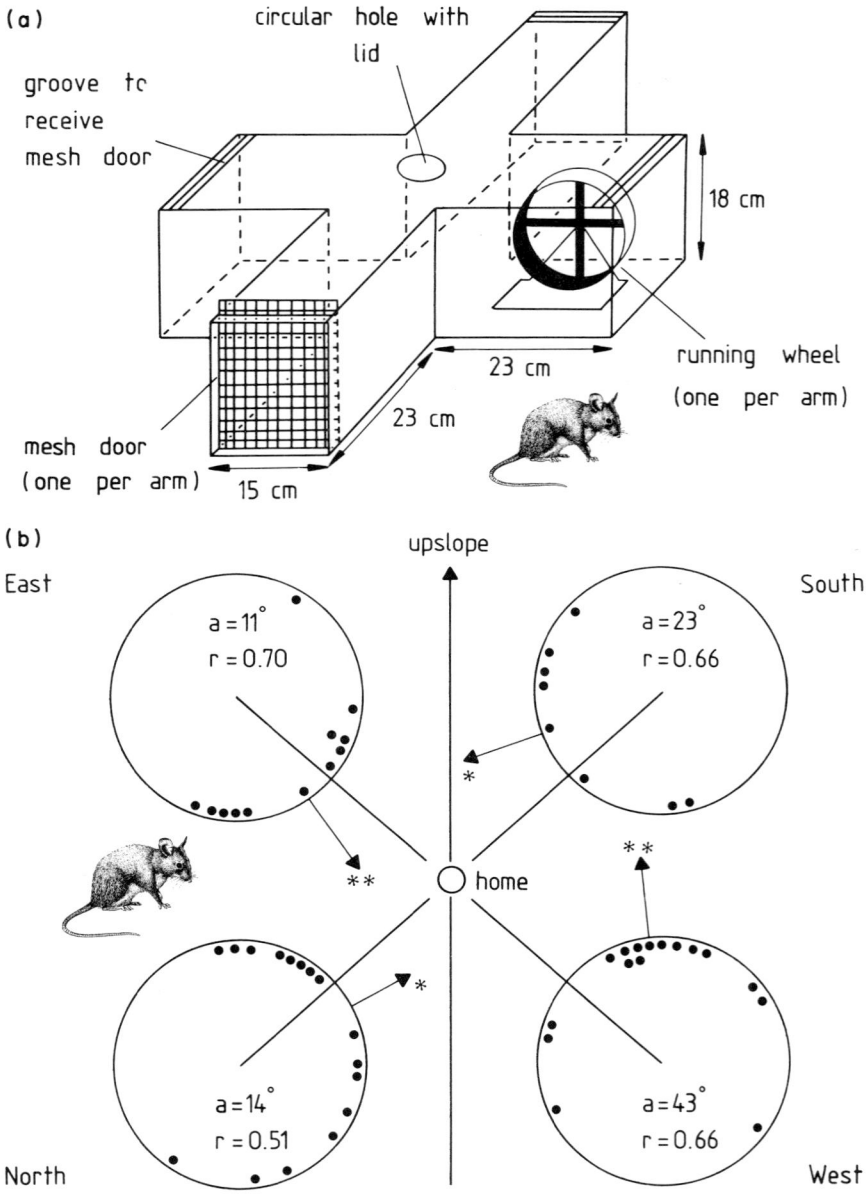

Fig. 3.3 The use of an orientation cage to study navigation by a small mammal

At the test site, an animal was introduced into the perspex cage (a) through the circular hole. The time spent in each arm was recorded and used to calculate a directional preference. When these preferences are plotted (b), they show that the mice are pointing more or less toward their place of capture ('home').

[From Baker (1981a), after Mather and Baker]

tests known collectively as circular statistics. All of the tests used most commonly in orientation research are described by Batschelet (1981). Here I shall give only the basic information required to be able to interpret the diagrams scattered through the rest of this book.

The first step in analysis of a set of directional estimates (e.g. Fig. 3.4) is to calculate their mean vector. It is usual when illustrating such a data set to show the individual estimates of direction as points around a circle relative to the direction being estimated (e.g. home). The mean vector is then shown as an arrow radiating from the centre of the circle.

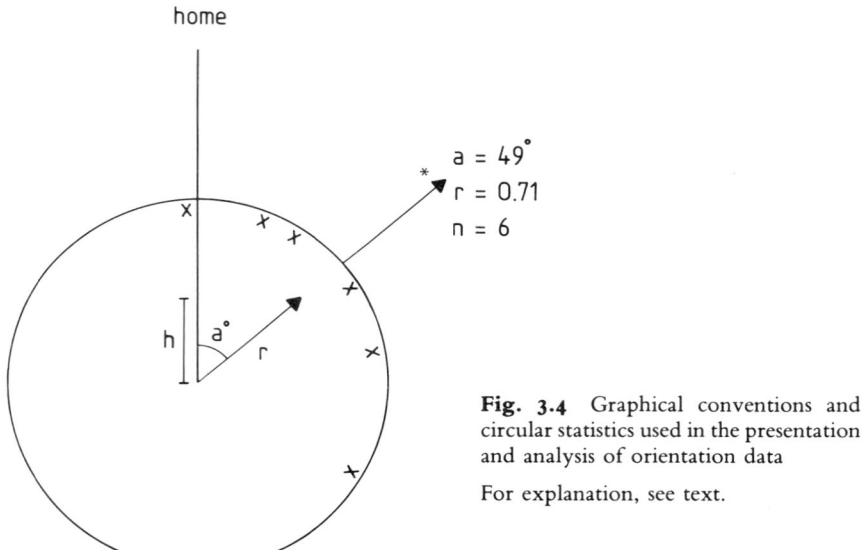

Fig. 3.4 Graphical conventions and circular statistics used in the presentation and analysis of orientation data

For explanation, see text.

A mean vector has both direction ($a°$) and length (r), where the length indicates the amount of agreement between the individual estimates. When all estimates are identical, agreement is total and $r = 1$. When estimates are uniformly distributed throughout 360 degrees so that there is no mean vector, there is no agreement and $r = 0$. By convention, the arrow is drawn to reflect both the direction and length of the mean vector such that, when $r = 1$, the length of the arrow equals the radius of the circle.

The direction of the mean vector is often given as a compass bearing with north $= 0°$, east $= 90°$, south $= 180°$, west $= 270°$, and so on. At other times, all estimates and the mean direction are given as an error ($e°$) relative to the direction being estimated. An accurate estimate of, say, home direction is thus given as $e = 0°$. An estimate ninety degrees clockwise of the correct direction would be $e = +90°$, or simply $e = 90°$, whereas an estimate ninety degrees anticlockwise of the correct direction would be $e = -90°$. An estimate precisely opposite to the correct direction would be $e = 180°$.

If a uniform set of estimates of direction gives a mean vector with $r = 0$, we can check for non-uniformity in the data by evaluating, in effect, whether the r that we calculate for our data is significantly greater than zero. The statistical test for this is known as Rayleigh's z-test which, to be precise, tells us whether our data show a unimodal departure from uniformity. There are other tests which allow for other, such as multimodal, departures, but these are less commonly used.

The z-test tells us whether there is significant agreement among the estimates of direction in a given experiment, but it does not tell us whether agreement is over the correct direction. After all, it would be possible for a group of pigeons, for example, all to depart, one after the other in precisely the same direction and for us to calculate a mean vector of $r = 1$. These data would be highly significant by the z-test, yet their mean error could be $e = 180°$. Obviously the z-test by itself is not enough to tell us whether a given set of estimates indicates an acceptable ability to recognize home direction among our experimental group. To do this, we need a test that takes into account the mean error.

The first step in testing for homeward orientation is to apply a modification of the Rayleigh test, known as the V-test. Effectively, this test examines the length of that component of the mean vector that lies in the home direction. This component, usually known as the homeward component, is represented in Fig. 3.4 by h and can vary from $+1$ (when all estimates are exactly in the home direction) to -1 (when all estimates are exactly in the opposite direction).

The V-test tells us whether there is a significant vector in the expected direction. It does not, however, tell us whether the mean error is significantly different from $0°$ (Aneshansley and Larkin 1981). The final step in evaluating homeward orientation, therefore, after establishing significant non-uniformity by either z- or V-tests, is to determine the 95 per cent confidence limits of the mean error. If these include $0°$ (e.g. when $e = 30° \pm 35°$), an ability to recognize home direction by the group being tested can be accepted. If not (e.g. when $e = -30° \pm 15°$), then homeward orientation has not been established.

3.6. Separating goal orientation from orientation

So far we have considered the evaluation of homeward orientation in purely statistical terms, once the data have been collected. However, even if such data are consistent with homeward orientation from a particular release site, this is not enough to demonstrate that goal orientation is in operation.

The earliest experiments on homing pigeons often used release sites on a narrow range of compass bearings from the home loft. From the loft used by Kramer at Wilhelmshaven on the north German coast, for example, it was much more convenient to release birds from the south than from the north. As pointed out by Matthews (1968), however, such experiments do not separate goal orientation from simple orientation. Suppose that, for some unknown reason, homing pigeons from a particular loft always fly to the northwest upon release. If such pigeons were only ever released from sites at different distances to the southeast of their loft, the impression gained would be of accurate homeward orientation at all distances. Not until the birds were released from sites to the northwest would it become clear that they were employing orientation, not goal orientation. It has become standard practice, therefore, in experiments on goal orientation, to release groups from at least two (opposite) directions from home, and preferably from four.

It might seem unlikely that any animal, when displaced from home and released, would ever simply orientate on a particular compass bearing, unrelated to the direction of home, unless trained to do so as in the case of racing pigeons. Yet this is precisely how some birds do behave. When mallard (*Anas platyrhynchos*), for example, are displaced from Slimbridge, England, and released, they fly almost always to the northwest (Matthews 1968), no matter what the direction of 'home' (Fig. 3.5). Mallard from other areas also show a preferred compass direction, though not always to the northwest (Matthews and Cook 1982). Wallraff (1978b) examined data from a variety of pigeon lofts in the United States and Europe and found that birds from each loft had their own, apparently loft-specific, preferred compass direction, or PCD as he termed it. When released the birds showed mean vanishing bearings that were a compromise between their PCD and the direction of home.

In the case of mallard, the PCD was maintained beyond the vanishing point for not more than about 20 minutes and most ducks had landed or broken away from their original direction within 16 km of the release site. Even then, there was no evidence that orientation gave way to goal orientation, for subsequent recoveries were scattered at random (Matthews 1968). In the case of homing pigeons, orientation in the PCD does seem eventually to give way to goal orientation, for many birds still return home, albeit with reduced homing speeds (Matthews 1968). In the study by Michener and Walcott (1967), in which pigeons were trained from one direction and then released from another (e.g. Fig. 2.3), orientation in the training direction was sometimes maintained for up to 80 km before goal orientation became evident.

3.7. Testing for orientation

There is still no good explanation for why birds such as mallard and homing pigeons should sometimes orient in a PCD following displacement rather than towards home (but see Chapter 11). At first sight, such behaviour is inconvenient, requiring caution in the design and interpretation of experiments. In another sense, however, it is convenient because it

Major types of experiment 27

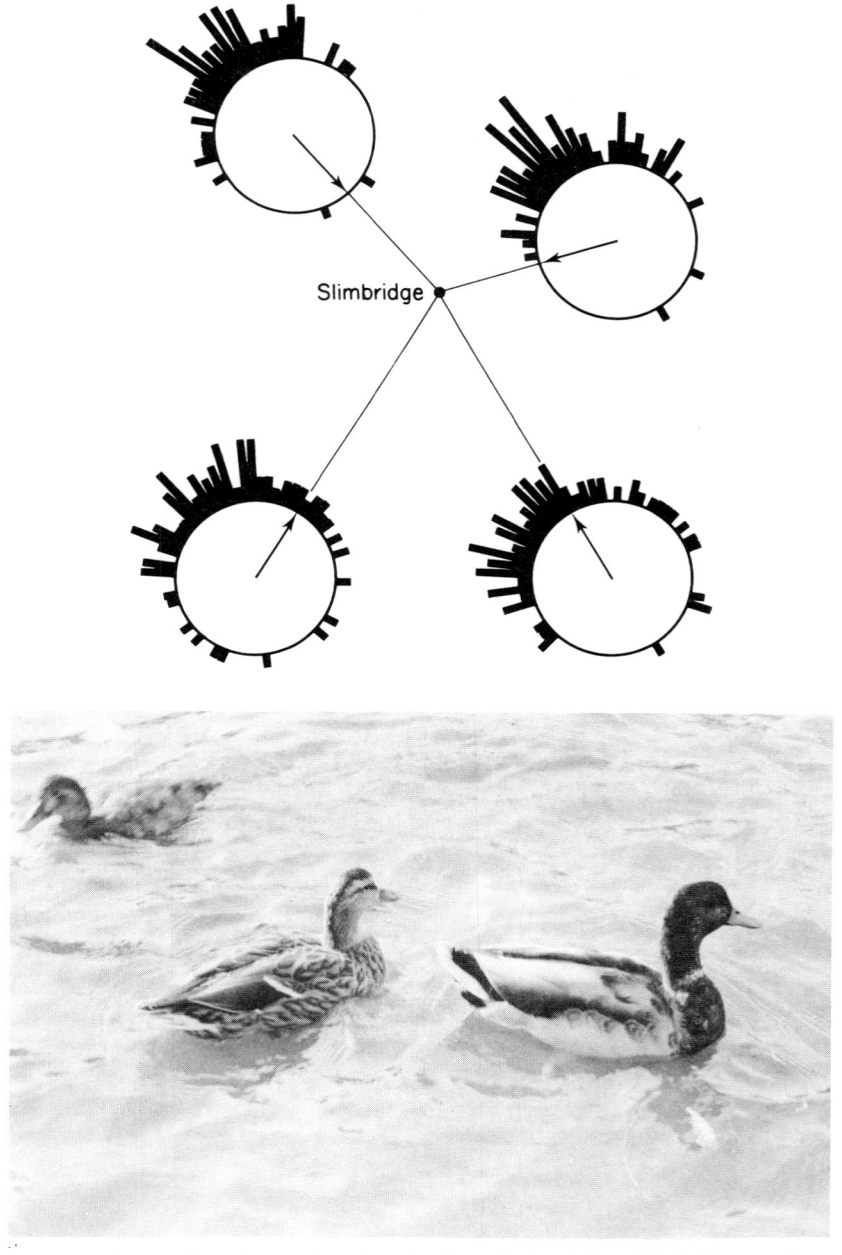

Fig. 3.5 When mallard (**Anas platyrhynchos**) are displaced from Slimbridge, southwest England, they fly to the northwest irrespective of the direction of 'home'

The shortest spoke represents the vanishing bearing of one bird.

(Re-drawn from Matthews 1968.)

(a)

(b)

(c)

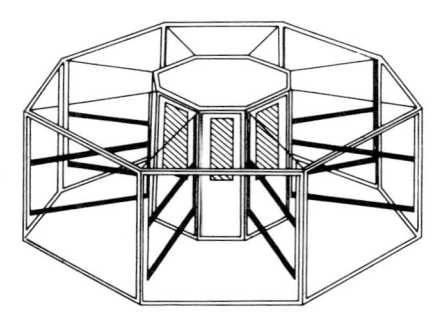

Fig. 3.6 The use of caged migrant birds to study migratory orientation.

Caged birds of migrant species, such as continental European starlings, **Sturnus vulgaris**, when housed in special orientation cages (as in (a)), may be used to study migratory orientation and timing. The caged starling in (b) becomes restless at about the same time as its free-living counterparts on the telephone wires are migrating. It will also orient in the same compass direction as they migrate. Orientation mechanisms may then be studied by, for example, altering the magnetic field through the cage by means of Helmholtz Coils as in (b). (c) the octagonal cage developed by Merkel and Fromme (1958) and later used extensively by Wiltschko.

Each time the bird lands on a perch the event is recorded automatically on paper tape. Data are later analysed by computer. (Photos by Gerlinde Orth)

offers an experimental situation in which compass orientation may be studied. As we shall see in the next chapter, there are many ways in which a bird could make use of an ability to recognize compass direction in solving problems of goal orientation. Indeed, paradoxical though it may seem, there are some forms of navigation that can be achieved entirely through detection of compass direction. A vital step, therefore, in trying to

30 *Bird navigation: the solution of a mystery?*

understand bird navigation is to study the ways in which birds detect compass direction. Experiments on how they manage to orient in their PCD is one method available, but there are others.

The most widely used of all methods of experimenting on compass orientation is to restrain a migrant bird in an orientation cage (Figs. 3.6 and 3.7). Throughout most of the year, such birds behave much like any other caged bird. At those seasons that the free-living members of their species are performing their annual migrations, however, caged birds become extremely restless. The pattern of this migratory restlessness, or *Zugunruhe* as it is often called, mimics in many respects the pattern of migratory activity in wild birds. For example, nocturnal migrants when caged are restless at night (Fig. 3.8). Longer-distance migrants show more restlessness than shorter-distance migrants (Chapter 12). Most important of all as far as orientation is concerned, and certainly the most unexpected, is that the restlessness is oriented in the migration direction traditional for the species concerned (Chapter 7).

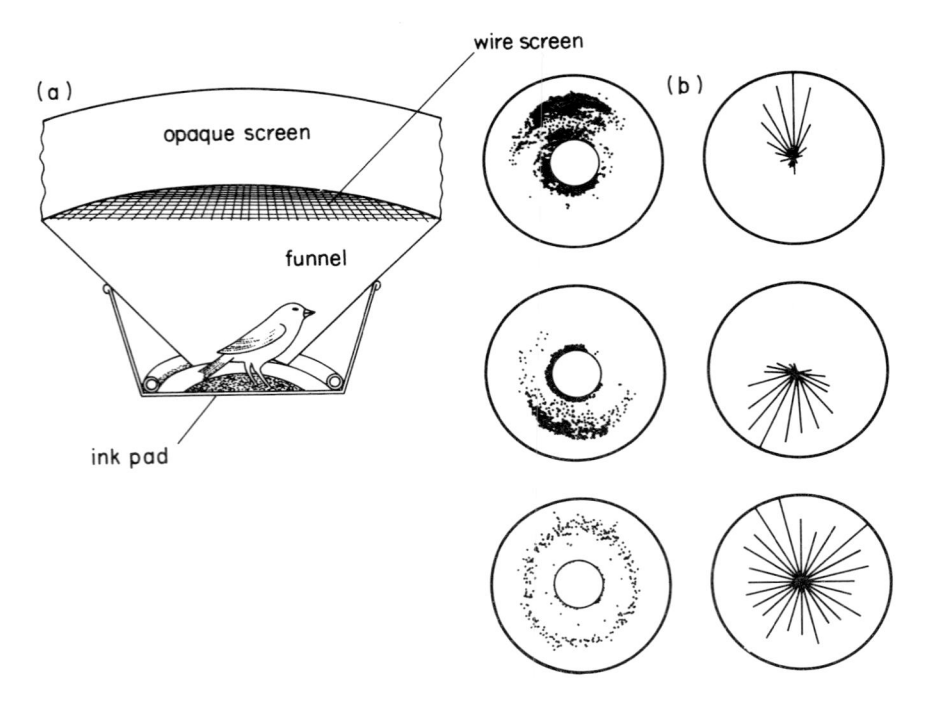

Fig. 3.7 The Emlen funnel introduced by Emlen and Emlen (1966)

When the bird jumps up the side of the cage (a), footprints are produced on the blotting paper lining the funnel. Footprints are then transcribed into vector diagrams as in (b). More recently, ink has been abandoned and the blotting paper replaced by 'Tippex' paper, which scratches easily when the bird jumps. Vector diagrams are produced by counting the number of scratches in each direction.   [Compiled and re-drawn from Schmidt-Koening (1979)]

Major types of experiment 31

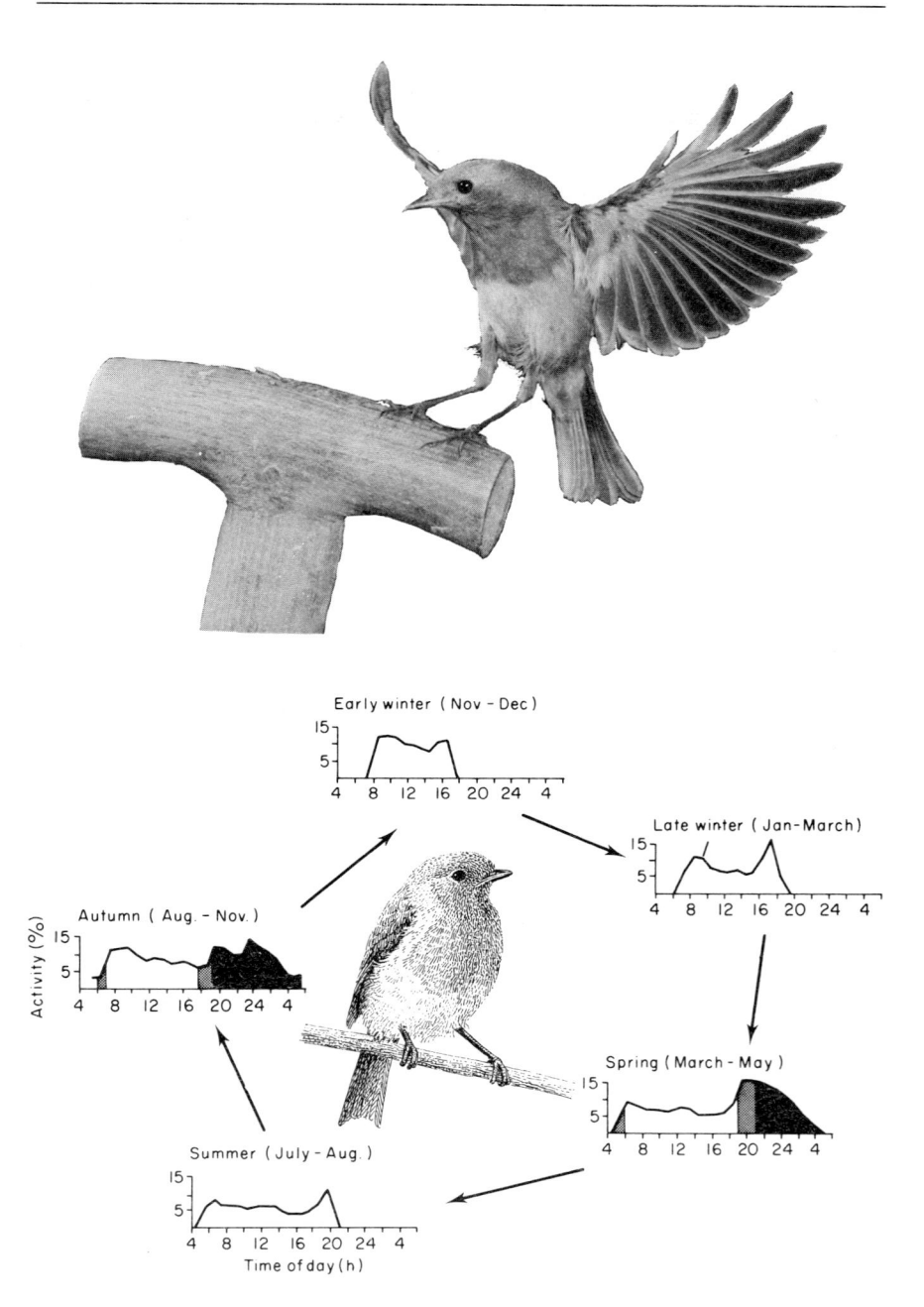

Fig. 3.8 Annual cycle of daily activity by caged European robins (**Erithacus rubecula**)

Diurnal activity is unshaded; nocturnal activity, black; dusk and dawn activity, stippled. During the migration seasons, caged robins are active at night when free-flying birds are migrating.

[From Baker (1978), after Merkel. Photo by Ronald Thompson, courtesy of Frank W. Lane]

This compass orientation by caged migrants during migratory restlessness was first noticed as long ago as 1949 by Kramer (1949) during experiments on caged passerines. His observation opened the way for subsequent generations of researchers to manipulate available compass cues and begin to unravel the compasses that are the basis of migratory orientation.

There are, however, disadvantages in being restricted to a study of compass orientation only during periods of migratory restlessness in caged birds. Not least, it means that there are long periods during the year when no experiments can be carried out. Kramer himself reduced this constraint (e.g. Kramer 1953) by devising an experiment to test for a sun compass in starlings (*Sturnus vulgaris*). For this, Kramer first trained a bird to take food from a pot in a particular compass direction. Then he began to manipulate the apparent position and movement of the sun (see Chapter 6). Training procedures similar in principle have been used to study the magnetic orientation of salamanders also (Phillips and Adler 1978).

In this chapter the major types of experiment used in the study of orientation and goal orientation have been surveyed. Such experiments have yielded a myriad of details. Sometimes results from different experiments are consistent with one another; at other times they are downright contradictory. If ever there were a danger of being unable to see the wood for the trees, it would be in the forest of orientation research. Now is the time, before we become engrossed in the detailed findings of particular experiments, to step back, to seek a few general principles, and to ask: how could birds navigate?

4 How could birds navigate?

When we think of navigation, we tend to think of it in its short-term context as the means by which an animal finds its way back to a familiar location without getting lost. Viewed in the context of exploration (Fig. 2.5), however, the main function of navigation is long-term. It is the means by which animals pioneer the most economical routes between places they will want to visit time and time again. It follows that the more efficient an individual is at navigation the more efficient will be its explorations, and the more efficient will be the future use of its familiar area and home range.

To gain a complete picture of goal orientation by birds, therefore, it seems we must examine how they find their way around their familiar area as well as how they form that familiar area in the first place through exploration. With the advantage of hindsight, I feel that understanding the former gives strong pointers to the nature of the latter.

4.1. Mental maps

It is almost axiomatic that in finding their way around their familiar area, birds make use of what is, in effect, a mental map of that area. Somewhere in the central nervous system must be stored details of: (1) which sites offer which resources and when; (2) the landmarks by which the different sites may be recognized; (3) the spatial relationships of the sites; and (4) how best to travel between them.

The concept of a mental map is familiar to humans because a moment's thought reveals that each of us has such a map and uses it almost continuously as we move from one place to another. Formal research into the nature of mental maps and which part of the central nervous system is responsible for their storage has made surprisingly little progress, even for mammals, though attention has recently concentrated on the hippocampus as a possible storage area (O'Keefe and Nadel 1979).

Perhaps the most important conceptual advance in understanding the mental maps of birds was made by Wallraff (1974) when he suggested that such maps take the form of a mosaic. This again should appeal to us as humans, for once more it describes nicely the probable form of our own mental map. In our mind's eye, the places with which we are familiar form a mosaic arranged roughly in accordance with their actual spatial

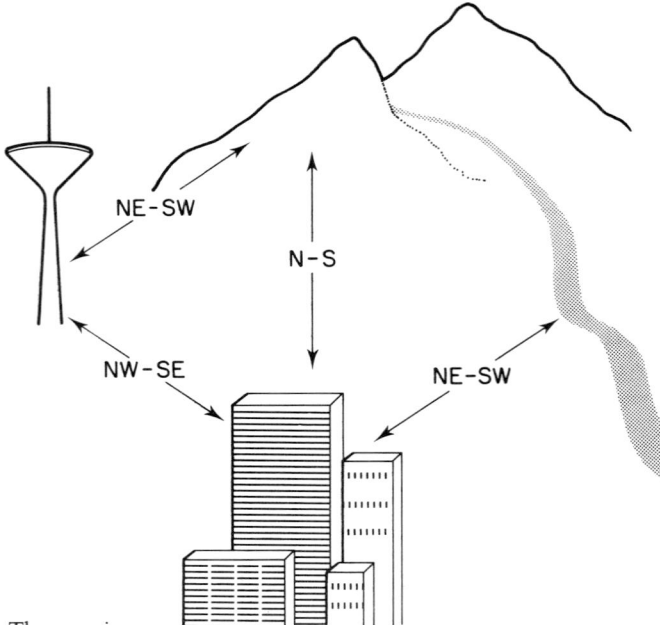

Fig. 4.1 The mosaic map

The spatial relationships of major landmarks are memorised in terms of their relative compass bearings.

[Inspired by Wiltschko and Wiltschko (1982)]

relationships. We may not be able to recall our entire map simultaneously, but from any one site we are able to indicate the approximate directions of other sites. Armed with pencil and paper, we could probably reconstruct our mosaic of familiar places, first entering those around our current home, then others around each of these, and so on until all familiar sites were exhausted. In addition to the sites themselves, we could also indicate the major landmarks that we associate with each site and by which we find our way around their immediate area. Finally, of course, we could also join the sites by the most familiar routes between them. The map remains a mosaic, however, because there are large tracts of land just to the sides of these routes and between the sites themselves with which we are totally unfamiliar.

Most modern city dwellers probably memorize the direction of one site from another in terms of major road systems. In the past, however, humans seem to have made much greater use of compass directions. Polynesians, for example, from any one island could point out the direction of many other islands, hidden beneath the horizon (Lewis 1972). Tupaia, the most famous of all Polynesian navigators, when met by Captain Cook in 1769, was able to construct a mosaic map of an area of ocean almost 4000 km in length at its longest axis and containing the positions of nearly 80 islands, many of which Cook then went on to 'discover'. Polynesians learn the directions between islands in terms of what are, in effect, compass bearings taken primarily from the stars. Australian aborigines, in common with many non-industrialists, use compass bearings not only to describe the directions of familiar places, but also in their everyday language (e.g. 'Do you wish me to sit to the north or south of you?') (Brandenstein 1972).

There is an economy in the use of compass bearings to memorize spatial relationships that is not found in other systems. For example, knowing that, from site A, the red building marks the direction of site B does not help in also remembering the direction of A from B unless the same red building can be seen from B. Knowing that B is north of A, however, is the same as knowing that A is south of B, no matter how far apart the two may be.

Directions memorized as compass bearings are particularly useful for animals that can travel in more or less straight lines, such as Polynesians, Australian aborigines and, of course, birds. We even have evidence that this is precisely how birds do store the direction of one site from another on their mosaic map.

At an early stage in the study of bird orientation, it was found that, if the sun were shining, it would be used to obtain compass bearings. To use the sun as a compass, however, a bird has to be able to allow for the sun's daily movement across the sky as it travels from eastern to western horizons. Compensation for such movement requires a sense of time and, in common with most animals that have been studied, including man, birds have a relatively efficient internal 'biological' clock. To work accurately, however, this internal clock needs to be reset each day to the correct local time.

The primary clue used to do this is the transition from dark to light at dawn and from light to dark at dusk. This use by birds of the light–dark transition can be exploited experimentally; its internal clock can be put out of phase with local time simply by placing the bird in an artificial light regime and having the lights turn on and off a certain number of hours before or after the local sun has risen or set.

One of the inevitable consequences of re-phasing a bird's internal clock is that, if it then tries to use the sun as a compass, it misinterprets direction. For example, consider a bird with an internal clock that has been re-phased to be six hours ahead of local time. At local noon, the bird's clock will tell it the time is 18.00 hours; at 06.00 hours, the bird will think it is noon. At real noon, therefore, when the sun is due south, the bird will interpret it to be due west; when the sun is east, it will be interpreted as being south, and so on. This technique, of phase-shifting a bird's internal clock (usually called 'clock-shifting'), has been extremely valuable in working out just when a sun compass is being used.

In an ingenious experiment, Graue (1963) released clock-shifted homing pigeons within one or two kilometres of their loft at sites they must have visited many times previously. When the pigeons could see the loft from the release site, they did the 'sensible' thing and flew straight home. When they could not see the loft from the release site, however, they did something quite extraordinary but very instructive. They flew at an angle relative to the sun that would have taken them home had their clocks not been phase-shifted. In other words, their vanishing bearings deviated from the direction of the loft by the amount predicted from the phase-shifting of their internal clock.

The most likely interpretation of these results is that the pigeons recognized their release site from the surrounding landscape, knew from their mental map that it was, say, south of their loft, and then proceeded to fly in the direction that they judged from the sun to be north. The implication is that when pigeons set off from one familiar site to another, they take their bearings from the sun rather than from the lie of familiar landmarks, even though the latter may be used to recognize the site in the first instance.

Many more experiments are needed before we can feel confident of the form of the mental map of birds, but for the moment we at least have a working hypothesis. The map is learned and stored in the form of a mosaic of familiar sites, identifiable by their surrounding landmarks. Spatial relationships of the different sites are stored as compass bearings: to travel from A to B fly north; from B to C, fly east; from C to A, fly southwest, and so on (see Fig. 4.1).

If this picture of the form of the mental map is correct, it has important implications for the way navigation might work during exploration. It suggests, for example, that the primary role for navigation, when a bird arrives at an unfamiliar location, is to work out the compass bearings of other familiar sites. There would seem to be two major ways that this could be achieved, whether during natural exploration or during homing experiments (Fig. 4.2). One way is to use 'route-based' navigation to keep track of the outward journey so that at all times during exploration or displacement the compass direction of home is known. The other is to use 'location-based' navigation and work out the compass direction of home from information available at the place from which navigation is to occur.

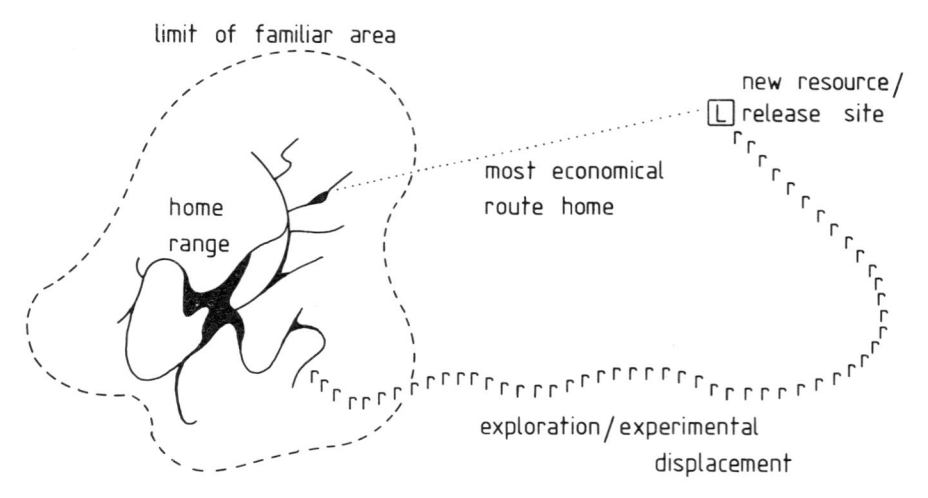

Fig. 4.2 Route-based and location-based navigation in both natural exploration and experimental displacement

[From Baker (1981a)]

4.2. Following the outward journey

Route-based navigation is most simple, of course, when the outward journey is a straight line. As long as the bird registers the compass direction of this straight line, either upon first leaving home or at some time during the journey, home direction is given simply by reversing that compass bearing. When the outward journey is not straight, however, it becomes necessary to register not only the compass direction of each stage or 'leg' of the journey but also the length, or at least relative length, of each leg. In Fig. 4.2, for example, the bird should register that it travelled to the east for about twice as far as it travelled north. It then follows that home direction is west-southwest. How accurately direction and distance need to be registered for successful navigation back to the original familiar area depends on the size of that familiar area relative to its distance from the unfamiliar site (Fig. 4.3).

Other forms of route-based navigation are possible which do not depend on judging compass direction during the journey, though some still give an answer for home direction in compass terms. As an example of one method, suppose that there are landmarks visible from the home site but some distance away. Suppose also that the bird learned the compass bearings of these landmarks from home. If the landmarks were passed during the outward journey, the compass bearing of home at that point in the journey would be known. For example, if a mountain range is visible from home 100 km to the north and the outward journey passes over that mountain range, then home lies to the south.

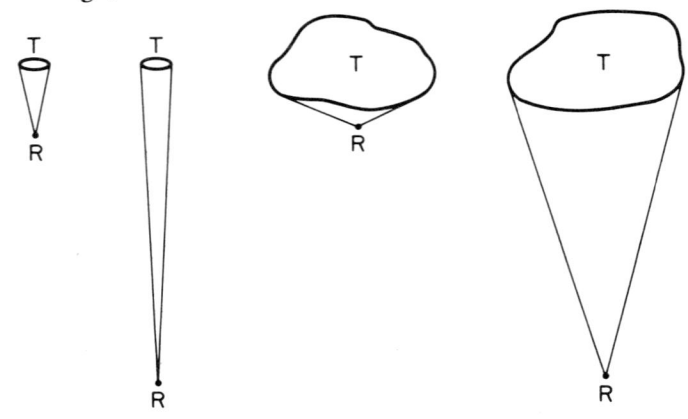

Fig. 4.3 The minimum accuracy of navigation depends on the size of the target area (T) and the distance of the release site (R)

Not all possible forms of route-based navigation give the answer for home direction in terms of its compass bearing. If birds, like modern jets and space-craft, were to use inertial navigation, they would end up with an answer relative to the orientation of their own body, rather than relative to a compass. Inertial navigation uses an internal gyroscope to register the twists, turns and speeds of the outward journey. The method would require the bird either to maintain continual awareness of home direction throughout the journey or, perhaps less likely, at the end of the journey to estimate home direction by integrating the accumulated information. Either way, the answer is obtained relative to the bird's body rather than relative to the environment. Having estimated home direction relative to its body, of course, the bird can then check with the environment to determine the compass bearing of home.

The different types of answer given by the different methods bears emphasis, for such differences could allow testable predictions to be made by which the methods could be separated by experiment. In effect, the methods of route-based navigation described earlier give the answer for home direction first in terms of compass direction. The bird then has to check with its compass to work out in which direction relative to its body to fly or 'point'. Inertial navigation allows a bird first to point towards home and only then, after checking with its compass, could it give the answer in compass terms.

The most likely site for an internal gyroscope would perhaps be the inner ear (Barlow 1964) though, as pointed out by Delius and Emmerton (1978), other parts of the body could give inertial information, as when the viscera move within the body cavity. However, there are means of following the twists and turns of a journey using the body as the axis for reference without using a gyroscope.

The angle of each turn during the outward journey could be judged relative to some environmental feature, such as the sun, shadows, magnetic field, wind direction or distant landmarks. The difference between this and earlier methods based on compass direction is that home direction would be updated after each turn relative to the bird's body. For example, suppose a bird has been flying away from home continuously, such that home is still directly behind it. Now suppose the bird makes a clockwise turn which it judges from the magnetic field or from the change in position of the sun or landmarks, etc., to be about 135 degrees. Home is now in front of it and about 45 degrees to the right.

The techniques described in the last two paragraphs, based as they are on the summation of angles, suffer from the enormous disadvantage that errors build up, one on the other. A gross mistake in judging a turn early on in the journey can never be corrected by accurate judgment thereafter, only by an equally gross mistake in the opposite sense. Techniques based on the judgement of compass directions, on the other hand, do at least allow some opportunity for partial correction of mistakes as the journey progresses.

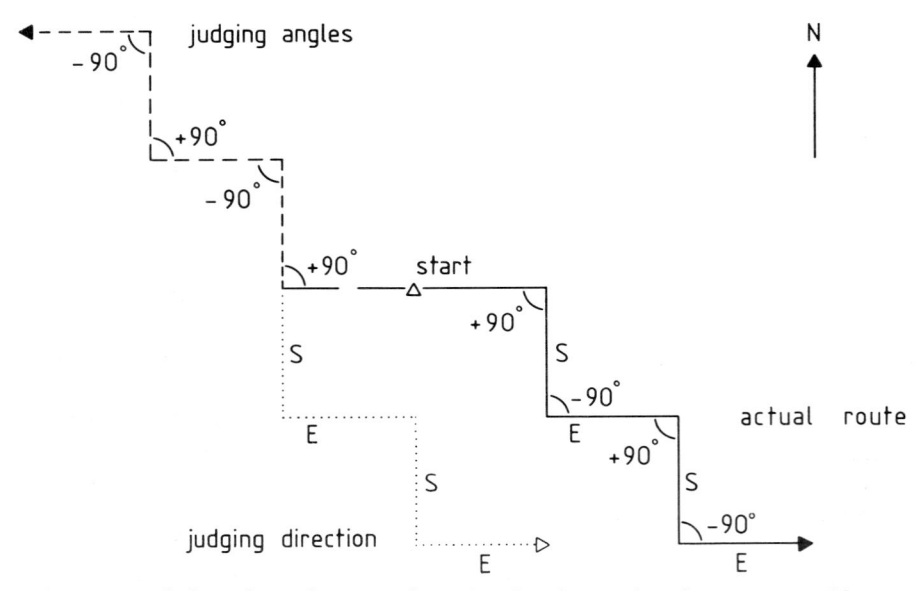

Fig. 4.4 Only by judging direction of travel, rather than angles of turn, is it possible to recover partially from initial disorientation during route-based navigation

[From Baker (1981a)]

The most simple form of route-based navigation has been left to last. If a bird memorized the sequence of landmarks it encountered during its outward journey, it could home simply by 'steeplechasing' back along that sequence. An apparent limitation to the steeplechase technique is that the birds must either restrict the outward journey to a straight line or return by an uneconomical route (Fig. 2.5). Some authors have rejected the possibility that homing pigeons might home by flying along sequences of landmarks because when a bird is released several times at the same site, it returns by a different route each time (Fig. 4.5). However, this limitation and criticism is more apparent than real, and loses its force if birds are suggested to use landmarks along the horizon. As shown in Fig. 4.5, a bird that registers the sequence of such landmarks during the outward journey, or on previous homing flights, can return home economically. Not only can a curved or tortuous outward journey be converted into a straight(er) flight, but a wide corridor of land could be available for the bird to use a variety of homeward routes on successive visits to the distant site.

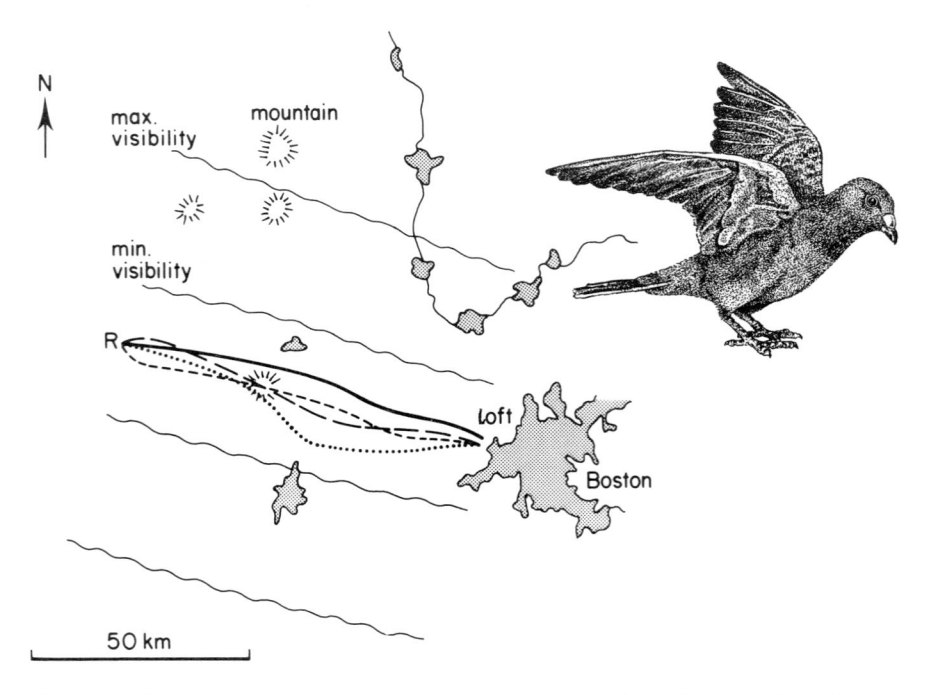

Fig. 4.5 When a pigeon is released on several occasions from the same site (R), its homeward track is rarely precisely the same. It flies, however, along a 'corridor' bounded by distant, potentially familiar, landmarks. In the diagram, the highest mountains in the region are shown, as are towns and cities (stippled). The distances of minimum and maximum visibility during the four homing flights illustrated are shown by a wavy line.

[Partly from Michener and Walcott (1967)]

Although birds could easily make use of these landmark-based techniques during natural exploration, researchers have until recently felt confident that such techniques were not available to birds taking part in navigation experiments. After all, in most experiments the birds cannot see the countryside through which they are passing. Recent work on the sensitivity of pigeons to smells and sounds, however, and their implications for the detection of olfactory and acoustic landmarks, require landmark-based techniques for following the outward journey to be taken much more seriously (Chapter 11).

4.3. Reading the landscape

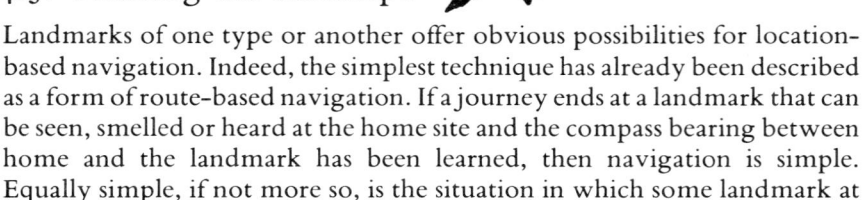

Landmarks of one type or another offer obvious possibilities for location-based navigation. Indeed, the simplest technique has already been described as a form of route-based navigation. If a journey ends at a landmark that can be seen, smelled or heard at the home site and the compass bearing between home and the landmark has been learned, then navigation is simple. Equally simple, if not more so, is the situation in which some landmark at home can be perceived from the unfamiliar site. Simple orientation in the direction of the landmark will take the bird home.

A bird does not have to arrive at the site of a familiar landmark to make use of such landmarks in location-based navigation. A range of distant landmarks can usually be detected from the home site, each one of which will lie on a particular compass bearing. As the animal moves away from home, the compass direction of each of the landmarks will change (Fig. 4.6). At any given location, the compass bearings of distant but familiar landmarks allow the bird to work out the direction it has to travel

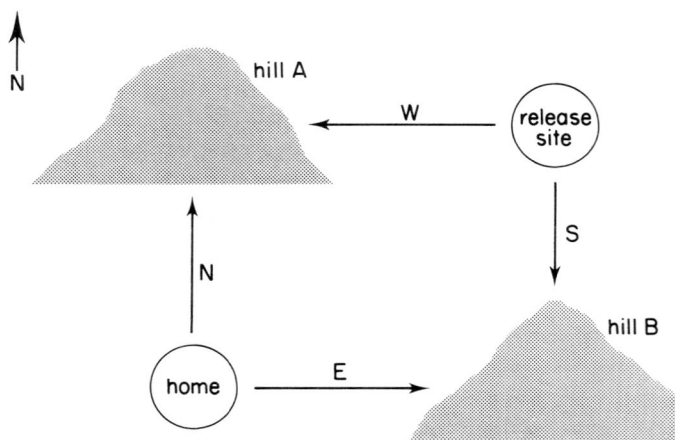

Fig. 4.6 When a bird explores or is displaced, the compass bearings or familiar landmarks change in a way that indicates the compass direction of home

[Simplified from Baker (1978)]

in order to return home and thus restore the landmarks to their proper bearings. For example, suppose that, from home, hill A is due north and hill B is due east, but from the unfamiliar location A is west and B is south. It follows that home lies to the southwest. This combined use of landmarks and compasses (Wallraff 1974; Baker 1978; Merkel 1978; Wiltschko and Wiltschko 1978) is similar to that used by surveyors and may be termed geodetic navigation.

Geodetic navigation works as shown in Fig. 4.6 only if the landmarks are visual or acoustic, being seen or heard from the home site. If they are olfactory, a different system has to be used that involves being able to detect the compass direction of the wind on which particular smells are carried to the home. The mechanisms that could be involved in this form of location-based navigation have been explored principally by Papi and his colleagues in Italy (e.g. Papi 1976, 1982).

Suppose birds learn while at home that particular smells arrive on winds from particular compass directions. Then suppose that on arrival at an unfamiliar location a particular smell is found to be much stronger than at home. It follows that home is in the compass direction in which winds that carry the smell to the home site are blowing (Fig. 4.7).

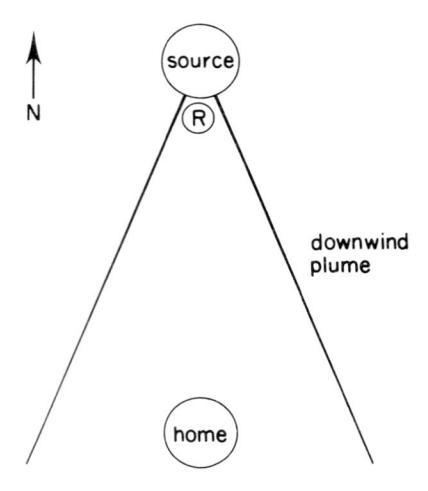

Fig. 4.7 If a bird learns while at home that a particular smell arrives with a north wind, upon release near to the source of the smell it knows that home lies to the south

Whereas in theory there is no limit to the distance over which a bird can use route-based navigation, there is a limit to the distance from which it can use a location-based system if the latter depends on landmarks. This limit is set by the furthest distance from home at which it can detect and recognize familiar features. How far this distance may be depends in part on the type of landmarks that are detected.

It may help at this stage to emphasize a difference between the familiar area, as we have conceived it so far, and what I shall call the familiar area map. The limits of the familiar area are set by the limits of the area that the bird has ever visited. Around the familiar area, however, is a zone the major landmarks of which are familiar because they can be detected at a distance. These landmarks thus have a place on the familiar area map but in one sense are not part of the familiar area. Thus when a person visits a village for the first time, the village becomes part of his familiar area, the mountain that he can see 50 km to the north becomes part of his familiar area map. The mental map, of course, accommodates both.

The width of this peripheral zone depends on a number of factors, not least of which is the range of senses by which the landmarks within the zone are detected. As humans, we are most familiar with visual landmarks, such as hills and mountains. The distance from which a hill can be seen depends on how high the hill projects above the surrounding countryside, how high the observer is, and, of course, visibility. This last factor is influenced both by atmospheric conditions and by the size and height of any obstacles

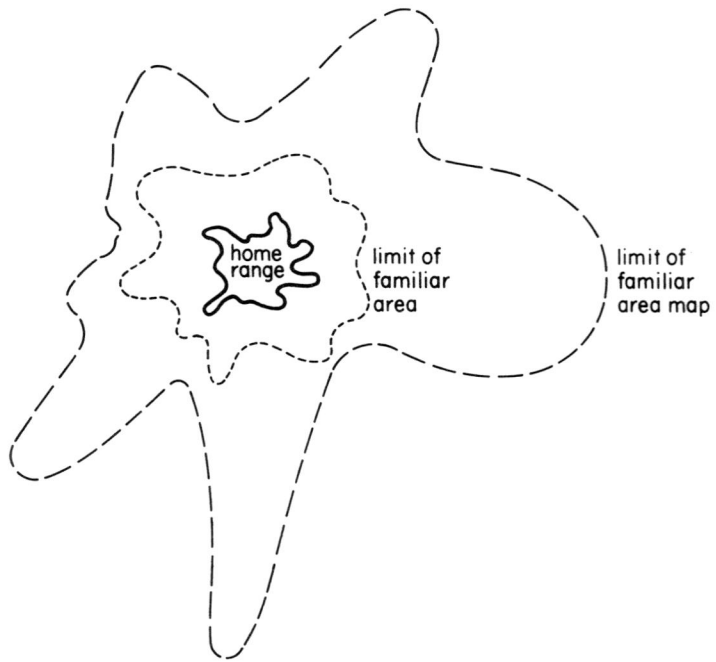

Fig. 4.8 Home range, familiar area and familiar area map

between the observer and the hill. The curve of the Earth's surface is such that, if it were smooth, the distance of the horizon in kilometres is given by $3.83 \sqrt{h}$ where h is the height of the observer in metres. Thus a hill projecting 300 m above surrounding 'flat' terrain can be seen from 80 km by a bird 10 m above ground level, from 100 km by a bird 100 m above ground level and from nearly 200 km by a bird 1000 m above ground level. The corresponding distances for a 1000 m hill are 130, 160 and 240 km. These calculations assume perfect visibility.

The distance from which sounds can be detected depends on the range of wave-lengths being used. Within the spectrum of sounds detectable by humans, the only sounds that could be used from a distance on a map are of fairly recent origin, such as the roar of aeroplanes arriving at and departing from a distant airfield. Perhaps people living near the sea could use the sound of surf as an acoustic landmark. In the study of birds, however, interest has recently turned to the possibility that they may be able to detect and make use of those low-frequency sounds known as infrasounds, the most useful of which would be those generated by landmarks such as coastlines and mountain ranges. The most interesting feature of infrasounds, however, is the long distance, 1000 km or so, from which they would allow such landmarks to be detected.

The distance from which smells can be detected is much less easy to decide in the absence of direct measurements. The sensitivity of the sense organ and the strength of the smell at source are obviously critical factors. So too is the rate of attenuation of the smell as it is carried from the source. This depends on air-flow patterns, including turbulence and convection, and on the chemical stability of the smell. Also important is the frequency with which the wind blows from different compass directions. A strong stable smell 10 km to the east will be far less valuable as a landmark than a weak smell 100 km to the west, if both can be detected but the wind only ever blows from the west. Finally, uniqueness has to be a factor. No matter how strong and stable may be a smell that is carried 100 km or so by winds from the north, it will be useless as a landmark if an identical smell is carried by winds from the south.

In theory, therefore, there seems nothing to prevent olfactory landmarks hundreds of kilometres away from the edge of the familiar area being incorporated into the familiar area map, but in practice the usefulness of such landmarks is likely to vary considerably from one area to another.

This brief survey of location-based navigation as based on landmarks has raised two points: (1) the range of landmarks that can usefully be included on the familiar area map is likely to vary from one area to another, depending on the topographical, acoustic and olfactory characteristics of each region; (2) the size of the familiar area map could be much larger than has often been assumed in the past. The first point is of interest later (Chapter 11) when we find that pigeons in, for example, Italy and Germany do seem to develop different navigational techniques. The second point is of interest immediately.

When a bird is displaced and released in a navigation experiment, the experimenter always tries to release the bird outside of its familiar area, for only by doing so can he study navigation and not pilotage. On most occasions he probably succeeds and the bird really is released beyond its familiar area. As most releases of pigeons are made between only 50 and 200 km from the home loft, however, and as few are made further than 400 km, we may wonder what proportion of releases are also beyond the limits of the familiar area map. Yet unless this is achieved, the results are impossible to evaluate in terms of the form of location-based navigation that in the past has excited most interest: navigation by grid map.

4.4. Reading a grid

The disadvantage of the familiar area map is that it does have limits. These are set, as we have seen, by the distance at which landmarks can be detected and recognized from the edge of the familiar area. Such limits are probably unimportant to a bird in its natural explorations since as it travels it can extend both its familiar area and its familiar area map. In a navigation experiment, however, such limits could matter if the experimenter, by design or accident, succeeded in (1) preventing the bird from detecting landmarks during the outward journey; and (2) releasing the bird beyond the limits of its familiar area map. Under such circumstances, location-based navigation could only be achieved if the birds were able to detect their position on some form of 'universal' grid.

The most familiar form of grid map to humans is one based on latitude and longitude. This is a 'universal' grid insofar as, given the coordinates of one's current location and with knowledge of the coordinates of home, it is possible to navigate from any point on the Earth's surface. There are a number of ways of determining latitude and longitude from environmental cues and, in theory at least, any one of these could be used by birds.

4.4.1. Finding 'latitude'

As one travels from the poles toward the equator, the height of the sun above the horizon gradually increases with decreasing latitude until, somewhere between the tropics of Cancer and Capricon, the sun is directly overhead at midday. This occurs actually on the equator only at the March and September equinoxes. On the northern tropic line it takes place at the June solstice and on the southern tropic line at the December solstice. If one travels from north pole to south pole, the noon height of the sun begins low in the southern sky, gradually rises to the overhead (zenith) position, and then slowly sinks through the northern sky. Outside of the tropics, therefore, a place where the sun at noon is lower than its height at home is further from the equator than home. A bird trying to return to the same latitude as home from such a place needs only, therefore, to fly towards the equator until the sun is the 'correct' height above the horizon at midday. Within the tropics, the situation is more complex. Navigation would require the bird to detect not only the height of the sun but also whether the sun at noon was in the southern or northern half of the sky and to know which half it should be in at home. Even in temperate regions the use of the sun to judge latitude is complicated by the fact that its height above the horizon changes with the seasons. Moreover, everywhere its height also changes continuously throughout the day.

46 *Bird navigation: the solution of a mystery?*

In many ways, it is easier to judge latitude by the stars than by the sun. The Earth's rotation makes the entire sky (i.e. the celestial sphere) appear to rotate (Fig. 4.9) about an axis. In the northern hemisphere, this axis of rotation of the sky is marked by *Polaris*, the pole star. In the southern hemisphere, however, there is no such convenient marker. The height of the axis of rotation above the horizon is a function of latitude. As one travels from pole to equator, the axis gradually sinks toward the horizon. In contrast to the situation when the sun is used for determining latitude, however, there are no complications of location or time. At each latitude the axis of rotation remains the same height above the horizon all night (and day) and at all seasons. At the equator, northern and southern axes just 'sit' on their respective horizons.

There is another way that stars can be used to determine latitude. This is the zenith star system, which may have been used by Polynesians and which David Lewis used so successfully when navigating his catamaran, without instruments, the 2000 km or so from Tahiti to New Zealand (Lewis 1972). As the celestial sphere rotates, a given star, as seen from a single location on Earth, always follows the same path across the sky. It always rises at the same point on the eastern horizon, reaches at its highest point the same height above the horizon, and always sets at the same point on the western horizon. The only change is that these three positions on its path are reached about four minutes earlier each day. At any one place, therefore, such as at 'home', the same few stars pass immediately overhead each night. Equally, each of these stars passes overhead at all points on the Earth's surface that share with 'home' the same latitude. Suppose that a bird could recognize and remember which stars pass overhead at the home site. If, at an unfamiliar place, these stars do not pass overhead, the bird has only to fly in the direction of each star at its highest point to arrive eventually back at the home latitude.

The movements of the sun and stars across the sky provide information about latitude because they are both caused by the rotation of the Earth about its geographic poles. There is a further effect of the Earth's rotation that birds have also been suggested to use to measure latitude: production of the so-called Coriolis force.

This effect can be illustrated if we imagine a particle situated on the edge of a disk of radius, r (Wilkinson 1949). If the disk rotates about a vertical axis with an angular velocity w, the sideways velocity of the particle will be rw. If, on the other hand, the particle moves towards the axis, say half-way along the radius, its sideways velocity decreases to $rw/2$. This reduction in

Fig. 4.9 Star maps of the northern and southern skies showing major constellations

Dotted outline indicates the Milky Way. In the northern sky, the axis of rotation is marked by **Polaris** which may be found from the pointer stars in **Ursa Major**. In the southern sky, the axis of rotation (X) has no such marker star, but is about mid-way between the Southern Cross and the bright star, **Achernar**.   [From Baker (1981a)]

How could birds navigate? 47

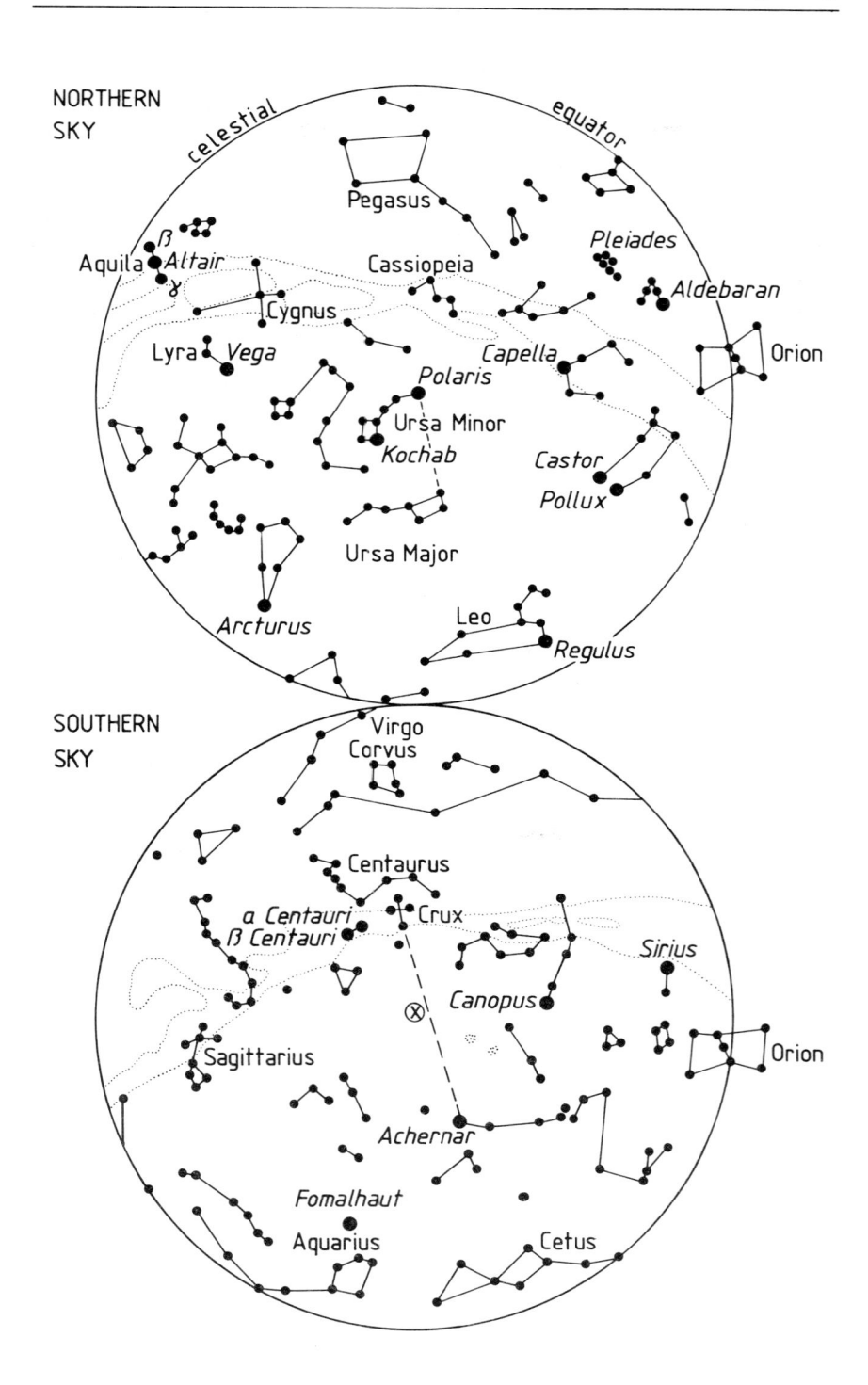

velocity is in effect produced by a sideways force acting in the direction opposite to the rotation of the disk. This sideways force is Coriolis force and its magnitude is a function not only of angular velocity but also of the mass of the particle and the speed with which it moves towards the axis.

If we now imagine the Earth to be made up of a series of concentric disks, decreasing in radius towards each pole, we can see that a body, such as a bird, as it moves from equator to pole is in effect moving towards the axis of rotation. The speed with which it does so depends in part, of course, on its own speed through the air, but in part also on its position on the Earth's surface; in other words, on its latitude. The Coriolis force that it experiences, therefore, is also a function of latitude. It follows that, if the bird could measure the Coriolis force acting upon it as it travels, it could in theory detect its latitude.

The Earth's magnetic field could also give information about 'latitude', except that in this case the poles to which it relates are the magnetic poles. These are not coincident with the geographic poles, which mark the axis of rotation of the Earth, and indeed are separated from them by over 2000 km. Lines of magnetic 'latitude' are also, therefore, not coincident with lines of geographic latitude.

The intensity, or strength, of the magnetic field is greatest at the magnetic poles and weakest at the equator. In addition, near the equator the lines of magnetic force are horizontal, running parallel to the Earth's surface with an angle of dip or inclination (I) of $0°$. At the magnetic poles, lines of force are vertical (I = $90°$). At intermediate latitudes, therefore, the angle of dip has a value somewhere between 0 and $90°$.

Given this global pattern, either total field intensity or angle of dip could perhaps be used to measure 'magnetic latitude'. However, there are difficulties. For example, over some areas of the Earth's surface there are large but gentle variations in field intensity totally obscuring the global patterns (Lednor 1982). Such variations, probably caused either by weak, large-scale differences in the permeability of the ground to magnetic fields, or by strong but very deep magnetic anomalies beneath the Earth's surface, would prevent any simple use of the intensity of the field to measure 'latitude'.

Angle of dip is perhaps a little more reliable, but even it is subject to distortion in some anomalous areas. Moreover, in order to measure dip, some other angle also has to be measured for reference. True vertical, measured presumably from gravity, seems the most likely candidate. It has been estimated, however, that to use angle of dip to determine 'latitude' a bird may need to measure true vertical to within about 0.03 degrees (Gould 1982).

4.4.2. Finding 'longitude'

There is no shortage of cues by which birds, or other animals, could work out whether they are north or south of the latitude of home. It is much more difficult to think of realistic cues that could be used to work out whether an unfamiliar location is east or west of the longitude of home.

As it happens, such a second coordinate is probably not essential during natural exploration. Normally, enough information can be collected during the outward journey to know roughly the longitudinal position (i.e. east or west) of the unfamiliar site relative to home. The technique is then deliberately to err during the return journey so as to be certain of arrival at the home latitude on, say, the east side of home. Final return is then achieved by travelling west along the line of home latitude. This trick has been used by human navigators for centuries (Gatty 1958; Lewis 1972). The same technique works if home can be recognized by smell. The trick is then to bias the return journey so as to be certain of arriving on the downwind side of home.

Although birds engaged in natural exploration may have no need for longitudinal coordinates, they would need such a system to solve the navigational problems encountered in homing experiments if (1) route-based navigation had been made impossible; and (2) the release site was beyond the limits of their familiar area map.

Human navigators have been able to measure accurately the longitude of their position only for the past century or so. The breakthrough came with the development of an accurate chronometer that could be set to home time for the duration of the voyage. Given both a knowledge of time of day at home and an ability to measure local time, it is possible to estimate longitude relative to home. This is because longitude and time of day are inseparable. Owing to rotation of the Earth, the further east one travels, the earlier the sun rises (relative to current time at home), the earlier it reaches its highest position, and the earlier it sets. Comparison of local time at an unfamiliar site with time of day at home would be one way, therefore, of working out their relative longitudes.

Matthews (1951, 1953, 1955b) suggested that birds might be able to use the sun's arc to determine longitude as well as latitude (Fig. 4.10). The proposed mechanism was for the bird to observe the sun's movement over a short period, say a minute, and then to extrapolate the arc to the noon position. In theory, the bird could then judge how far along its arc the sun had travelled relative to how far it would have travelled at home. If the sun had travelled too far along its arc, home lay to the west; if not far enough, to the east. This information, combined with north–south information from the height of the sun above the horizon, would give the bird the correct bearing of home.

Similar information could be obtained from the paths of individual stars. If, at an unfamiliar site, a particular star is observed to reach the highest point of its path across the sky earlier than at home, then home lies to the west; if later, home lies to the east. In one way, the use of individual stars to determine longitude would be easier than using the sun, in another way more difficult. Unlike that of the sun, the arc of an individual star does not

50   Bird navigation: the solution of a mystery?

rise and fall with the seasons. On the other hand, whereas the sun always reaches its highest point at local noon, individual stars reach their highest point about four minutes earlier each day. Moreover, stars well away from the axis of rotation are for part of each year above the horizon only during the day, when they cannot be seen.

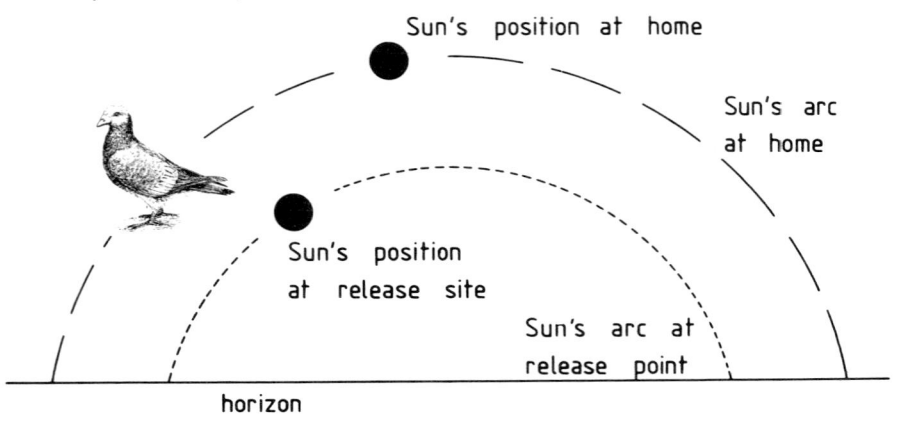

Fig. 4.10   Major elements of Matthews' sun-arc hypothesis

A displaced bird observes the sun's position at the release site and compares this with the position the sun would be in if the bird were at home. If the sun is lower than it should be, the release site is further from the equator than is home. If the sun has travelled less far along its daily arc than at home (i.e. local time is lagging behind home time), the release site is further west than is home. In the diagram, the release point, if in the northern hemisphere, is northwest of home.   [From Baker (1981a)]

Even the magnetic field shows variations with time of day that could be used to judge longitude. The total field strength of the Earth at about 50 degrees latitude is about 50 000 nT. This intensity, however, varies with time of day since it derives from several sources. By far the strongest field is generated by the Earth's core. A weaker field arises through induction caused by the movement of ions around the Earth in jet streams. The daily heating and cooling of the atmosphere displaces these jet streams north and south (Gould 1982). This gives a more or less regular variation in field strength of 30–50 nT as observed on the ground (Fig. 4.11) and is phased to local time. If a bird could detect how far along this curve the daily variation had reached relative to the stage it should have reached at home, it could again identify its position as east or west of home.

All of these methods allow a bird to detect an east or west displacement by measuring in effect, differences in time between the unfamiliar site and home. To do this, a bird, like a human mariner, would need a clock immutably set to home time against which it could compare the time of local events at an unfamiliar site. Such a clock, once set to home time, would need to remain accurate and unchanged no matter where the bird travelled and no matter what light/dark regimes it might encounter. This fixed clock would have to be separate from those other clocks that, as we have already seen, can be phase-shifted simply by placing the bird in an artificial light/dark regime.

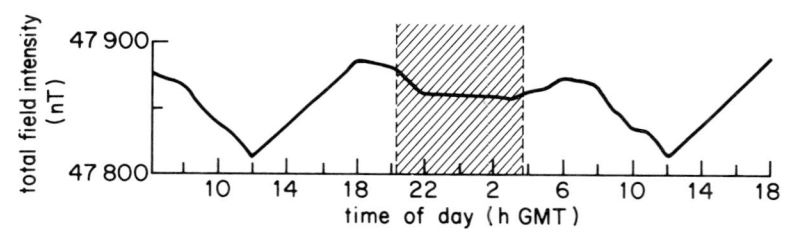

Fig. 4.11 Variation in intensity of the geomagnetic field with time of day

Drawn from data provided by Hartland magnetic observatory, southwest England, on a magnetically 'quiet' day in July, 1981. Stippling shows hours of darkness.

A 'map' based on latitude and longitude would be the best of all forms of bicoordinate grids because the gradients of the two axes cross each other at ninety degrees (i.e. north–south and east–west). This means that all sites have a unique pair of coordinates. However, other grid maps have been suggested. As the geographic and magnetic poles are separated, by about 2300 km in the north, geographic and magnetic lines of latitude are, as we have seen, not coincident (Fig. 4.12). The lines of these two types of latitude intersect, but in places at a fairly acute angle. In the northeastern United States, for example, the angle is about 30 degrees. Moreover, not all

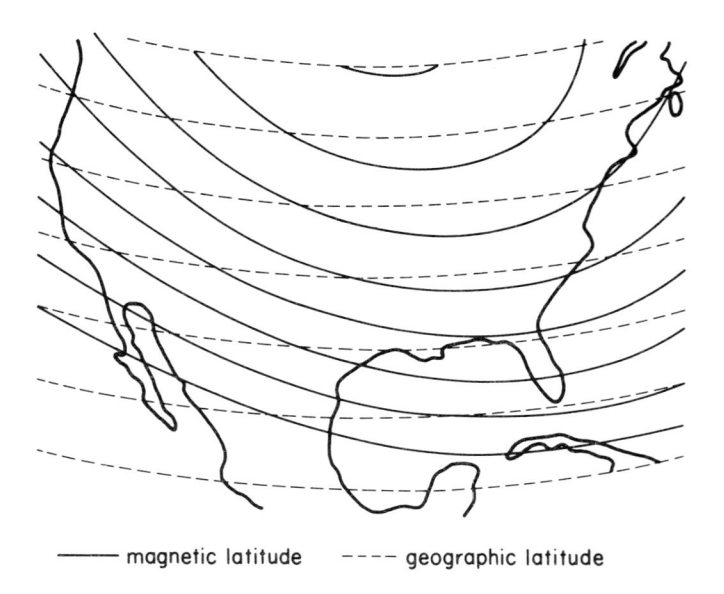

Fig. 4.12 Grid formed by magnetic and geographical 'latitude' over North America

locations are unique, different locations having the same pair of coordinates. Nevertheless, in theory the two axes could provide a usable grid map for navigation.

Another effect of the separation of geographic and magnetic poles is that in most places there is a difference between true north (as taken, say, from the sun at local noon) and magnetic north. The angular difference between true and magnetic poles is called the declination of the magnetic field. Declination shows a complex pattern over the surface of the Earth (Fig. 4.13). Although the grid it forms with either geographic or magnetic latitude is not neat, again in theory such a grid could serve to identify location. Indeed, both of the last two grids described have the great advantage that they are independent of time of day, unlike those involving true longitude.

4.5. Reading gradients

In the past, the temptation has been to suppose that birds are born with a pre-programmed ability to read gridmaps, though with the potential to refine this ability with training and experience (Matthews 1955b). Indeed, one theory of bird migration (Rabol 1970, 1978) is based on the supposition that seasonal migrants are born with the coordinates of their migratory track coded into their genetic material. More recently, with the travels of young birds being seen less as under the control of inflexible, pre-programmed instructions and more as explorations (Baker 1978, 1982; Wiltschko and Wiltschko 1978), emphasis has shifted toward the possibility that birds learn environmental gradients around their home during their early travels.

Wallraff (1974) and Wiltschko and Wiltschko (1978) have suggested that young birds, as they explore in different directions around their original home, notice that different features of their environment gradually change in an orderly way. Examples would be that in one direction, magnetic field intensity increases in strength while the sun's arc becomes lower in the sky whereas, in the opposite direction, the trends are reversed. Having noticed these gradients, the birds can extrapolate the trends beyond the area in which they have travelled. If a bird is then displaced to an unfamiliar site and observes that, say, field intensity is lower than at home, it flies in the compass direction that it has learned previously leads to an increase in field intensity. Viewed in this way, the distinction between a familiar area map and grid map becomes less clear cut. It becomes even less clearcut if we then extend the idea of gradients to include gradual changes in the landscape (Baker 1978).

Suppose a bird living near a coast observes that all rivers run to the sea. On navigating from an unfamiliar site further inland, the bird is able to orient towards the coast simply by observing the direction of flow of the rivers beneath it. The technique is not foolproof but has been used by many

How could birds navigate? 53

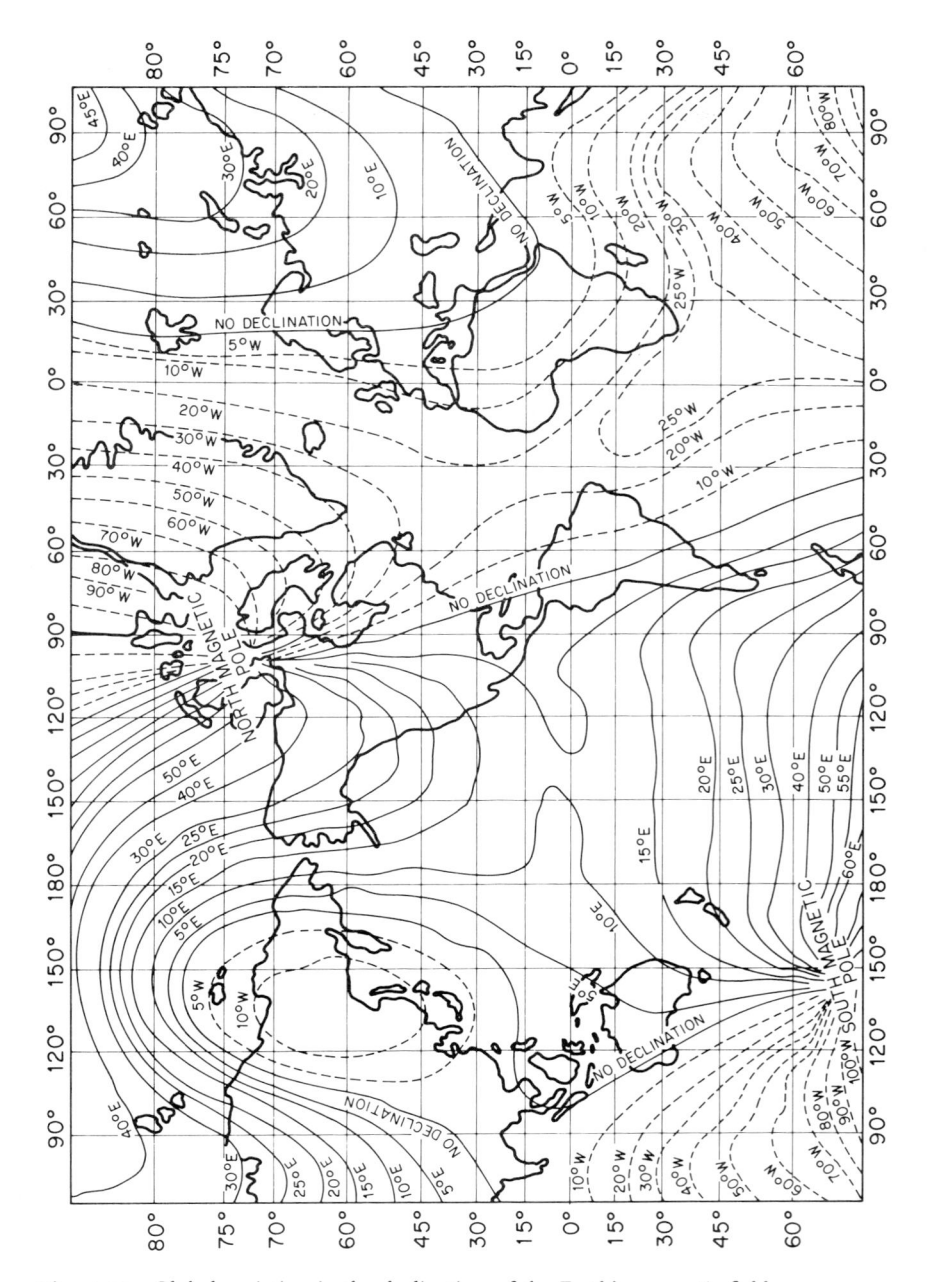

Fig. 4.13 Global variation in the declination of the Earth's magnetic field

human explorers (Gatty 1958). Similarly a bird living near a major river may observe that all smaller streams eventually join that river. As a final example, a bird living on a large flat plain may observe that the land slopes upwards to the southeast. On navigating from high ground, previously beyond the limits of its familiar area map, the bird may be inclined to fly northwest. Such landscape gradients can often provide reliable navigational information far beyond the area in which the bird has ever travelled.

In the chapters that follow, we encounter a potentially bewildering array of experimental data. At first sight these data appear a mess of often contradictory information. The aim of this chapter was to provide a framework within which each piece of evidence might find a place.

At this point, the framework itself might seem muddled and complex. However, if we brush aside the variety of compasses, landmarks, and grid coordinates that a bird might sense, we are left with a simple and limited range of navigational possibilities. All methods resolve into the combined use of route-based and location-based navigation. These, in turn, resolve eventually into an ability to recognize and use landmarks, compass directions and geographical coordinates. A good place to start, therefore, is with evidence for the range of landmarks birds can detect. Chapter 5 examines the avian landscape.

5 The avian landscape

The human landscape is dominated by topographical features that we detect by sight. Hills, valleys, mountain ranges, lakes, coasts, oceans, forests, isolated trees, cities, individual houses, roads, railways, rivers, coloured soil, rock formations, etc. all find a place on the mental map of our familiar surroundings. In only a minor way is this visual tapestry embroidered further with sounds and smells. Nevertheless, the noise of aeroplanes taking off or landing at a nearby airfield or of foghorns from ships near the coast, the smell of gasworks, brewery, or fish dock when the wind is blowing from a particular direction, may all find some small niche in our predominantly visual landscape.

Although familiar with such a visually-biased landscape, we do not expect all animals to share our view of the world, even on land. It seems likely that dogs, cats, mice, deer and probably most frogs, toads and salamanders, for example, have a landscape dominated by familiar smells, with visual landmarks much less prominent. Equally, most bats probably have a landscape dominated by sounds, albeit produced through echolocation, the patterned reflection of noises emitted by the bats themselves.

Increasingly, however, there is evidence that even for bats visual landmarks are important at long distances (reviewed by Baker 1978).

Until recently, most people expected the landscape of the majority of birds to be very similar to that of humans and to be dominated by visual landmarks. This could still be true, but it now seems that smells, and even sounds, may well have a more prominent position on the avian than the human landscape. Even so, it is convenient to begin with the surprisingly scanty evidence that birds detect and make use of visual landmarks.

5.1. Visual landmarks

At a pigeon loft in wooded country, young, untrained racing pigeons had been loath to fly any distance from the loft (Whitney 1963). At the same time, trained birds suffered heavy losses during races, apparently because they failed to break away at the right place from the stream of racing birds flying along the coast 10 km away. Then, a galvanized steel tower, about 30 m high and with a large golden ball on top, was built near the loft. Thereafter, young birds explored further and racing losses fell sharply.

The implied recognition by pigeons of familiar landmarks around their loft can actually be seen when individual birds are tracked from their release site (Michener and Walcott 1967). For example, birds homing to Cambridge, Massachusetts, seem to turn toward the loft as soon as the tall buildings of Cambridge come into view (Fig. 5.1). Similarly, it has been observed that pigeons are reluctant to home at night (Lipp and Frei 1982) but can be trained to do so more readily if conspicuous lights are erected near the loft.

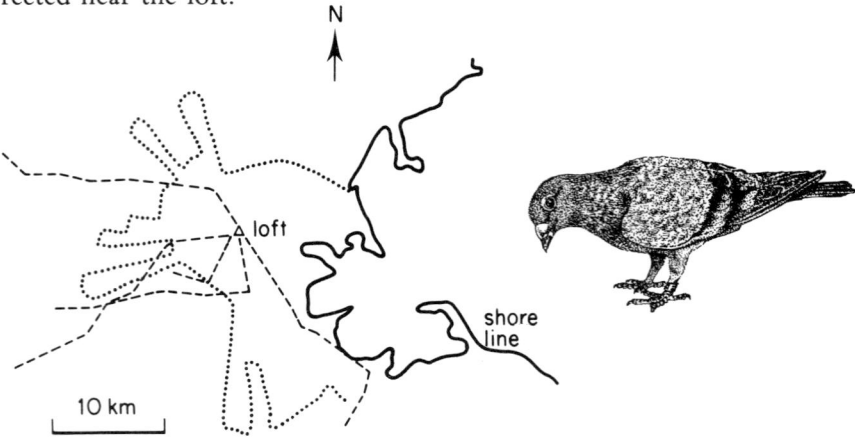

Fig. 5.1 The final approach of a single bird to its loft at Cambridge, Massachusetts, on 5 separate homing flights

Outside of the area bounded by the heavy line an observer at treetop level cannot see the top of the tallest building in the city. The final turn directly toward the loft occurs within 3 km after crossing this line.

[Simplified from Michener and Walcott (1967)]

Bank swallows (*Riparia riparia*) restrained in an orientation cage can orient toward their nesting site from distances of up to 16 km. Orientation ceases, however, when the cage has opaque sides (Sargent 1962).

It seems fairly clear, therefore, that birds are familiar with visual landmarks in the immediate vicinity of home. It is worth reminding ourselves, however, of the use they make of this familiarity. To judge from the clock-shift experiments carried out by Graue (1963) in which the birds were released only 1 km or so from the loft, the landmarks are used by the birds to work out the compass direction of home from that point. They do not seem to orient to the landmarks themselves, unless they can see the loft.

Wagner (1972, 1978) has shown that pigeons do notice, and are influenced by, the topography of the release site. Releases opposite the home loft on the far shore of Lake Constance produced vanishing bearings along the shore or inland, rather than across the water to home. Similarly, the pigeons flew along valleys upon release rather than crossing mountain ridges, and when birds were released above low-lying fog in mountainous regions, they flew toward distant mountain peaks projecting above the fog rather than downwards to penetrate the layer of reduced visibility.

A more direct study of the scanning of landmarks by pigeons at the release site was carried out by Köhler (1978). In an ingenious experiment, a simple camera was attached to the top of a bird's head (Fig. 5.2). The sun, shining through an aperture on top of the camera, formed a spot on the

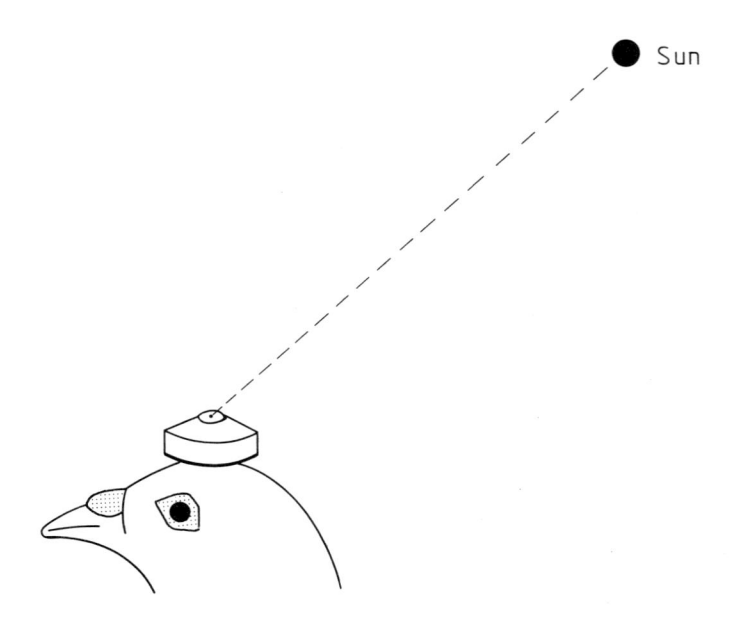

Fig. 5.2 Simple camera attached to a pigeon's head used by Köhler to record the direction in which a pigeon fixes its attention during initial orientation

[From Baker (1981a), after Köhler and Schmidt-Koenig]

horizontal film each time the bird held its head steady for a few seconds. With the position of the sun known to the experimenter, the direction of fixation whenever the head was steady could be calculated. Köhler interpreted his results as showing that, upon release, pigeons systematically scan the horizon and temporarily fixate on horizon features. Scanning is more intense when visibility is only 12 km than it is when visibility is 30 km.

We can probably accept, therefore, that following experimental displacement, pigeons take note of the visual landmarks that surround them when they are released. This is not the same as accepting that they use such landmarks to reach a decision about the compass direction of home. On this point, there is little by way of direct evidence for any bird.

Herring gulls (*Larus argentatus*), when displaced from breeding colonies on the shores of Lake Huron and released on the shore of another lake, spend their time flying along the shore-line until their urge to home wanes (Southern 1970), thus suggesting that they are confused by the similarity in topography. Gulls released from other sites and followed by radio-tracking on their return flights were found to follow erratic courses and to be influenced by large topographical features. Williams et al. (1974) released herring gulls in fog and over the sea within 50 km of their home area. The birds seemed to adopt a search pattern of more or less straight flights, 3 – 10 km long, until they reached the edge of the fog bank or encountered some familiar landmark. In good visibility, the birds then appeared to choose not the shortest route home but one requiring the least effort, taking into account wind direction and topography.

Starlings (*Sturnus vulgaris*), taught to discriminate between locations 200 km apart, could only continue to do so as long as both the sun and distant landmarks were visible (Cavè et al. 1974).

Finally, Dornfeldt (1982) carried out an analysis of the vanishing bearings of over 2000 individual releases of homing pigeons on 153 journeys, mainly from the loft at Göttingen, Germany. He found that the deviation of pigeons from the true home direction could be attributed to topographical features, not only such factors as the slope of the land, but also, and in particular, the direction of structures, such as power lines and railway stations, that were reminiscent of similar structures near the home loft.

Just because birds seem to observe and interpret the visual landmarks around their home and at more distant sites, it does not follow automatically that they also notice the visual landscape while travelling between such locations.

The evidence that birds do observe the topography beneath them as they fly comes almost entirely from studies of individuals on seasonal migration. Most often, such data come from radar observations, but some are obtained by watching birds flying through the beam of a vertical shaft of light or flying across the face of the Moon.

Day-flying migrants are so often and so obviously seen to follow coastlines or ranges of hills and to fly along valleys or the edges of forests and deserts that there has never been any doubt they are aware of the landscape beneath them. For nocturnal migrants, however, the situation has been different. At night, radar screens show migration over a broad front so consistently that it has been very tempting to think that the birds concerned are relying exclusively on compass orientation. In consequence, various authors have concluded that at night birds ignore even the most obvious landscape features beneath them (Emlen 1975; Wallraff 1977).

It is indeed commonplace for night-migrants to pass over coastlines, islands or other major landmarks without changing course. However, this does not mean that the landscape passes unnoticed. Observations in which night migrants turn or land apparently in response to the shorelines of large bodies of water (Lowery and Newman 1966; Richardson 1978; Gauthreaux in Able 1980) rather suggest that where no response to visual landmarks is observed it reflects simply a lack of response rather than a lack of awareness.

Strong support for this view is provided in two recent papers (Bingman et al. 1982; Bruderer 1982) that make out a powerful case for accepting that even night migrants are aware of the topography over which they are passing. Bruderer, for example, found that along the northern edge of the Alps, changes in direction in relation to topographical features were more frequent than had previously been reported. Bruderer also saw indications

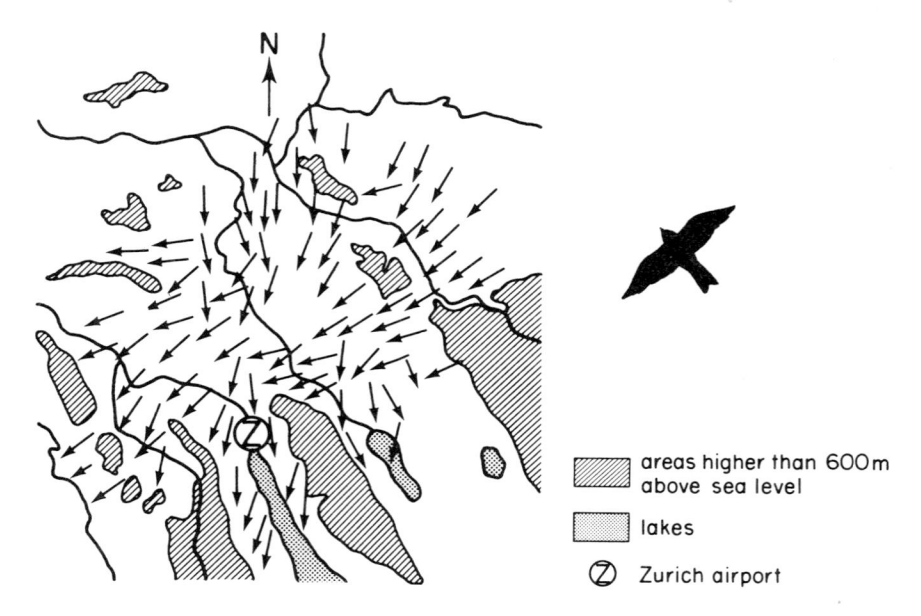

Fig. 5.3 Effect of topography on the migration direction of nocturnal migrants as observed by radar [Compiled and simplified from Bruderer (1982)]

in his results that, with favourable winds and good visibility, it was normal for birds to orient with respect to topographical features ahead, perhaps by lining up middle and far distance landmarks.

Bruderer argues that when consistent changes in direction are shown by migrating birds over the course of just a few hundred kilometres, such changes are likely to be responses to topography. Thus, along the northern border of the Alps, west-southwest migration directions predominate instead of the southwest or south-southwest directions expected of birds passing through southern Europe. Towards the west of the Alpine chain the extent of the deflection decreases whereas towards the east, south-bound and south-southeast-bound migrations are more common.

Reasons for a bird to scan the visual landscape as it migrates are discussed further in Chapter 12, but two reasons are relevant here as we search for evidence that such scanning occurs. Firstly, it would allow a bird to maintain its migration direction, perhaps by point-to-point use of a succession of landmarks (Bergman 1964), if the bird is forced to fly under a layer of cloud thick enough to prevent it from seeing the sky. Secondly, it allows the bird to monitor and perhaps correct and compensate for any tendency to be drifted off-course by winds from the side.

It is now clear that when nocturnal migrants fly under a layer of cloud thick enough to obscure the Moon and stars, they are still able to maintain their migration direction over land quite as well as they can under clear skies (Able 1982a). Visual observations in mist of low-flying birds over the sea suggest that they, also, can maintain a heading, perhaps by reference to wave patterns (King 1959; Drury 1959).

Most authors agree that the best way for a bird to judge the extent to which the wind is tending to drift it off-course is to maintain visual contact with landmarks (Griffin 1969; Emlen 1975; Able 1980). Radar observations of wood pigeons (*Columba palumbus*) and cranes (*Grus grus*) migrating over Sweden and the Baltic Sea during the day suggested that the birds compensated totally for wind drift while flying over land. Over the sea they drifted slightly, as if they were using the moving waves below them as stationary landmarks (Alerstam and Pettersson 1976). Radar studies of nocturnal migrants by both Bingman *et al.* (1982) and Bruderer (1982) found that an ability to avoid being drifted off-course when the wind blew from the side coincided with flight over linear structures (e.g. rivers) below. The study by Bingman and his colleagues in particular highlights the way that under some conditions migrating birds seem to ignore topography while under other conditions they react to it.

We can assume from the above data that travelling birds are always aware of the visual landscape passing beneath them, but that by no means do they always show an obvious reaction to each major feature they encounter. At the same time, there is no shortage of evidence that visual landmarks are not essential for birds to travel or even to return to the vicinity of home. Migrating birds, for example, can maintain their direction even when flying above fog or low cloud that obscures all view of the ground (Griffin 1973). Moreover, pigeons wearing contact lenses made of frosted glass that, at the very least, impair the ability to recognize or even detect topographical landmarks, can set off towards home and maintain their direction as well as birds wearing clear lenses (Schmidt-Koenig and Schlichte 1972).

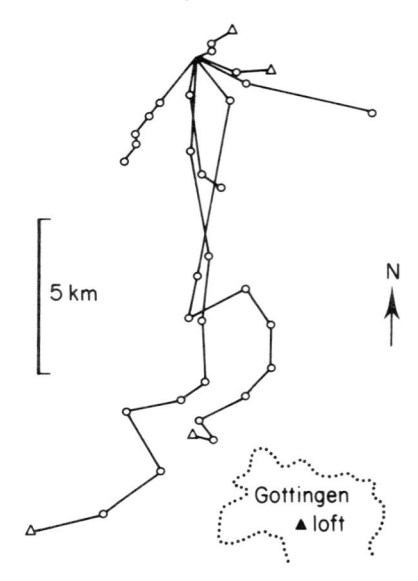

Fig. 5.4 Tracks of pigeons homing while wearing frosted-glass contact lenses

Each dot shows one fix of a pigeon's position. All releases under sunny skies.

[Simplified from Schmidt-Koenig and Walcott (1978)]

This last experiment illustrates an important principle which we encounter time and again in research into bird navigation. Evidence that birds can navigate in the absence of a particular type of information is not evidence that normally they do not use it. It is clear that birds do detect, recognize and react to visual landmarks. It is clear, also, that they do not monitor such landmarks continuously and, particularly when travelling between sites, often prefer to maintain direction using celestial or other cues when these are available. Equally clearly, from the frosted lens experiment, a detailed view of landmarks is not essential for a bird to set off in the home direction or to maintain that direction, even to within a few kilometres of the loft. The last two conclusions, however, do not contradict the first, that normally birds do use visual landmarks.

The point is so important, so self-evident, yet so easily forgotten, that it bears repeating in yet another way. Once it has been demonstrated that birds can detect and react to a particular environmental cue in solving navigational problems, then use of that information has to be accepted as part of the navigational armoury. Subsequent experiments which show that a bird can still navigate even if denied that particular information only serve to identify the range of situations in which the bird can manage without it. Such experiments do not show that under normal circumstances the information is not used. A human, for example, normally finds his way out of a familiar room through the use of visual landmarks. He avoids chairs and desks and takes the most economical route to the door. Switch off all the lights and he still finds his way out, albeit more slowly and with more bruises, by a combination of kinaesthesis (i.e. judging steps, angles, distances), touch, and perhaps even, to judge from the studies of blind people, crude forms of echolocation. Such an experiment does not prove that these methods are used when visual cues are present, nor does it prove that visual cues are not used when they are present.

Normally, birds detect and react to visual landmarks at all stages during migration and homing. When such landmarks are unclear or absent, as when wearing frosted lenses, pigeons can still return to within a few kilometres of home. They do so more slowly, and with more bruises through crash-landing on trees. They also suffer more casualties through predation by hawks, but many do return. The secret of how they manage to adopt and maintain home direction seems to be that they fly at the correct angle relative to the sun, which can be seen through the frosted glass. Clock-shifted birds set off in the wrong direction in accordance with their faulty sun compass (Fig. 5.5). This still leaves unanswered the question of how they work out the compass direction of home in the first place. Of all the possibilities, route-based navigation seems the most likely, but the critical experiments have not yet been carried out.

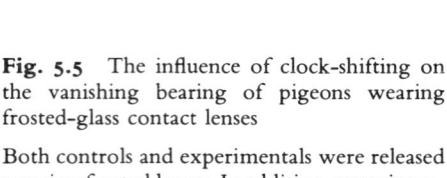

Fig. 5.5 The influence of clock-shifting on the vanishing bearing of pigeons wearing frosted-glass contact lenses

Both controls and experimentals were released wearing frosted lenses. In addition, experimentals were exposed to an artificial day−night cycle 6 hours in advance of the natural cycle. The mean bearing for experimentals is significantly different from that for controls.

[Simplified from Schmidt-Koenig and Keeton (1977)]

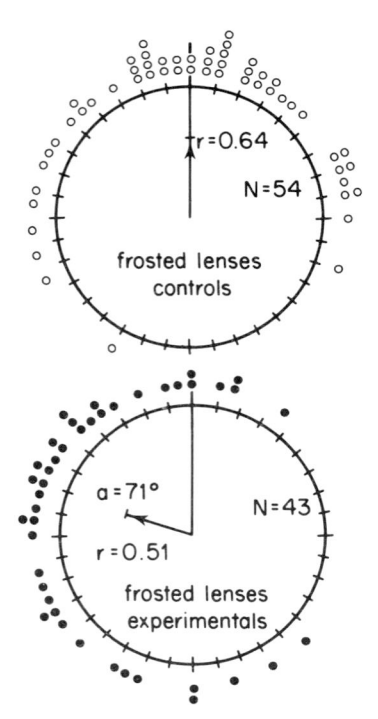

Lacking a detailed view of landmarks, few birds with frosted lenses get nearer to the loft than one or two kilometres. Some, however, actually return to the loft itself. Whether these are just lucky or can see just enough detail to pick out the loft is unknown. Perhaps they find the loft by smell!

The principle highlighted by the frosted lens experiment has to be remembered at almost every step in the remainder of this book, not least when we consider experiments to determine whether there is any place on the avian landscape for familiar smells.

5.2. Smells

In the mid-1970s, I attended a conference organized by a society that meets every year to discuss bird migration. The star speaker on that occasion was an eminent zoologist, who was there to enlighten the audience on current advances in the study of bird navigation. Towards the end of his lecture, with more than a hint of humour in his voice, the speaker drew his audience's attention to the possibility that, in Italy, pigeons might use smells within their environment to find their way home. He stressed, however, that in other parts of the world, this did not seem to be the case. The audience laughed. Since then, the determined efforts of Papi and his colleagues in Italy have produced enough evidence to convince all but the most hardened sceptics.

The most elegant and convincing experiment so far involved two groups of pigeons, each raised in aviaries walled with plastic and bamboo so that air could enter diffusely. The birds also had access on top of the aviary to two glass-walled corridors through which air was blown. One group had the smell of olive oil added when the air was blown from the south and of synthetic turpentine when blown from the north (Papi *et al.* 1974). The other group received the same treatment in reverse. At the release point, east of the loft, a drop of olive oil was applied to the bill of half the experimental birds and turpentine to the other half. Birds with olive oil on their bills flew to the south if they had received olive oil at the loft on north winds and to the north if they had received it on south winds. The same effect was found with the turpentine birds. Homing speed was unaffected, however, suggesting the birds soon realized their mistake, presumably through detecting other cues.

Except that the experiment has not yet been replicated by other workers, the results give convincing support to Papi's theory (Papi *et al.* 1972) that pigeons notice the smells arriving at the home site and associate each smell with the compass direction of the winds on which the smell arrives. Put more elegantly, pigeons at the loft learn the chemical signature of winds from different compass directions. The pigeon landscape has olfactory landmarks.

Almost as convincing is an experiment in which pigeons were raised in lofts rigged with fans to reverse the natural direction of air flow (Ioalé *et al.*

1978). Vanishing bearings of these birds were either roughly reversed or highly scattered (Fig. 5.6). This experiment is less convincing than the previous one because in this case, we have no proof that the critical influence of the fans was on the direction of arrival of smells. Unless experimental design is such that only the direction from which particular smells arrive is altered, it is difficult to be sure that the experiment is testing what it is meant to test. This principle is clearly highlighted in another series of experiments designed to test for the use of olfactory landmarks, the deflector loft experiments.

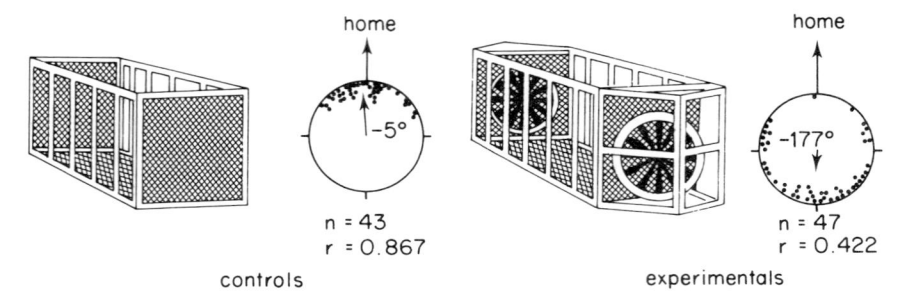

Fig. 5.6 The effect of reversing the direction of air-flow through the loft on the vanishing bearings of homing pigeons

[Re-drawn from Papi (1982), after Ioalé]

A deflector loft is an aviary fitted with baffles made of wood and glass designed to deflect the direction of winds, along with their associated smells, as detected by the birds inside (Baldaccini et al. 1975). The baffles are arranged such that some lofts deflect the winds in a clockwise direction and some counterclockwise.

Deflector lofts have given consistent results in the hands of a wide variety of workers: in Italy (Baldaccini et al. 1975); Germany (Kiepenheuer 1978a, 1979, 1982b); and the United States (Waldvogel et al. 1978). When pigeons raised in such lofts are released, their vanishing bearings are deflected in the same sense as the winds at the loft (Fig. 5.7). It is as if the baffles really have changed the direction from which a particular smell seems to arrive at the loft. For example, when a west wind blows, bringing with it a characteristic distant smell, the birds in a 'clockwise' loft (Fig. 5.7) should perceive the wind as coming from the north. If the bird is then released near to that smell, it flies south instead of east.

All replications of the deflector loft experiment obtained results that fitted beautifully this interpretation, until Kiepenheuer (1979) tried releasing birds that could not smell. Even with their nostrils plugged or anaesthetized, the birds still showed the usual deflection. Yet if these pigeons were unable to recognize their release site from its smells, they should have been in no position to use the false information they had

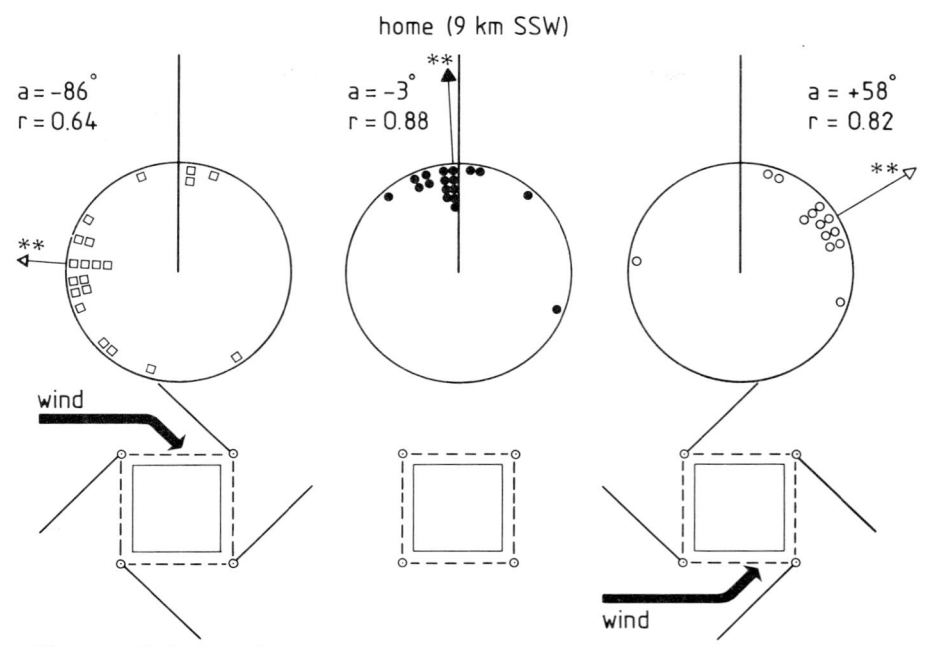

Fig. 5.7 Deflection of wind direction at the home loft influences the initial orientation of pigeons when they are displaced and released

[From Baker (1981a), after Baldaccini, Benvenuti, Fiaschi and Papi]

acquired while in the deflector loft. Suddenly, it seemed possible that the detection of smells may have nothing to do with the deflector loft effect.

Since 1979, various authors have experimented with lofts that deflect the wind in one direction and other cues in another direction. Waldvogel and Phillips have been impressed by the light that is reflected by the glass louvres of many designs of deflector loft and by the pattern of reflections, particularly of polarization patterns, that are produced. In the conventional deflector cage, these light cues are deflected in the same sense as the wind. In an ingenious feat of design, the two Americans built a loft that deflected the wind in one direction and light cues in another. They found that birds that were permanent residents in the deflector lofts responded to the deflected wind rather than to the reflected light (Waldvogel and Phillips 1982). In another series, however, they took birds from one loft and housed them in one of their modified deflector lofts for a few days. When released, these birds homed to their original loft, not their new one. Nevertheless, upon release they still showed the deflector loft effect. Moreover, in the modified loft they were deflected in accordance with the light cues, not the wind (Phillips and Waldvogel 1982).

Kiepenheuer (1982b) is unimpressed by the suggestion that reflected light may be the basis of the deflector loft effect. He points out that even with non-reflecting material for the louvres, the effect is still shown. He looked instead to some other design of loft, one that would provide no information on wind direction yet allow other factors to be deflected as in conventional deflector lofts. By slightly altering the design of the loft walls, he managed to produce what he terms a 'whirlwind' loft, in which the wind swirls round within the loft and gives little information concerning the direction from which it originated. Nevertheless, other cues, such as sounds, are deflected as in the normal deflector. Kiepenheuer found that first-flight birds showed the same deflection as in a conventional deflector loft but experienced birds did not. Even more puzzling, Kiepenheuer noticed that the deflection was shown even if the birds were released at only very short distances (50–400 m) from the loft.

The basis of the deflector loft effect still requires explanation. The hope is that it will tell us something important about the way a pigeon builds up its map of smells while at its home. If Phillips and Waldvogel are right, that the effect is due to polarized light reflection, or if Kiepenheuer is right, that it is due to the deflection of sounds, or even if the original olfactory interpretation turns out to be right after all, then this hope will have been realized. Kiepenheuer's observation that the effect shows immediately upon release even over short distances, however, suggests to me that the real explanation may yet prove to be fairly trivial (as far as navigation is concerned). It may even be something as simple as perhaps the way in which the bird learns to correct for wind drift upon first taking flight. We do not know how birds learn to do this, but if it involves a particular relationship between wind direction, as perceived on the body, wind direction, as perceived from the movement of clouds, and some additional reference system, such as a compass, then life upon release would be very confusing indeed for a bird housed in a deflector loft.

There is an important moral to the deflector loft saga. When designing an experiment to test for the effect of a particular factor, always manipulate that factor directly, never indirectly. This rule was followed in the olive oil/turpentine experiment, but not in the fan or deflector loft experiments. Nor was it followed in experiments that attempted to prevent pigeons from being able to smell their surroundings upon release.

Plugging the nostrils or anaesthetizing with xylocain the critical tissues in the nose are two of the methods by which pigeons have temporarily been prevented from smelling their surroundings upon release. Birds treated with xylocain are unable in the laboratory to respond to strong artificial odours until about 90 minutes after treatment (Schmidt-Koenig and Phillips 1978).

The third method used to prevent birds from being able to smell is more permanent and involves surgical bisection of the olfactory nerve. Ethical considerations apart, such surgery rarely leads to convincing results. Indeed, it can never do so if all it produces is disorientation. Only when such experimental treatment is found to have no effect are the results obtained meaningful, as is shown by experiments on pigeon olfaction.

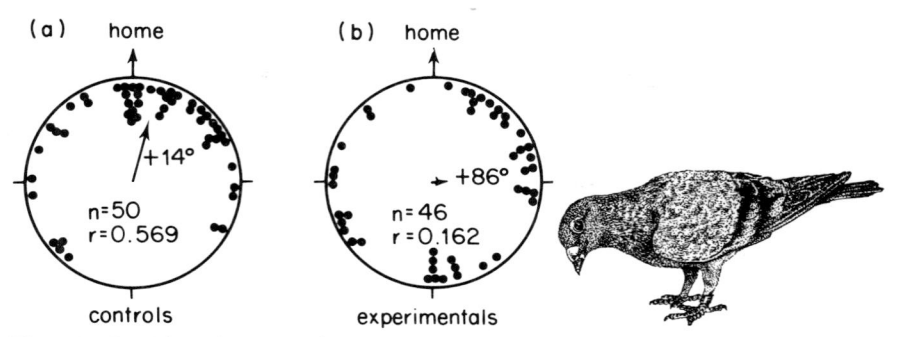

Fig. 5.8 Vanishing bearings of pigeons that are unable to smell are randomly oriented

In experiments on pigeons from Florence, Italy, controls are well oriented (a), whereas birds unable to smell disappear in random directions (b).

[Re-drawn from Papi (1982)]

Pigeons whose nostrils have been plugged (Papi et al. 1972) or whose olfactory nerves have been cut (Benvenuti et al. 1973; Papi et al. 1973) are poorly oriented at release and home slowly. When attempts were made to replicate the surgical experiments by collaboration between Papi's and Keeton's research groups, no effect on initial orientation was found though homing speed was slower (Papi et al. 1978b). Wallraff (1980a), however, successfully replicated the original results when he found that birds with cut olfactory nerves were less well oriented and slower to home than untreated birds.

In the face of such inconsistent results almost any interpretation is possible. Experiments on impairment of the sense of smell of pigeons have been reviewed by Gould (1982) and Papi (1982).

Gould stressed that in none of the experiments by the various groups was homeward orientation abolished in the treated birds. Illustrating this with results from the experiments by Wallraff (1980a) he points out that whereas the length (r) of the mean vector (see Fig. 3.4) for control birds was an impressive 0.8, it was still a respectable 0.4 for experimentals. Moreover, the mean vanishing bearing for birds with sectioned nerves differed from that of the untreated controls by only 38 degrees while different groups of control birds at the same site differed from each other by 41 degrees.

Papi, on the other hand, argued forcefully that in no experiment so far have birds incapable of smelling their surroundings been well oriented toward home. Experiments that found no difference between experimental birds and controls did so because of the poor performance of controls not because the experimental birds were well oriented.

Gould (1982) favours the view that the results of experiments on birds with an impaired sense of smell may not be due to an inability to detect olfactory landmarks but be due instead to a generalized and distracting trauma as a result of their treatment. In support of this possibility are two observations. First, experimental birds in these tests usually land and seem to spend extended periods grounded (Papi *et al.* 1978b). Secondly, when Keeton and his colleagues (Keeton *et al.* 1977), instead of plugging the nostrils, used hollow tubes so that, although the birds could still not smell, they could at least breathe, no disorientation was evident.

Papi (1982) has pointed out, however, that in these experiments with nasal tubes, most releases were from the south. As Keeton's pigeons have a PCD (preferred compass direction–Chapter 3) of north-northwest (Windsor 1975), the homeward orientation of birds that could not smell may have owed little to the recognition of home direction. Moreover, Papi stresses that when the results for this experiment are pooled, the birds with nose tubes did significantly worse than those without. Even so, it could then be argued that tubes, even hollow tubes, inserted in the nostrils may be traumatic enough to reduce performance.

There is yet another possible explanation for the poor performance of birds that have undergone surgery. Perhaps, when the olfactory nerve is sectioned, disorientation is due to interference, not with the sense of smell, but with the sense of magnetism (Baker and Mather 1982b). We still do not know the location of the magnetic sense organ for any animal (see Chapter 8) but for humans (Baker *et al.* 1982, 1983), mice (Mather *et al.* 1982), tuna fish (Walker and Dizon 1981) and even homing pigeons (Walcott and Walcott 1982) an intimate association between the magnetic sense organ and the nasal region is a possibility.

The evidence against both of these suggestions, trauma and interference with the magnetic sense, derives from experiments in which only one nostril is impaired through surgery (Wallraff 1980a). When the same nostril is also plugged, leaving the other unblocked and functional, homeward orientation is unaffected. When the other nostril is plugged, so that one nostril does not function because its nerve has been sectioned and the other does not function because it is blocked, the birds are disoriented. Although this may be some defence against the suggestion that the magnetic sense is impaired by surgery on the olfactory nerve, it is little defence against the suggestion that birds are traumatized by such treatment. It can still be argued that trauma is only severe enough to influence homeward orientation when both nostrils are incapacitated.

The fault with all of these experiments involving impairment of the sense of smell is that they are designed to generate disorientation. This is very weak, for it is almost always possible to think of reasons why a given procedure could cause disorientation through influences other than those being tested. It is always preferable to design an experiment to produce a rotation of orientation. Indeed, I should go further and suggest that only when an experiment produces a predictable rotation of orientation can the results be considered to be at all critical.

A few experiments have been carried out in which the sense of smell of pigeons should have been disrupted yet in which the birds seemed to be able to orient normally. Experiments with nasal tubes to bypass the olfactory chamber (Keeton *et al.* 1977) have already been noted. Schmidt-Koenig and Phillips (1978) and Kiepenheuer (1979) reported normal vanishing bearings in pigeons with nasal linings anaesthetized with xylocain. Keeton *et al.* (1977), Papi *et al.* (1978b) and Hartwick *et al.* (1978) found that strong masking odours applied to the beaks and nostrils of pigeons failed to have any predictable effect. This was in contrast to Benvenuti *et al.* (1977), who found that application of a strong smell to the nostrils to block detection of natural smells disoriented the pigeons at release. Finally, Schmidt-Koenig and Phillips (1978) found no ability of pigeons in the laboratory to distinguish natural air from pure, filtered air.

These experiments are sometimes cited (e.g. Gould 1982) to indicate problems in accepting that pigeons use olfactory landmarks in navigation. Conditioning experiments, however, of the type used by Schmidt-Koenig and Phillips and which are described in Section 5.3, are only useful if they produce positive results. Failure to show that an animal can be conditioned to a particular factor shows only that it cannot be conditioned to that factor, not that it cannot detect it. Similarly, as we have already discussed in relation to visual landmarks, experiments in which a pigeon can still navigate even when unable to smell tell us only that olfactory landmarks are not essential. This does not mean they are unimportant when available. In the face of evidence that such landmarks are used when available, evidence that they are not essential tells us little about the normal navigational mechanism. As we have seen, however, of all the experimental series that might seem to provide positive evidence that pigeons detect and use olfactory landmarks, only three have the strength of predicting a rotation of orientation. Even two of these, the fan and deflector loft experiments, had the fault that they deflected factors other than smells.

Of all the experiments that have been carried out to test for the presence of smells on the avian landscape, only one, the adding of olive oil and turpentine to the winds arriving at the home loft, has provided unequivocal positive evidence. Nevertheless, with the proviso that the experiment has not yet been repeated by other workers, this one experiment is enough. On the basis of that single set of results, we are forced to consider that the pigeon landscape has olfactory, as well as visual, landmarks. The same does not yet have to be said about acoustic landmarks.

5.3. Sounds

In addition to our subjective human experience that sounds can sometimes make useful landmarks, there is one piece of experimental evidence that other animals can also use such information. Chorus frogs (*Pseudacris triseriata*), caged for several days near a breeding chorus of their own species, then displaced and released in an arena, selected a compass direction that would have taken them to the chorus, and hence the breeding pond, from the original site but not from the site of testing (in Ferguson 1971).

Evidently the frogs had learned the compass direction of the sound.

There is no such comparable evidence for birds, though few people would be surprised if the avian landscape were also dotted with acoustic landmarks in a similar way. Most such sounds, however, do not extend the familiar landscape beyond the limits set by visual and olfactory landmarks. On the other hand, if infrasound landmarks could be detected, the situation would be altogether different.

The world is a very noisy place at infrasound wavelengths of 1 to 10 Hz. Not only do local winds and turbulence generate noise but also thunderstorms, magnetic storms, earthquakes, ocean waves, jetstreams and mountain ranges produce coherent sounds that can be identified thousands of kilometres from their source (Kreithen 1978). Griffin (1969) suggested that birds might use geographical and meteorological sources of infrasound for orientation, navigation and weather forecasting.

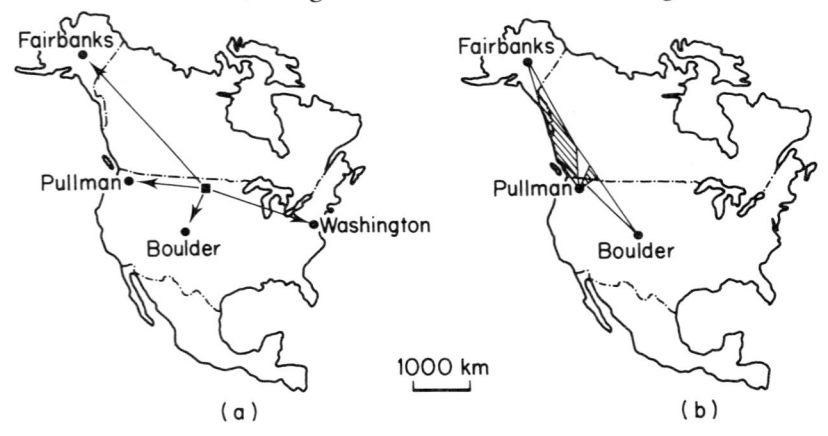

Fig. 5.9 Infrasounds as long-distance landmarks (a) A severe thunderstorm in central North America (solid square) can be tracked infrasonically by stations several thousand kilometres away.

(b) Shading shows areas of continuous infrasound production due to the interaction of winds and mountains in northwest America.   [Re-drawn from Kreithen (1978)]

Humans can detect infrasounds but only if the sounds are very loud (e.g. about 100 dB at 10 Hz) (Fig. 5.10). Our threshold for detection is above that of natural infrasound levels with the result that we can go through life more or less oblivious to the noise that surrounds us. Until the late 1970s, the same was thought also to be true for birds. Then, Yodlowski et al. (1977) found that homing pigeons were sensitive to sounds below 10 Hz and Kreithen and Quine (1979) found that this detection extends as low as 0.04 Hz. Between 1 Hz and 10 Hz, the pigeon's thresholds are 50 dB more sensitive than thresholds reported for human hearing (Yeowart and Evans 1974) and should allow pigeons to detect many of those natural infrasounds to which humans are oblivious.

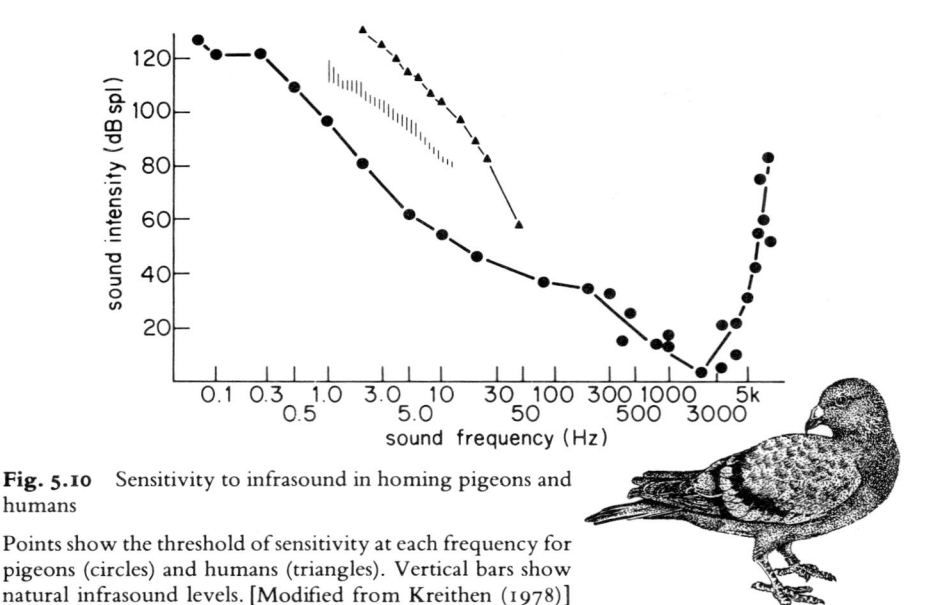

Fig. 5.10 Sensitivity to infrasound in homing pigeons and humans

Points show the threshold of sensitivity at each frequency for pigeons (circles) and humans (triangles). Vertical bars show natural infrasound levels. [Modified from Kreithen (1978)]

The technique used in these studies on pigeons was cardiac conditioning (Kreithen and Keeton 1974a). The pigeon was fitted with two electro-cardiogram (ECG) surface electrodes and two shock electrodes (implanted under the skin of the breast). The bird was then restrained by a leather harness and placed in a sound-insulated conditioning-chamber (Yodlowski et al. 1977). Various mechanical means of generating infrasounds artificially within the chamber were used, each infrasound being presented to the bird as a stimulus at random intervals. The heart rate of the pigeon was monitored for 5 s before each presentation which consisted of a 10 s stimulus. This was in turn followed immediately by a shock of 1 mA for 350 ms. After a few trials, a conditioned response developed and the heart rate increased, before the shock, each time the bird detected the stimulus.

The site of detection, where the infrasound signals are converted into nerve impulses, is not known with certainty but the mechanism is probably located in the inner ear organ system. Pigeons with surgically removed cochleas and lagenas do not respond at all to infrasounds (Yodlowski et al. 1977). Other individuals, with their columellas removed, responded to infrasounds but with a 50 dB reduction in sensitivity (Kreithen and Quine 1979).

In the laboratory, therefore, pigeons show a sensitivity to infrasounds that should allow them to hear natural sources when free-living. There are, however, two major obstacles to our taking the step from accepting that pigeons can detect infrasounds to accepting that infrasound landmarks have a place on the avian landscape. Firstly, birds have to be able to separate geographical infrasounds from those 'pseudosounds' generated by local turbulence, not least by the turbulence generated by the bird itself as it flies.

Secondly, birds have to be able to detect, identify and locate the source, particularly the direction, of a distant sound.

The recognition of useful landmarks from true infrasonic signals, when these are imbedded in the complex array of wind noises and the sounds generated by the bird's own flight movements, seems a formidable task. There is as yet no evidence that birds can separate these signals, though Quine (1979) has evidence that pigeons can detect man-made infrasounds outdoors. Some progress has been made, however, in determining whether the direction of distant sounds could be detected (Quine and Kreithen 1981; Quine 1982).

Previously, all known mechanisms for localizing the direction of sounds relied on the difference in phase, intensity or time with which sound waves arrived at the ears on each side of the head or body. None of these methods is useful at infrasound wavelengths. For example, a sound of 0.1 Hz has a wavelength of 3.4 km in air (Kreithen 1978). A pair of ears would need to be very far apart to detect differences in the signal being received. However, as birds are able to move with some speed through the air, a different method of localizing sounds is available to them, based on the Doppler shift principle, that is not available to land-based animals such as humans.

Regardless of its frequency, a sound will appear higher in pitch if a bird flies toward the source and lower in pitch if it flies away. A pigeon flying at 20 m/s should encounter a 14 per cent shift in all sound frequencies as it flies toward or away from the source. This shift may even allow the bird to separate infrasounds generated at a distance from those generated by its own flight movements.

Quine and Kreithen (1981) have shown that homing pigeons can detect small shifts in sound frequency at 1, 2, 5, 10 and 20 Hz. Thresholds range from a 1 per cent shift at 20 Hz to a 7 per cent shift at 1 Hz. Such sensitivity makes it feasible that natural Doppler shifts can be detected and that a bird in flight could detect the direction of distant infrasonic landmarks, as long, that is, as the sounds from such landmarks can be separated from local pseudosounds.

Clearly we cannot yet be certain that the avian landscape supports infrasonic landmarks. Equally clearly, we should no longer ignore the possibility. Indeed, there is one feature of this study that persuades me that birds probably can detect and do use infrasound landmarks in navigation, using the Doppler shift system for detection.

Humans, in their natural movements, would rarely, if ever, have been able to move fast enough to detect the direction of infrasonic landmarks by means of a change in pitch. Unable to detect direction, and thus to use the sounds they hear for navigation, the detection of atmospheric infrasounds would have been a liability, leading, as it would, to a great deal of apparently useless noise. The human threshold would thus have been set by evolution above environmental levels. Birds, however, able as they were to move toward and away from sources of infrasounds, had the potential to locate the direction of those sources.

Only if the detection of infrasounds confers some advantage, as for navigation or weather forecasting, is it possible to make sense of a threshold set by evolution below environmental levels, for this must increase the background noise through which a bird has to detect and discriminate all those other sounds, such as songs, that are so important in its everyday life. It would be interesting to know if the threshold of sensitivity for flightless birds is more similar to the threshold for humans or homing pigeons.

5.4. Fields of magnetism and gravity

Over the Earth's surface, fields of magnetism and gravity show gradual changes in intensity. Superimposed on these global fields are regional fields, perhaps thousands of kilometres in extent, that in places mask the larger fields. These variations in the magnetic field have already been noted in Chapter 4. In addition to these fields, however, the Earth's surface is littered with smaller, more intense, magnetic anomalies, many of which are also gravity anomalies.

Evidence is mounting, though most of it circumstantial, that pigeons are influenced by magnetic anomalies (see Chapter 10). Gould (1980), using data provided by Walcott, has plotted a number of pigeon tracks in relation to magnetic topography in the region of anomalies. The result is visually impressive (Fig. 5.11) and could well be taken as evidence that pigeons are aware of the magnetic topography over which they fly.

Keeton (1981), encouraged by further circumstantial evidence that pigeons may be able to detect small changes in the strength of gravity (Larkin and Keeton 1978), suggested that the topography of gravity fields might also be part of the bird's map of its environment.

At present, the evidence is meagre that features such as magnetic and gravitational topography might have some place on the avian landscape. All we can do is await further information.

5.5. Landscape differences between species

The image of an avian landscape that we have built up in this chapter is so far heavily biased in terms of a pigeon flying by day over land. Not all birds have the same sensory balance as pigeons, fly solely by day, or live primarily over land.

As far as sensory balance is concerned, we can say little at present concerning the importance of infrasounds to different species. If infrasonic landmarks do prove to be important to birds, and if their direction is detected by Doppler shift mechanisms, then, apart from flightless species, all birds would seem to gain an advantage from including such landmarks on their mental landscape.

74 Bird navigation: the solution of a mystery?

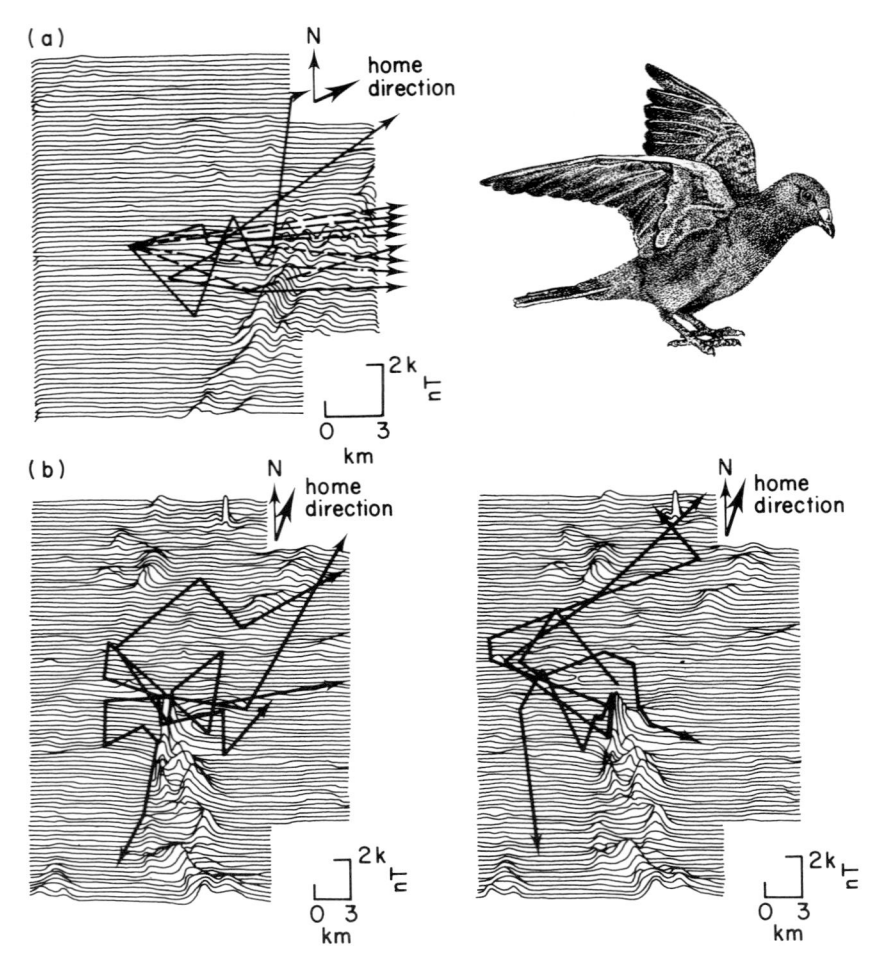

Fig. 5.11 Flight paths of homing pigeons in relation to magnetic topography

Contours show magnetic topography, peaks indicating magnetic anomalies at which the total magnetic intensity is elevated by up to 3000 nT. (a) Pigeons released on magnetic 'plains' near Worchester, Massachusetts, usually fly straight over magnetic anomalies. (b) Pigeons released at the magnetic anomaly at Iron Mine Hill, Rhode Island, are disoriented and remain so for some time after leaving the vicinity of the anomaly.

[Simplified from Gould (1980, 1982), from data by Walcott]

As far as smell is concerned, pigeons, along with other seed-eating birds, are quite low on the league table of development of the relevant part of the brain and of the nasal apparatus (Bang 1971; Wenzel 1982; Hutchinson and Wenzel 1982). Pride of place goes to fish-eaters such as the petrels, shearwaters and albatrosses which have greatly enlarged nose tubes. Other birds, species that live on water, are carnivorous, colonial nesters or nest in the ground, also rank quite highly on this league table. Olfactory landmarks could well dominate the landscape for such birds, especially as many of them, such as storm petrels and shearwaters, come to nesting sites only in the hours of darkness.

There seems no good reason to doubt at present, however, that most birds have a landscape dominated by visual landmarks. This is probably true even for birds that fly by night. There are relatively few occasions when, away from the lights of civilization, the form of distant topography cannot be seen against the night sky. There is circumstantial evidence that even bats, with their poorly developed eyes, use distant topography in navigation (Williams and Williams 1970). Rivers, darker patches of woodland, and nowadays, of course, the lights of towns and cities, will on most nights be visible to an airborne bird. Griffin and Buchler (1978) have suggested that the call notes of birds migrating at night may reflect from the ground in such a way that the birds could detect the type of area (e.g. woodland, open country) over which they are flying. They could thus supplement their visual landscape by a crude form of echolocation.

Sounds and smells, of course, continue to be available at night. Griffin and Hopkins (1974) have suggested that the distribution of lakes and ponds as marked by the breeding choruses of frogs and toads could be useful landmarks to birds migrating at night in spring.

Most humans think of oceans and seas as featureless voids, totally bereft of landmarks. This is not true, even for humans, once the senses have become attuned to the important features. The oceanic navigators of Polynesia were adept at learning cloud and wave patterns, the colour, taste, and temperature of different currents of water, and the different animal communities associated with different regions (Lewis 1972). More subtle cues, such as the colours that reflect from the underside of clouds, indicating the colour of the water beneath, extended their visual landscape beyond the limits of the horizon. Smells and sounds were also used. Bellrose (1972) has pointed out that all of these cues are also available to birds flying over the sea. Indeed, whereas Polynesians are hindered by being confined to the water's surface, birds have the aviator's advantage of being able to rise well above the ocean surface, thus extending their visual horizon, and obtaining a more coherent landscape. In a more global sense, winds blowing primarily from a particular direction are often characteristic of a particular area of the sea and would be a useful landmark to wide-ranging species such as albatrosses and shearwaters. Olfactory and perhaps infrasonic landmarks would also be invaluable additions to the 'seascape'.

Whether a bird flies by day or by night, over land or over sea, short-distances or far, it seems that it will carry around within its head a mental image of the landscape with which it has become familiar during its earlier explorations. Depending on whether it feeds on seeds, green leaves, fruit, insects, meat, fish or carrion and thus depending on the balance of its senses, this landscape will be dominated by either sights or smells and perhaps embroidered by local sounds and distant, unseen, infrasonic landmarks.

Which landmarks are most conspicuous and important on the landscape of any individual bird will depend in part on its sensory balance, in part on the extent of its own explorations, and in part on which landmarks are most conspicuous in the region in which it is born and lives. Whichever landmarks are used, however, the individual's familiar area will be surrounded by features tens, hundreds or thousands of kilometres beyond the limits of the places that it has ever visited.

A familiar landscape of such dimensions might at first seem to require a prodigious memory. We have already seen, however, in Chapter 4, that the avian landscape is not continuous. Instead it seems to take the form of a mosaic consisting of only the most important landmarks and places. The spatial relationships of these major landmarks seem to be memorized, not in terms of continuous landscapes, but rather in terms of their relative compass bearings. Such a system has a pleasing economy. A landscape containing a few major, widely-spaced landmarks that can be detected over long-distances with movement between any pair of landmarks being achieved by orientation on a learned compass bearing, requires relatively few details to be memorized for mental maps nevertheless to encompass large areas.

Mosaic maps, therefore, make relatively few demands on a bird's memory and in consequence make the mental retention of information on areas stretching over thousands of kilometres, as required of long-distance migrants, far more feasible. It follows, however, that an understanding of mosaic landscapes requires more than an understanding of just the sights, sounds and smells that birds use as landmarks on such a landscape. It requires in addition an understanding of the compasses a bird could use to detect and then memorize the compass bearings of familiar sites, one from another.

Our next step, therefore, is to look at the evidence that birds can detect compass direction and the range of information they might use to do so. It bears testimony both to their importance and to the experimental effort that has been put into understanding them, that our examination of bird compasses needs to be spread over four chapters. We begin by looking at the compasses birds have been found to use during the day.

6 Bird compasses: by day

6.1. The sun compass

The first bird compass to be studied experimentally was the sun compass. Kramer (1950) retained a starling (*Sturnus vulgaris*) in an orientation cage during the normal period of spring migration and found that, under sunny skies, the bird oriented to the west-northwest. When the sun was deflected by a mirror, the bird's orientation was deflected by a corresponding angle.

Dependence on birds with an urge to migrate limits experimentation to the relatively short migration seasons. Kramer and Saint Paul (1950) relieved this constraint when they succeeded in training starlings in a circular cage to look for food in a certain compass direction. All visual landmarks were excluded by a screen and only the sun and sky were visible. Their birds maintained the same learned compass preference throughout the day. Kramer concluded that the starlings must be making allowance for the sun's movement across the sky during the day. The critical proof, however, was provided by Hoffmann (1954) who demonstrated that clock-shifting a bird by six hours (see Chapter 4) produced the predicted 90 degrees shift of the bird's compass orientation.

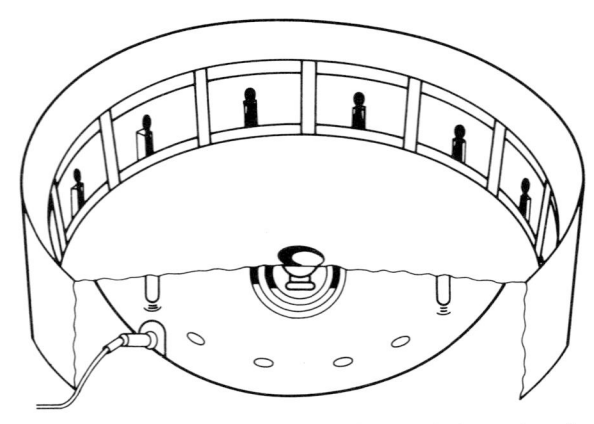

Fig. 6.1 Automatic training cage used by Schmidt-Koenig in arctic and transequatorial sun-compass experiments

The section through the cage shows the pecking discs at the periphery of the upper platform. If the bird's choice was correct, food appeared automatically in the food cup in the centre of the arena. Electrical impulses from the pecking disc passed through brushes from the upper rotatable platform to contacts in the lower fixed platform.

[Re-drawn from Schmidt-Koenig (1979)]

Similar training cages have been used to demonstrate sun–compass orientation in homing pigeons, meadow larks (*Sturnella neglecta*) and others (Kramer and Riese 1952; Saint Paul 1956). Schmidt-Koenig (1958, 1960, 1961) showed that clock-shifts of six hours slow, of six hours fast and of 12 hours all affected the orientation of homing pigeons in training cages (Fig. 6.2). The effect was not immediate, however, and the expected shifts in compass orientation (90 degrees following a 6 h shift, 180 degrees following a 12 h shift) took four days to appear in full.

After the discovery that birds possessed a time-compensated sun compass, interest shifted to how this compass coped with the latitudinal peculiarities of the sun's arc across the sky described in Chapter 4. Hoffmann (1959) transported starlings from their latitude of training in cages at Wilhelmshaven in Germany (53° 30′ N) to a testing latitude at Abisco, Sweden (68° 22′ N). Schmidt-Koenig (1963a) transported homing pigeons from Durham, North Carolina, USA (36° 00′ N) to Barrow, Alaska (71° 20′ N). Both groups of birds had received directional sun-compass training at their mid-northern latitudes and were being tested north of the Arctic Circle during the local season of continuous daylight.

The results showed that both starlings and pigeons could orient reasonably well by the local sun during the part of the local day that was also day at their home latitude. Starlings were also well-oriented under the midnight sun. Less clear was the pigeon's use of the local sun during what would have been night at home. This difference, however, may have had less to do with differences between the species than with experimental conditions (Matthews 1968). The starlings had several weeks of exposure

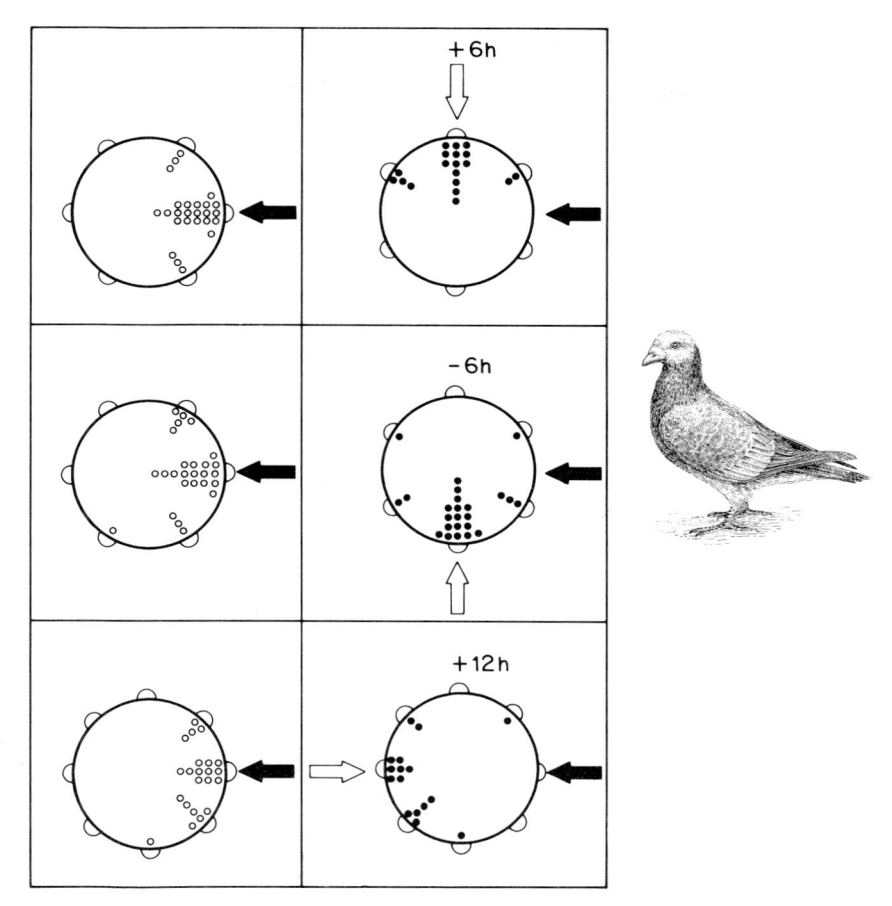

Fig. 6.2 The influence of clock-shifting on the directional choice of homing pigeons in a cylindrical training cage

Each dot shows one peck of a bird. Controls (natural day) are shown in the left column; experimentals (clock-shifted) are shown in the right column. Black arrows show the preference expected if clock-shifting has no effect; white arrows the preference expected if the bird is using a time-compensated sun compass.

[Modified from Schmidt-Koenig (1979)]

to the continuous day before being tested whereas the pigeons were tested immediately.

When wolf spiders (*Arctosa*) were taken from Italy to Finland at 69°N but retained on an Italian day/night schedule until being tested, they were well-oriented at midday but quite disoriented at midnight (Papi and Syrjämäki 1963). Finnish spiders of the same species, however, were well oriented at both midday and midnight.

Starlings and homing pigeons have also been subjected (by Hoffmann and Schmidt-Koenig respectively; see Schmidt-Koenig 1979) to trans-equatorial displacements. Again the experimental technique was to use birds trained to feed in a particular compass direction in a circular orientation cage. Birds thus trained to the north in the northern hemisphere should orient 90 degrees to the left of the sun at 06.00 h, directly away from the sun at noon, and 90 degrees to the right of the sun at 18.00 h. Such birds, transported to the southern hemisphere (where the sun still moves from east to west but across the northern sky), might be expected to orient to the southern sun as they had learned to orient to the northern sun. In which case, they would orient correctly to the north at 06.00 and 18.00 h but would orient to the south at noon. This is just how the experimental starlings and homing pigeons behaved. So, too, in cases where they are not simply disoriented, do bees and fish (e.g. Kalmus 1956; Hasler and Schwassmann 1960).

Unfortunately, starlings do not cross the equator in their normal movements and the homing pigeons tested were from temperate regions. In fact, experiments on homing pigeons have been more or less confined to stocks from lofts in north temperate regions, though experiments are about to begin on birds from Brazilian lofts (Ranvaud 1982).

We do not know, therefore, whether birds that are transequatorial migrants or birds that live permanently between the tropics have any special ability to cope with those vagaries of the sun's movements to which they are exposed. The only direct attempt to study the sun compass of a transequatorial migrant was when Schmidt-Koenig trained bobolinks (*Dolichonyx oryzivorus*) in an orientation cage in Durham, North Carolina. The birds were to have been transported to South America for testing but all either died or escaped before leaving North America (Schmidt-Koeing 1979).

Homing pigeons are raced successfully from lofts at São Luis do Maranhão and Recife in Brazil, within eight degrees of the equator. Yet here the seasonal compass reversal of the sun's position at midday (Chapter 4) is striking (Ranvaud 1982). This does not mean, of course, that the pigeons are observing and allowing for the vagaries of the sun's motion, although it seems likely.

Even if pigeons and other birds in the tropics do observe and allow for the sun's complexities, it does not follow that they have any special ability to do so; an ability denied to temperate birds. All birds, as all other animals, seem to have to learn the characteristics of the sun's movements. No animal of which we know is born with an innate ability to compensate for these movement patterns, only an innate predisposition to learn how to do so. The complexities of the sun's arc are greater in the tropics, so there is more to learn. The question we need to answer is whether birds that visit or live in the tropics are more predisposed to learn these complexities, or learn them faster, than more temperate birds.

The fact that birds learn their sun compass was first indicated in experiments by Hoffmann (1954) and then confirmed by Wiltschko, Wiltschko and Keeton (1976).

Hoffmann took six 12-day-old starlings from a nest box and raised them with no view of the natural sun. He then trained them to a feeding direction at just one time of day using an artificial sun. With only six birds, no firm conclusion could be drawn but, when tested at a time of day different from their training time, there was little indication of any innate ability to compensate for the movement of the natural sun.

The Wiltschkos and Keeton used the elegant technique of raising pigeons in an artificial photoperiod six hours behind the natural day. The birds were allowed to fly short-distances from the loft in the afternoon (their morning). When eventually released on a homing experiment, these pigeons were as well oriented as controls raised in a normal photoperiod. The experimental birds were then placed in natural conditions, whereupon they oriented upon release as if their clocks had been set six hours ahead of normal time. On still later releases, the birds could orient correctly. Their sun compass had readjusted to normal time.

These experiments show that birds learn the association between time, sun and geographical direction during their first few weeks of life. They also show that if this association changes, it can be relearned. They do not show what range of information the birds need for a complete sun compass to be achieved. Experiments on lizards showed that if the animals are trained to orient by the sun at several times of day, they can orient at intermediate times, as if by extrapolation. The need to see the sun at different times of day has also been demonstrated with pigeons. The Wiltschkos showed that if young pigeons are allowed to see only the descending (afternoon) part of the sun's arc, the birds do not have a functional sun compass in the morning. This follows from the result that clock shifting such birds produces only the expected change in orientation in the afternoon, not in the bird's 'morning' (Wiltschko and Wiltschko 1980).

The accuracy with which birds learn to use their sun compass has proved extremely difficult to measure. Largely, this has been due to difficulties in experimentation. The bane of all experiments on orientation and navigation is that birds are not single-minded enough to do only what is required of them. Train them that food is provided from one direction only and they insist from time to time on exploring other directions, apparently just in case food does suddenly appear elsewhere.

These visits to 'unprofitable' areas are not mistakes by the bird. As shown by Smith and Sweatman (1974), the behaviour is adaptive. Birds collect information by such visits. When the best foraging site disappears they are then able to go immediately to the previously next-best site. To experimenters on orientation, however, such exploration by the birds is a disadvantage for it maintains a level of scatter in the results that confounds attempts to measure accuracy.

Early attempts to measure the accuracy of the sun compass (reviewed by Schmidt-Koenig 1979) were thus frustrated. The perhaps most successful attempt was by McDonald (1972), using operant conditioning methods. A pigeon sat on a box on a turntable and was presented with an artificial sun (a projector light presented to the bird by a system of mirrors). In front of the bird were two keys and a cup in which food appeared. If the bird pecked one key, the table was turned anti-clockwise, slightly and automatically, by a motor. If the bird pecked the other key, food appeared but only if the table, and thus the bird's long axis, were aligned, say, 15 degrees anti-clockwise of the sun. If the bird pecked the reward key when the table was not so aligned, or continued to peck the rotation key after the table had reached correct alignment, then the table returned to the starting position. In this way, the pigeon trained itself to judge angles relative to the artificial sun.

The accuracy achieved by the birds in this apparatus was between 3.4 and 5.1 degrees. An interesting discovery, however, was that the birds judged the angle between the table and the light, not by looking at the light itself but by watching the shadows cast on the table. McDonald argued that the use of shadows in natural conditions could be advantageous, because the sun's movement can be magnified by a factor of up to nearly six. Whether birds in their normal life use shadows for their sun compass rather than, or as well as, the sun's disk is unknown. We may wonder, also, whether accuracy of orientation to an artificial light by a sitting bird is likely to indicate, or be better or worse than, accuracy of orientation to the real sun by a free-living bird.

One of the factors which influences the accuracy of the sun compass of a free-living bird is the bird's ability to measure time of day; in other words, the accuracy of its internal clock. Matthews (1968) reviewed early experiments aimed at answering this question. Evidence that birds can measure short time-intervals of a few seconds or less can be gleaned from the highly synchronized duets sung by species such as the black-headed gonolek (*Laniarius erythrogaster*). Experiments in which birds are trained to feed at a particular hour each day or in which some behaviour, such as daily onset of locomotor activity, is monitored under constant light conditions, suggest that accuracy of time-keeping is no better than a few minutes in each 24 hours. The greatest accuracy so far recorded still seems to be that of a mammal, the flying squirrel (*Glaucomys volans*), which showed an error of between two and nine minutes from one day to the next (DeCoursey 1962).

Evidently we are some way from being able to say with any confidence how accurately birds can learn to use the sun as a compass. Accuracy, however, is not the only consideration in evaluating how useful such a compass may be to a bird. As everybody knows, even on days that are only partially cloudy, the sun itself may spend a great deal of time hidden behind clouds, despite there being large areas of blue sky. It has been known for some time that under such conditions bees use polarized light for orientation (von Frisch 1967) and over 90 species of invertebrates have now been found to respond to such light. Vertebrates have been little studied from the point of view of sensitivity to polarization, man being the species for which such sensitivity is most convincingly documented, though a salamander and perhaps some fish are also sensitive (Kreithen and Keeton 1974b). Now it seems that birds also are sensitive to polarized light.

6.2. Polarization patterns

Due to atmospheric scattering, the light from blue sky is linearly polarized. Moreover, the plane of polarization is closely related to the position of the sun. The so-called E-vector of sky polarization, for example, in a given region of sky is perpendicular to a line drawn by the observer between that point in the sky and the sun. In addition there is a band of maximum polarization (Brines 1980), that parallels the E-vector and forms a ring around the Earth 90 degrees away from the sun. This band moves across the sky during the day maintaining a fixed relationship to the position of the sun. As a result, the detection of polarization patterns in blue sky would provide a very useful compass, either in its own right or as a back-up to the sun compass on occasions when the sun itself is either just below the horizon or obscured by patchy cloud. Early experiments, however, suggested that homing pigeons were insensitive to polarized light. (Montgomery and Heinemann 1952).

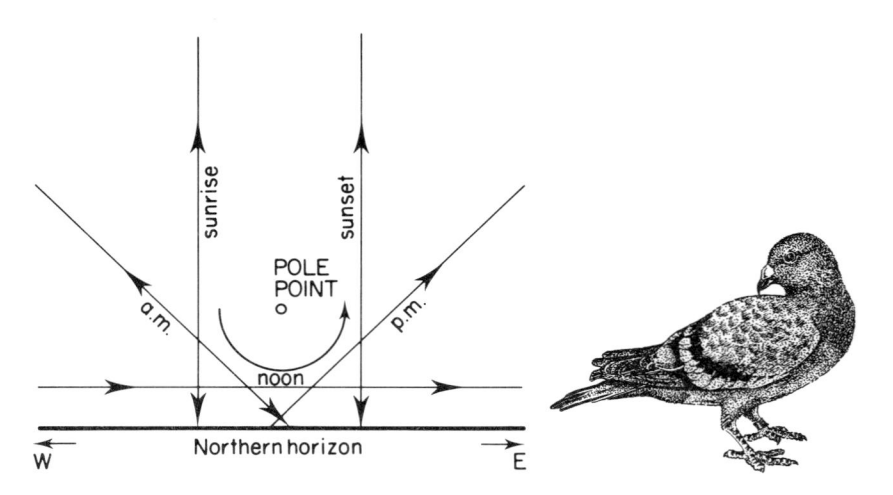

Fig. 6.3 Rotation of the band of maximum polarization visible to an observer when looking north in the northern hemisphere during the summer

Straight lines show the band of maximum polarization at different times of day, with double-headed arrows showing the E-vector of polarization. Single-headed curved arrow shows the rotation of the band from sunrise to sunset.

[Simplified from Phillips and Waldvogel (1982)]

These early tests used a light source close up to the pigeon, the significance of which was missed until Kreithen read a series of papers by Blough, Catania and others (see Kreithen 1978). These studies showed that the pigeon's retina has specialized regions, each with its own peculiar organization of nerves, distribution of coloured oil droplets, and spectral and spatial sensitivity. Whereas humans focus all images on the fovea, regardless of the distance of the object, pigeons use the peripheral retina, with red oil droplets, for near objects, and the central and lower part of the retina, with yellow oil droplets, for distant objects. Binocular vision is used only for objects that are nearby. For distant objects, pigeons use each eye separately with independent, monocular fields of view. The polarization patterns of the sky, being both distant and overhead, fall on the region of the yellow retinal field. In contrast to Montgomery and Heinemann, therefore, who had presented their light source onto the binocular red field of the retina, Kreithen and Keeton (1974b) positioned their polarized light source at a distance of 1.9 metres from the bird to illuminate the yellow retinal field (Fig. 6.4). The birds responded to the polarizers.

Two years after Kreithen and Keeton had demonstrated the sensitivity of pigeons to polarized light, Delius et al. (1976) also obtained positive responses, this time to polarized targets presented overhead. Targets

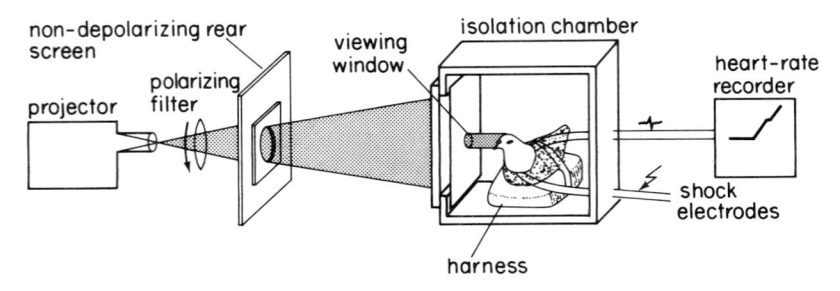

Fig. 6.4 Experimental set-up during cardiac conditioning tests of the type carried out by Kreithen, in this case for testing perception of linearly polarized light

[Simplified from Kreithen (1978)]

presented to the binocular field as in the early experiments failed to produce a response. As in Kreithen's experiments on infrasound, the experimental technique used in these experiments was cardiac conditioning.

There is no direct evidence as yet that free-flying birds make use of the polarization patterns of the sky for compass orientation. Phillips and Waldvogel (1982), however, are investigating the possibility that at least part of the deflector loft effect on homing pigeons (Chapter 5) is due to deflection of polarization patterns rather than to deflection of wind direction.

Able (personal communication) studied the orientation of the white-throated sparrow (*Zonotrichia albicollis*) during twilight, the period between the time that the sun's disk disappeared below the horizon and the time the first bright stars became visible, about 50 minutes later. White-throated sparrows are nocturnal migrants but, like many such migrants, they frequently initiate migration during twilight. Moreover, when restrained in orientation cages, they begin to hop in seasonally appropriate directions at this time of day.

Able carried out tests in Albany County, New York, using birds caught locally during spring migration and orientation cages with tops made from a plexiglass polarizer. Rotation of the polarizer had a clear effect on the orientation of the sparrows. Some role for polarization patterns in orientation during twilight is therefore indicated. The nature of this role, however, is not clear. The results showed that the birds were not using polarized light to locate the position of the sun below the horizon. On the other hand, they may have been orienting along the axis of the E-vector. During the migration seasons in north temperate latitudes, the twilight E-vector at the zenith provides a roughly north–south axis. If the birds tested had been orienting along the E-vector axis, however, they were being biased for some reason toward that end of the axis nearer sunset. This may mean only that some other factor is also involved in compass orientation during twilight. In which case Able's experiment may have provided the first evidence that polarization patterns are used by birds for orientation, at least once the sun is below the horizon.

6.3. A role for ultraviolet light

The combined use of sun and polarization patterns would allow compass orientation from dawn to dusk on all days with clear skies or partial cloud. Such orientation would be impossible, however, on overcast days. How often the sun and polarization patterns would be totally obscured depends on the thickness of the cloud and the wavelength of light used to detect them.

Many insects, including ants and bees, use ultraviolet wavelengths to detect polarization patterns since shorter wavelengths penetrate thin cloud layers and thus are available when other cues are masked. An ability to see ultraviolet light would therefore increase the amount of time that a sun compass and polarization patterns would be available to a bird. Humans have pigments in the lens of the eye that absorb ultraviolet light. These prevent short wavelengths from reaching the retina. The lenses of pigeons and other birds, however, are very clear in the ultraviolet range.

We now know that hummingbirds (Huth and Burkhardt 1972) and pigeons (Kreithen 1979) among birds have a peak of spectral sensitivity in the ultraviolet range (325–360 nm) (Fig. 6.5). Among other vertebrates, toads, newts and lizards also show sensitivity to ultraviolet light.

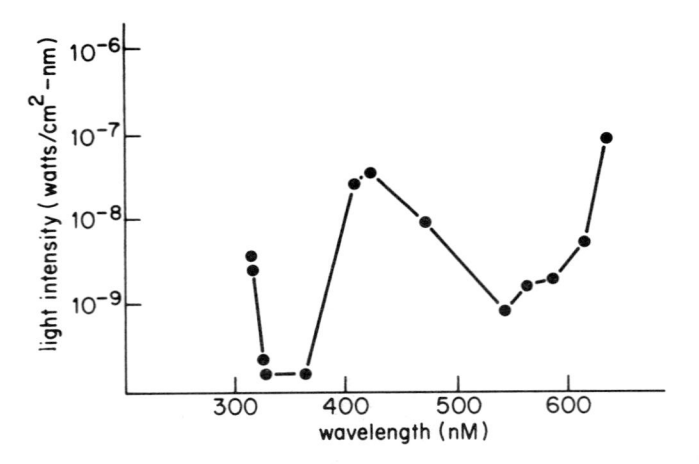

Fig. 6.5 Sensitivity of a pigeon to ultraviolet and visible light
Each dot is a behavioural threshold obtained by cardiac conditioning.
[Re-drawn from Kreithen and Eisner (1978)]

During the day, therefore, aided by a sensitivity to ultraviolet light, birds have compass information from the sun and/or polarization patterns at all times except under thick overcast. Neither of these compasses, however, is available once it is dark. Any bird needing to orient at night needs other compass cues, and these are the subject of the next chapter.

7 Bird compasses: by night

At night, the sky offers two major compass cues, the stars and the Moon, and, of the two, stars have received by far the greatest attention from researchers. The major characteristics of the night sky were described in Chapter 4.

7.1. Stars

First experiments on the star compass of birds were carried out on sylviid warblers in Europe (Sauer and Sauer 1955, 1960; Sauer 1957, 1961) and a few years later on indigo buntings (*Passerina cyanea*) in the United States (Emlen 1967a, b). In both cases, birds were tested in orientation cages. Experiments by the Sauers have since received a great deal of criticism, largely because the data were recorded by direct observation. The experimenter watched the birds from below and recorded the time they fluttered in a certain direction during migratory restlessness. Emlen and

Emlen (1966) overcame this difficulty and obtained a permanent record of a bird's orientation by introducing the 'Emlen funnel' (Fig. 3.7). Despite the shortcomings of their experimental design, however, the Sauers were the first to provide evidence that birds do have a star compass.

The main facts to emerge from these early experiments by the Sauers and Emlen were: (1) birds in an orientation cage at night are able to recognize their normal direction for migration, both under the natural starry sky and under a planetarium sky; and (2) if a planetarium sky is rotated horizontally through 180 degrees, the direction of orientation of the birds reverses accordingly. At the time, it was also concluded that when the stars are turned off and the planetarium dome diffusely illuminated with dim light, the direction of activity of the birds becomes random or activity ceases altogether. Whether the birds in those early experiments really hopped randomly under these conditions is now less clear (Emlen 1975). The conclusion that the birds were relying heavily on information from the stars, however, is still justified.

Sauer suggested that a view of the stars when the birds are young is unnecessary for the development of normal migratory orientation by warblers. Moreover, his results gave some indication, as for the sun compass, of a compensation for the rotation of the sky as the night progressed. These suggestions have not been supported by later workers and, unless they receive support from other studies, must be considered doubtful. For the moment, the general concensus is that, as with the sun compass, birds have to learn how to use the stars for orientation but that, having learned this, unlike with the sun compass, compensation for time of night is not involved.

Evidence against any compensation for time of night in the use of the star compass comes from experiments on both free-flying and caged birds.

Experiments on free-flying birds have exploited the 'preferred compass direction' (PCD) reaction described in Chapter 3. The flight of wildfowl at night can be followed by the ingenious technique of attaching a small electric lamp to their leg. By this means, Bellrose (1958, 1963) demonstrated that a variety of North American wildfowl adopt the same PCD upon release at night as they do during the day. Matthews (1963) confirmed this with European mallard (*Anas platyrhynchos*). Both authors agreed that orientation did not occur under overcast skies. The results were inconclusive, however, with regard to orientation under partial cloud or high, thin overcast. Star orientation was thus indicated, albeit circumstantially, but what of time-compensation?

When mallard are clock-shifted and released by day with a view of the sun, their PCD is shifted as expected for a time-compensated sun compass. When clock-shifted and released under starry skies at night, however, their PCD is unaffected (Matthews 1963). If they are orienting by the stars, time-compensation is not involved.

Confirmation that birds can learn complex star patterns and evidence

Sky overcast

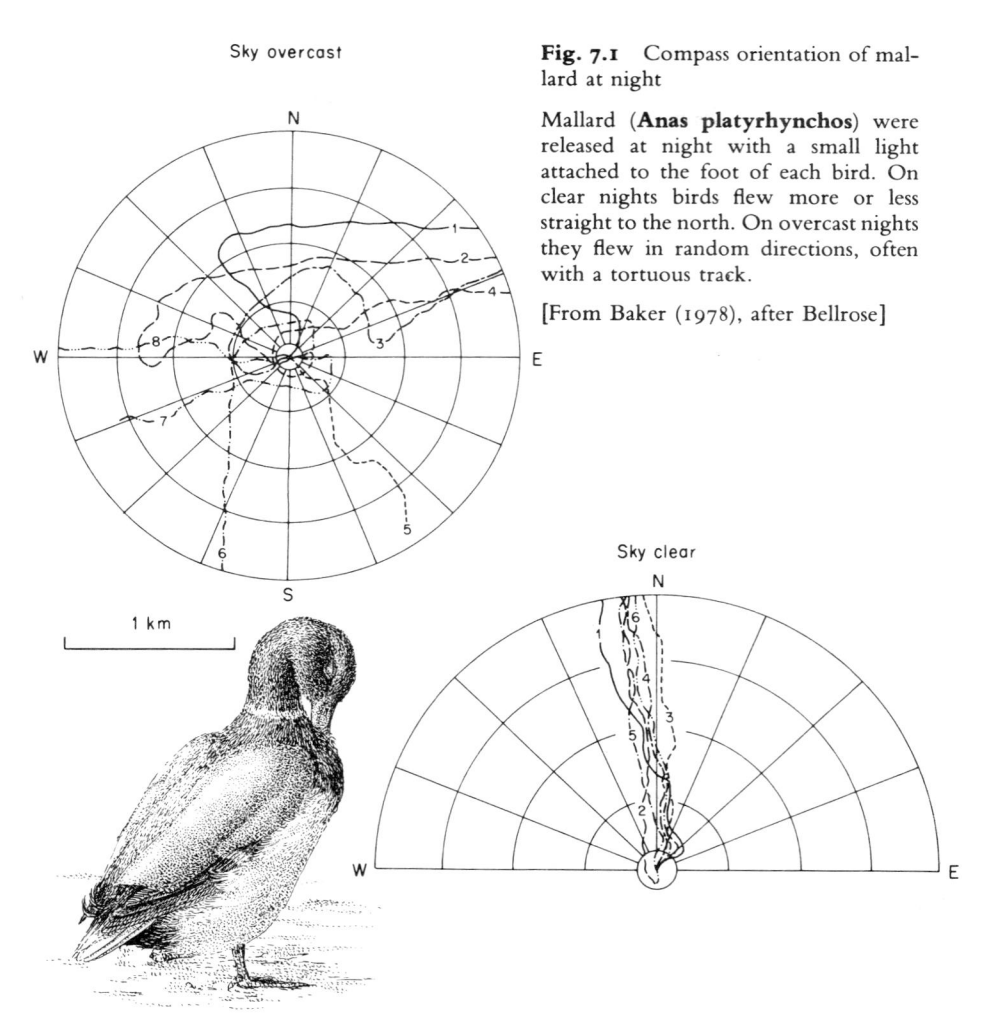

Fig. 7.1 Compass orientation of mallard at night

Mallard (**Anas platyrhynchos**) were released at night with a small light attached to the foot of each bird. On clear nights birds flew more or less straight to the north. On overcast nights they flew in random directions, often with a tortuous track.

[From Baker (1978), after Bellrose]

Sky clear

that they have no instinctive ability to compensate for the rotation of the night sky came from cardiac conditioning experiments on caged ducks. Wallraff (1969, 1972) conditioned teal (*Anas crecca*) and mallard to a particular direction under a planetarium sky. When the stars were familiar from previous training sessions, the birds reacted as expected. If they were exposed to a sky that had been rotated about the pole star so that it was appropriate to the same location but for a few hours later in the night, the birds seemed unable to recognize their training direction. To do so, they needed to receive separate training at the new time of night or under the new starry sky. Once they had learned a particular star pattern, however, the ducks were able to remember that pattern for long periods. The longest interval between training and testing was six months, during which time other additional star patterns had been learned. Training was achieved just as readily under totally unnatural star patterns as it was under simulated natural patterns. These birds at least, therefore, are able to learn to distinguish between different patterns composed of more than a thousand star-like points of light and to memorize at least some characteristics of the patterns over long periods.

Wallraff's results implied that his ducks were learning fairly gross star patterns. Emlen (1969, 1970, 1972) has shown for indigo buntings that, although star patterns over large areas of the sky may be learned, they are used specifically to locate a particular and important point in the sky, the axis of rotation. In the northern hemisphere, this axis is marked by the pole star. In the southern hemisphere, there is no such marker star (Fig. 4.9).

Emlen began his extremely elegant series of experiments with a group of 10 nestling buntings brought into the laboratory and reared without sight of either stars or any other sources of points of light. When the birds came into a migratory state for the first time, they were tested in orientation cages in a planetarium with a stationary but otherwise natural star pattern. All showed nocturnal migratory restlessness but with no indication that they could adopt the southerly direction that is normal for the species during autumn migration. Indeed, over many nights their orientation was random. There was no support for Sauer's suggestion that previous experience with stars is unnecessary for orientation using a star compass.

The experiment continued with two other groups of young buntings. During their early development, from three weeks before their first moult, both groups were allowed to see planetarium skies with star patterns normal for the season. The two skies to which the different groups were exposed both rotated at the normal speed, but differed in one major respect. One group observed a sky with a normal axis of rotation, centred in the northern sky on *Polaris* (Fig. 4.9). The other group observed a sky manipulated to rotate around a different star, this time in the southern half of the sky: *Betelgeuse* in the constellation *Orion*. When the birds began migratory restlessness, both groups were tested simultaneously in the planetarium under the same, stationary sky. The two groups oriented in quite different directions, but in both cases the preferred orientation was toward what would have been south had the axis of rotation of their particular sky been north.

Once birds have developed their star compass, they do not need to see the pole star to know its position. In the same way that humans in the Northern Hemisphere can tell roughly the position of the pole star so long as they can see the two 'pointer' stars that point towards it in the constellation *Ursa Major* (the plough or big dipper; Fig. 4.9), so too, it seems, could Emlen's buntings, except that they did not necessarily use *Ursa Major*. By blocking out parts of the planetarium sky, Emlen was able to show that different individuals used different star patterns as 'pointers' to fix the position of the axis of rotation. As long as each bird could see its own 'pointers' among the star patterns visible on a particular night, it could

Bird compasses: by night 91

Fig. 7.2 Vector diagrams for an indigo bunting (**Passerina cyanea**) tested in an Emlen funnel in spring

(a) under a normal spring planetarium sky
(b) under a spring planetarium sky rotated horizontally by 180 degrees

[Simplified from Emlen (1967a). Photo by G. Ronald Austing, courtesy of Frank W. Lane]

judge the location of the axis of rotation and hence use a star compass. Finally, it seemed that most individuals learned 'pointers' among star patterns within about 15 degrees of the pole star, a preference that would seem to be advantageous. Such stars never disappear below the northern horizon unless, or until, the bird's migrations carry it near to the equator.

It seems, therefore, that young indigo buntings, between leaving the nest and beginning autumn migration, watch the night sky and observe which part rotates least. They then learn how to judge the position of this point in the sky from the surrounding star patterns. Having done these two things, they have a compass with a fixed, unmoving north for as long as they can see the reference patterns of the natural night sky. With such a compass there is no need to compensate for time in any way. The same is far from true, however, for that much more conspicuous but also more occasional inhabitant of the night sky, the Moon.

7.2. Moon

In a review of animal orientation, Able (1980) concluded there was only one group of animals that had been shown to have a Moon compass. This is a group of beach-living amphipods (e.g. *Talitrus, Orchestoides* and *Talorchestia*) known as sandhoppers.

Able implies that relatively few organisms have evolved a lunar compass because there are a number of difficulties that have to be overcome for the Moon to be useful. Most of these difficulties involve time compensation and result from the vagaries of the Moon's movement through the sky during the course of each lunar month. As a result of this movement, the Moon is visible at night for only a part of each lunar cycle. Even while it is visible in the night sky, the size of its image and position against the backdrop of stars changes from night to night. Of course, this nightly change is predictable, the Moon rising about an hour later each night and its image waxing in a characteristic way. Nevertheless, the lunar precession means that to use the Moon as a compass, an animal needs an internal clock in phase with lunar time. In effect, this would mean that most organisms would require two independent timing systems: a sun clock in phase with the day–night cycle; and a Moon clock that operates at a slightly slower rate and is synchronized by other unknown stimuli.

Most recent authors (e.g. Able 1980; Wallraff 1981b) now seem to accept, though often reluctantly (e.g. Enright 1972), that, on balance, the evidence for sandhoppers supports the hypothesis originally favoured by Papi and Pardi (1959, 1963). This was that these tiny crustaceans do indeed have sun and Moon compasses, both of which are time-compensated, which they use to orient their feeding excursions up and down the beach between tides.

It seems unlikely, in fact, that the paucity of examples of a Moon compass reflects difficulty in the evolution of such a compass. Rather than a lack of use of the Moon by animals, so few examples almost certainly reflects a lack of attention from zoologists. As far as birds are concerned, there is direct evidence for a time-compensated Moon compass only for the mallard (Matthews 1973). Indirect evidence, however, comes from an unusual source and derives from recent work on the response of insects to light traps.

The large yellow underwing moth (*Noctua pronuba*) has been shown to use both lunar and stellar orientation in straightening out its cross-country migration (Sotthibandhu and Baker 1979), though in neither case does time-compensation seem to be involved. Moreover, individuals of the same species adopt the same angle of orientation relative to an artificial light source, such as on a light trap, as they do to the Moon. This evidence supports a suggestion first made by von Buddenbrock (1937) that moths are attracted to a light trap because they mistake the artificial light for the Moon. The nearer the artificial light mimics the Moon with respect to

Bird compasses: by night 93

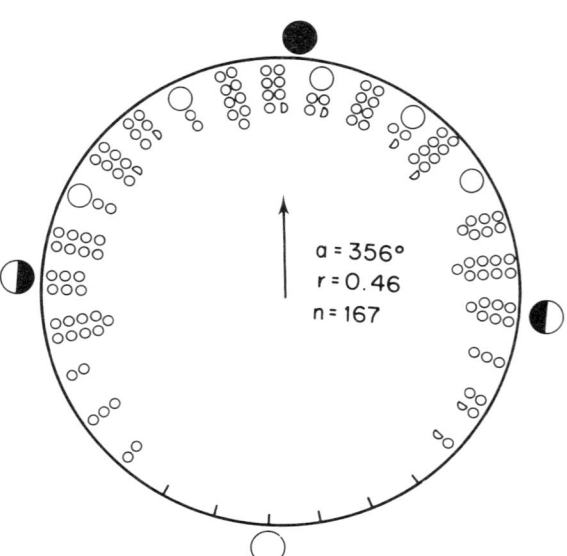

Fig. 7.3 Frequency of bird kills at a Dutch lighthouse in relation to day of lunar month

Large circles = 10 nights; small circles = 1 night. Most collisions occur around New Moon and none around Full Moon.

[Re-drawn from Verheijen (1981)]

certain important characteristics, the more likely a moth is to orient to it (Baker and Sadovy 1978, Baker in press). Moreover, it seems that light traps may catch selectively moths that have a preferred compass direction to the south (i.e. towards the Moon $\pm \angle 90$ degrees).

Each year, large numbers of nocturnal migrants among birds are attracted and killed by tall structures with a light on top. The chief culprits are lighthouses, tall buildings, ceilometers, and radio and television towers. It has been estimated that television towers in the USA take an annual toll of more than a million birds (Aldrich et al. 1966). An extensive analysis of the weather conditions associated with mass kills of birds at lighthouses showed they were characterized by dark, starless nights, particularly with rain, haze or overcast (Clarke 1912).

Is it possible that these kills are evidence of migrants attempting to use lunar orientation on nights when the natural Moon is not visible? If so, we might expect two further characteristics. Collisions should be: (1) greatest around the New Moon phase of each lunar month when the real Moon is not visible for comparison; and (2) greater during autumn migration, when preferred orientation should be toward the Moon, than during spring migration, when preferred orientation should be away from the Moon (for transequatorial migrants, the converse should be true once the birds have passed through the area where the Moon is directly overhead at its highest point).

Verheijen (1980, 1981) has carried out a detailed analysis of bird-kills with respect to the phase of the Moon. His results show convincingly that collisions are clustered around the phase of New Moon, with a noticeable lack around the phase of Full Moon. My own impression from Verheijen's data is also that bird-kills at lights are almost entirely a feature of autumn migration. If such collisions with lights really are evidence for lunar orientation, they suggest that a large proportion of nocturnal migrants may make use of a Moon compass.

Observation of the sun and polarized light by day, particularly if based on perception of ultraviolet wavelengths, and of stars and the Moon by night, should enable a bird to have access to compass information for a large part of its time. There are relatively few parts of the world, however, where such celestial compasses are available continuously, for most places suffer total overcast at some time. If continuous information is advantageous, therefore, some further source of compass information is needed. In Chapter 8, we examine recent evidence for von Middendorf's suggestion that birds are able to detect compass direction by reference to the Earth's magnetic field.

8 Bird compasses: magnetism

It is more than a century since von Middendorf (1855) suggested that birds may be able to obtain compass directions from the Earth's magnetic field. Ever since, the possibility has been a subject for continual controversy and argument (see reviews by Matthews 1968; Keeton 1974a; Baker 1981a). Some people still totally reject evidence for a magnetic compass in birds (e.g. Gerrard 1981). Others, although open-minded, remain sceptical and require more convincing evidence than at present exists (e.g. Griffin 1981). The general concensus, however, is that we already have enough evidence to accept that birds have a magnetic compass, even though we are still a long way from understanding when and how magnetic fields are used and perceived.

8.1. Evidence for a magnetic compass

The first evidence that birds could detect compass direction in the absence of celestial information was provided by Merkel and his colleagues (Merkel and Fromme 1958; Merkel and Wiltschko 1965). Working in Germany with European robins (*Erithacus rubecula*) in orientation cages, Merkel reported that migratory restlessness was still oriented in seasonally appropriate directions even in enclosed rooms. These reports were met by considerable scepticism (see Matthews 1968), largely because the orientation of the birds was so variable and only showed significant directionality if subjected to a particular form of statistical manipulation, known as second-order analysis.

The orientation of a bird in an orientation cage over the course of a single night rarely departs significantly from randomness (or, more precisely, uniformity). On the other hand, if for each night for each bird the mean direction (Chapter 3) of all hops is calculated, and then all of these means are

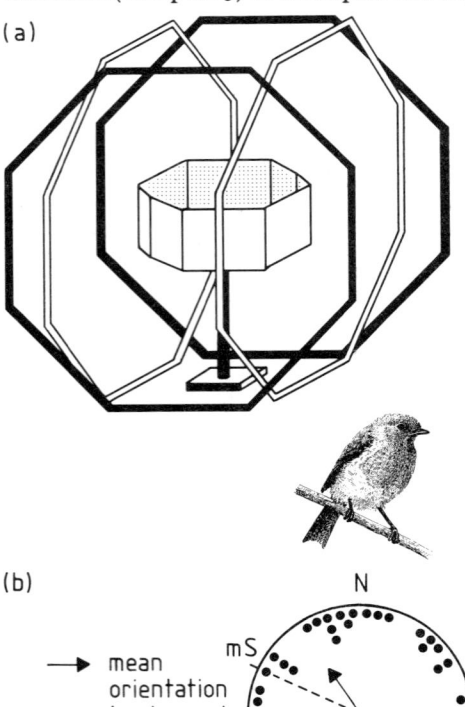

Fig. 8.1 Influence of rotation of the magnetic field on the orientation of the European robin (**Erithacus rubecula**) during migratory restlessness

(a) Orientation cage (see Fig. 3.6) and Helmholtz coil system used by Wiltschko.
(b) Each dot is the mean bearing for an individual as calculated from one night's exposure in an orientation cage. Vertical line shows the traditional migration direction at each time of year. mN, mS show magnetic north and magnetic south of the changed magnetic field.

[From Baker (1981a), after Wiltschko]

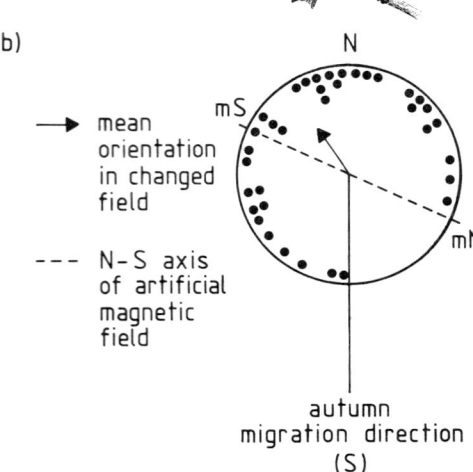

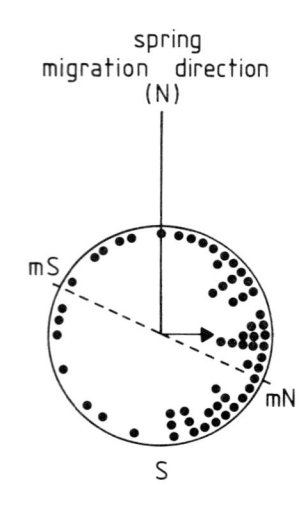

gathered together and analyzed, statistical significance often emerges.

Such statistical manipulation is perfectly justified (Batschelet 1981), and when the Frankfurt data were analyzed in this way, several encouraging trends emerged. Firstly, spring and autumn data gave mean directions that were nearly opposite and corresponded well with the traditional migration direction for the species as known from ringing recoveries. Secondly, the birds seemed able to orientate only if the magnetic field intensity through the cage was very close to that of the Earth (about 46 000 nT at Frankfurt). Finally, but most importantly, when the cages were enclosed by Helmholtz coils and the direction of the horizontal component of the resultant magnetic field was changed, the orientation of the birds shifted accordingly (Merkel and Wiltschko 1965; Wiltschko and Merkel 1966; Wiltschko (1968).

Since these pioneer experiments on robins, similar data have been collected on a variety of nocturnal migrants. Whitethroats (*Sylvia communis*), garden warblers (*S. borin*), subalpine warblers (*S. cantillans*) and blackcaps (*S. atricapilla*) in Europe and indigo buntings (*Passerina cyanea*) and savannah sparrows (*Passerculus sandwichensis*) in the United States have all been studied (Wiltschko and Merkel 1971; Wiltschko 1974; Wiltschko and Wiltschko 1975a; Emlen et al. 1976; Viehmann 1979; Bingman 1981) and when the data have been subjected to second-order analysis the results have been consistent in their support for the existence of a magnetic compass sense.

8.2. The magnetic compass: polarity or inclination?

It is not always easy in the mind's eye to envisage the form of the Earth's magnetic field and to picture the way it is being manipulated during experiments. The two most important points to keep in mind are, first, that the lines of force have polarity (i.e. north−south) and, secondly, that they are only horizontal at the equator. Elsewhere on the Earth's surface, the lines have an angle of dip (or inclination). They become steeper at higher latitudes and are more or less vertical at the geomagnetic poles. Numerically, angle of dip is zero at the equator and 90 degrees at the poles.

The combined product of polarity and angle of dip can be illustrated in the following way. Imagine yourself in the Northern Hemisphere at about the latitude of Britain, where angle of dip is about 70 degrees. If you stand and face north, the magnetic lines of force that pass through your head have their more southerly part going up into the sky behind you while their more northerly part disappears into the ground a short distance in front. If you face west, the lines travel up and southwards to your left and down and northwards to your right whereas, if you face south, the lines come down from the southern sky in front of you and disappear into the ground behind you. If you stand in the Southern Hemisphere at about the latitude of New Zealand and face south, the lines of force descend from the northern sky behind you and disappear into the ground to the south in front of you.

Everywhere, therefore, the lines of force are polarized north to south. In addition, except at the equator, the lines of force descend into the ground towards the poles and ascend into the sky towards the equator. These characteristics in theory offer birds two alternative ways of using the magnetic field as a compass: (1) as a north–south 'polarity' compass; or (2) as a pole–equator 'inclination' or 'dip' compass.

8.2.1. Alternatives

The most familiar way of 'reading' the magnetic field, as when using a hand-held compass, is to make use of the polarity of the lines of force. If a dipole magnet, such as a compass needle, is suspended in the geomagnetic field, it will align itself with the north-seeking arm pointing, naturally, to the north. Most such needles are suspended so that they can only rotate in the horizontal plane. Such a compass will always point to the north whether it is used in the Northern Hemisphere, where lines of force dive into the ground to the north, in the Southern Hemisphere, where they disappear up into the air, or at the equator, where they are horizontal. In fact, the only places it will fail to work will be at the geomagnetic poles where, because the lines of force are vertical, there is no horizontal component of the field with which the compass needle can align.

The second way to obtain compass information from the magnetic field is to make use of the angle of inclination. In essence, the animal needs to turn until the lines of force make their smallest angle with respect to gravity. Put another way, the animal needs to turn until the lines of force go into the ground most steeply. When facing in this direction, the animal will be facing the nearer of the two geomagnetic poles, the north pole in the Northern Hemisphere, the south pole in the Southern. Such a system, however, provides no compass information near the equator where the lines of force are horizontal.

8.2.2. Evidence

Experiments to test whether birds used a magnetic compass based on inclination or polarity were initiated by Wiltschko (1972), again using robins in a state of migratory restlessness. A system of electromagnetic coils around the orientation cage (Fig. 8.1) allowed the horizontal and vertical components of the magnetic field to be manipulated independently.

If the vertical component of the geomagnetic field is opposed and reversed by a stronger field produced by coils, the polarity in the horizontal plane remains the same but the lines of force disappear upwards towards the pole instead of descending into the ground. A compass needle still points to the north but an inclination compass now finds the smallest angle with gravity towards the equator instead of towards the pole. In other words, reversing the vertical component reverses an inclination compass but not a polarity compass.

If the horizontal component of the geomagnetic field is reversed, the lines of force are now bent so that their more northern end disappears into the ground towards the south pole of the Earth rather than towards the north. Both a polarity and an inclination compass would be reversed, the former pointing south instead of north, the latter pointing towards the equator instead of the pole.

Finally, if both the horizontal and vertical components are reversed, the polarity of the lines are simply reversed. They still dive into the ground toward the pole, as do the unaltered lines of force. Now, however, it is their more southerly end that dives into the ground in the northern hemisphere and their more northerly end in the southern hemisphere. A polarity compass shows reversal but an inclination compass is unaffected, still finding the smallest angle to gravity towards the pole (see Fig. 8.2).

Wiltschko carried out all of these manipulations and came to the surprising conclusion that, although robins use the alignment of the lines of force to provide themselves with a north-south axis, they determine which is its poleward end, not by using polarity but by using angle of inclination. The clearest reversal was obtained by reversing either the vertical or the horizontal component (Fig. 8.3). If both components were reversed, the robins' orientation was unaltered. Finally, in a horizontal field with polarity and alignment but with an inclination of zero degrees, such as would be experienced at the equator, orientation was uniform. Similar results have been obtained for the blackcap (*Sylvia atricapilla*) (Viehmann 1979).

8.2.3. Advantages and disadvantages

An inability to use a magnetic compass under equatorial conditions is no disadvantage to a robin which travels no further than Mediterranean latitudes in its autumn migration. The same is probably true for blackcaps. Although many travel to North Africa, few reach the magnetic equator. Such an inability would be more of a disadvantage to a transequatorial migrant.

To test whether transequatorial migrants also use an inclination compass, Wiltschko (1974) repeated his experiments, this time using the garden warbler (*Sylvia borin*). This species breeds through most of Europe between 40° and 65° N and winters in tropical and southern Africa between the latitudes of 10° N and 30° S, with by far the largest part of its winter range lying to the south of the magnetic equator. Most garden warblers, therefore, during the course of their annual migrations, have to cross a zone in which the geomagnetic field is horizontal and in which, therefore, an inclination compass of the type possessed by robins and blackcaps would be useless. Wiltschko found that garden warblers, which are nocturnal migrants, would show their normal migration direction in an orientation cage and that they could do so even in the absence of stars. Experimental rotation of the magnetic field produced a corresponding shift in their

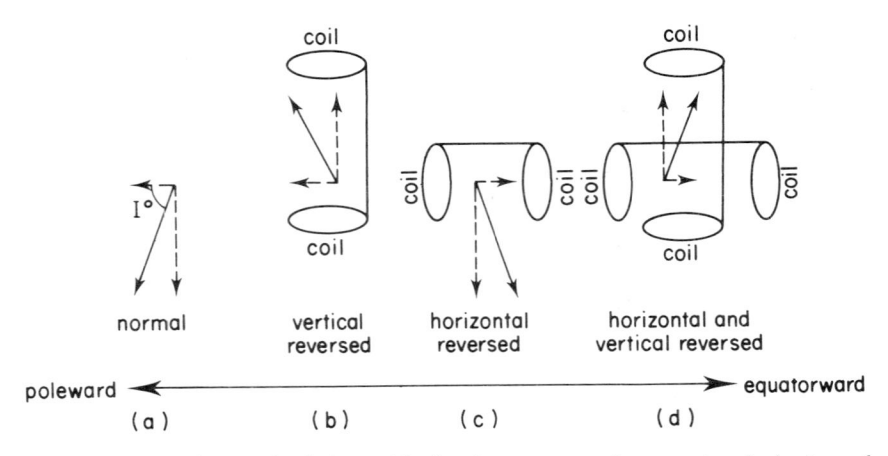

Fig. 8.2 Manipulation of polarity and inclination compasses by reversing the horizontal and vertical components of the geomagnetic field.

In mid-temperate latitudes the lines of force of the geomagnetic field (solid line) are not horizontal (a) but have an angle of inclination (I). These lines of force, however, can be resolved into horizontal and vertical components (dashed lines). By opposing either the horizontal or vertical component with a double-strength field generated by appropriate electromagnetic coils (b–d), one or other (b, c) or both (d) of these components can be reversed. When only the vertical component is reversed (b), an inclination compass is reversed but a polarity compass is unaffected. When only the horizontal component is reversed (c), both polarity and inclination compasses are reversed. When both horizontal and vertical components are reversed (d), a polarity compass is reversed but an inclination compass is unaffected. (An inclination compass is unaffected if the lines of force slope from top right to bottom left, irrespective of the direction the arrow is pointing, but is reversed if the lines slope from top left to bottom right. A polarity compass is reversed whenever the arrow points to the right instead of to the left.)

orientation . . . and they were *unable* to orient in a horizontal magnetic field.

The available evidence, therefore, albeit for only three species, is that birds have evolved to use a magnetic compass based on inclination rather than polarity. In so doing, they seem to have sacrificed an ability to use a magnetic compass near to the magnetic equator. We might suppose that, for such an obvious disadvantage to have evolved, it must be more than offset by some compensatory advantage. One such advantage comes to mind that would apply with particular force to seasonal migrants such as the three species so far tested.

The question of whether migrant birds are born with an instinctive ability to read one of their compasses is discussed in Chapter 9. Suppose, for the moment, that migrant birds are born with an instinctive urge to migrate towards magnetic south in their first autumn. Any advantage conferred on a bird by such a predisposition would, of course, be completely negated if the Earth's magnetic field were suddenly to change in any significant way. Yet, periodically, this is precisely what does occur or,

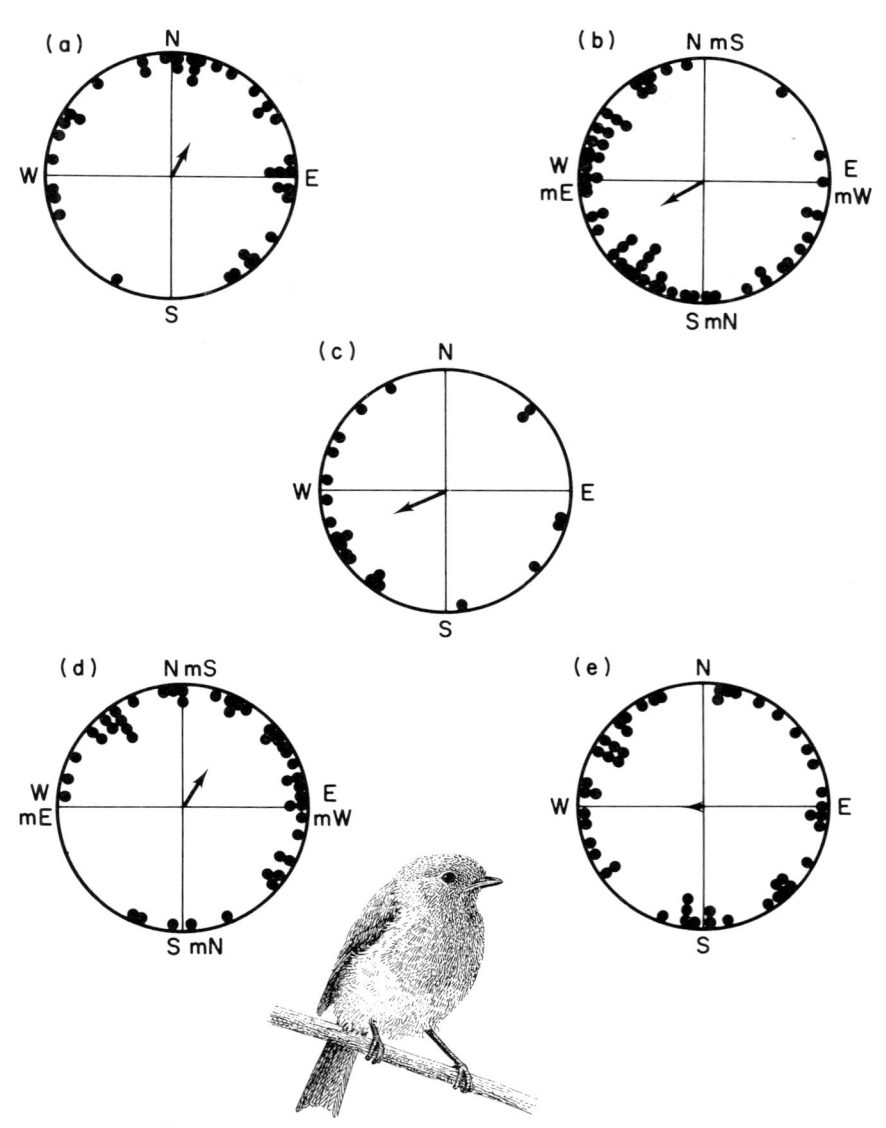

Fig. 8.3 Evidence for an inclination compass in the European robin (**Erithacus rubecula**)

Each dot shows mean bearing for one bird in an orientation cage over the course of a single night during the period of migratory restlessness in spring. In the normal geomagnetic field, robins orient to the northeast (a). They continue to do so (d) when both vertical and horizontal components of the field are reversed (Fig. 8.2d). When either (b) the vertical component is reversed (Fig. 8.2b) or (c) the horizontal component is reversed (Fig. 8.2c), the robins orient to the southwest. If the lines of force are made horizontal, orientation is random with a tendency toward bimodality (e).

[Modified from Wiltschko and Wiltschko (1972)]

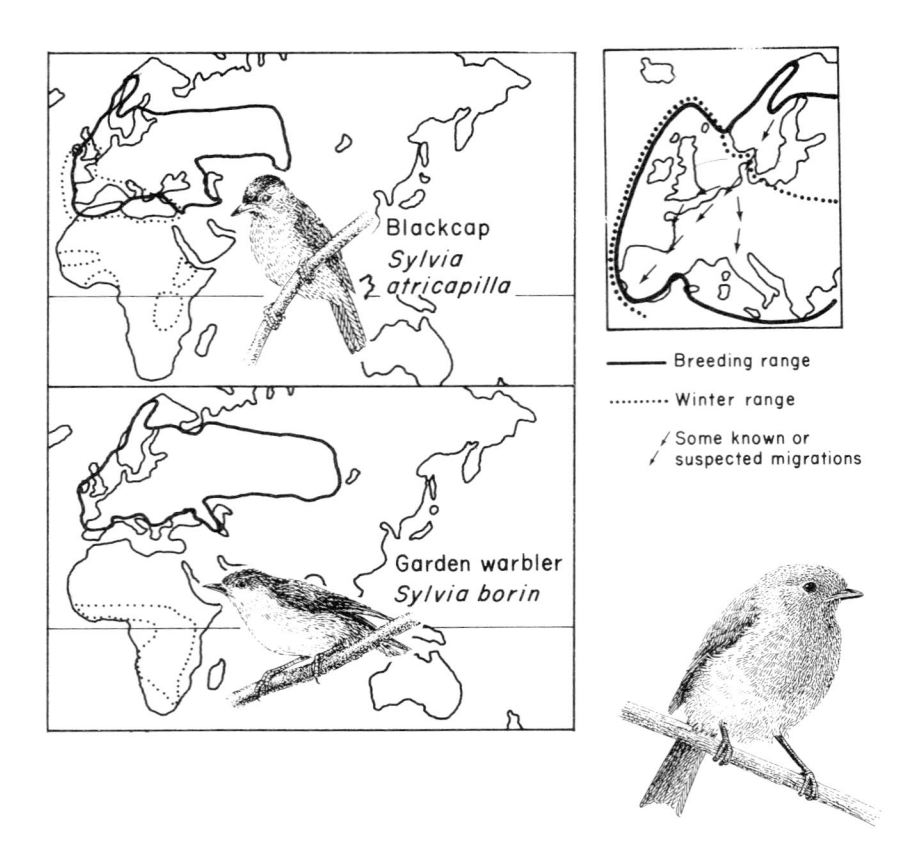

Fig. 8.4 Breeding and wintering grounds of the European robin (**Erithacus rubecula**), blackcap (**Sylvia atricapilla**) and garden warbler (**S. borin**)

[Re-drawn from Baker (1978), after several authors]

at least, it does if the bird happens to be using a magnetic compass based on polarity!

Every few thousand years, the polarity of the Earth's magnetic field reverses. Consider a young bird of a species in the northern hemisphere that has evolved an instinctive urge to fly towards magnetic south on its first autumn migration but that uses a polarity compass by which to recognize south. After the next reversal, which is predicted to be nearly upon us, such a bird would head north to spend the winter in the Arctic. A bird that uses an inclination compass, on the other hand, would still be predisposed to fly toward the equator. What is not clear is to what extent, and for how long, even an inclination compass might give false information during the reversal period.

The progress we are making toward an understanding of the magnetic compass of birds is positive but slow. One of the reasons for this slowness is that so many people fail to obtain positive results when they attempt to repeat experiments that seem to work perfectly well in another laboratory. This ease with which experiments on the magnetic sense can produce negative results is worth further consideration.

8.3. Critical factors

Wallraff and Gelderloos (1978) summarized the position well when they found that the orientation of various caged passerines tracked the horizontal rotation of a planetarium sky perfectly well but failed to track a similar rotation of the magnetic field. They stressed, as Wallraff (e.g. 1978a) had done elsewhere, that magnetic orientation in a cage seems to be a very weak phenomenon that, at best, can only just be detected. As a result, even small changes in experimental conditions may cause any magnetic orientation that still occurs to be masked completely by the background 'noise'. They then suggest various factors in their experiment that may have prevented detection of magnetic orientation, either because the birds were prevented from detecting the magnetic field or because the experimenters could no longer detect magnetic orientation among the birds' other behaviour.

Of the factors which Wallraff and Gelderloos suggested may have been critical, one was that the orientation cages used were of the Emlen-funnel type rather than the Merkel-cage type (see Chapter 3) used so successfully by the Wiltschkos. The concensus of opinion is that non-visual orientation is less pronounced in funnels than cages (though for some species, e.g. pied flycatchers, *Ficedula hypoleuca*, the converse is true; Beck, personal communication). Perhaps some feature of a bird's movements make detection of the magnetic field (i.e. 'magnetoreception') more difficult in a funnel than in a cage. Another possibility is that the total intensity of the magnetic field in the testing funnels, although the same as in the birds' living quarters, was rather low (39 000 nT) compared to the field intensity outside the buildings (46 000 nT). The magnetic field inside the room was also distorted by the steel reinforcements of the building and the distortion was not fully corrected by the artificial field, which itself was not strictly homogenous. Yet another factor that may be critical is that the room was permeated by the artificial 50 Hz noise which penetrates all modern buildings. This might disturb magnetoreception if the birds measure the magnetic field by some electric rather than magnetic means, e.g. electromagnetic induction. Finally, the field to which the birds were exposed did, after all, rotate, thereby introducing yet other variation.

Any one of these factors could have prevented magnetoreception (or the detection of magnetoreception). Until we know which factors are important and which are not, experiments will continue to be designed more or less in the dark. It seems most important that a search begins in earnest for those factors that are critical in the success or failure of experiments on magnetoreception.

8.3.1. Field intensity

Only one of these potentially critical factors has so far received attention in direct experiments on birds. In the 1950s and '60s, many authors were impressed by the weakness of the Earth's magnetic field and suspected that this fact above all else made its detection by birds unlikely. Any experiments on magnetoreception, therefore, tended to use field strengths greatly in excess of that of the natural geomagnetic field. Such experiments consistently obtained negative results.

One of the breakthroughs came when Wiltschko (1968, 1972, 1978) demonstrated how absolutely critical it was to experimental success for field intensities to be near to the strength of the geomagnetic field and to be relatively steady over short time intervals. European robins, adapted to the 46 000 nT of the Frankfurt region, only showed orientation to the ambient magnetic field within the range of intensity from 40 000 to 57 000 nT. When field intensity was taken outside of these limits, the birds were temporarily disoriented. Within three days of adaptation to a new intensity, however, the birds were once again able to orient. Robins adapted to extremes of 16 000 and 150 000 nT at the same time could orient at these intensities but not in the intermediate intensity range of 60 000 to almost 150 000 nT.

Several interesting conclusions emerge from this study. One, of course, is that a bird would be capable of adapting to the gradual change in intensity of the magnetic field that it would experience if it migrated over any distance between pole and equator. Another is that the magnetic sense organ (i.e. the 'magnetoreceptor') needs a period of time to adapt to, or recover from, any change in its magnetic environment. The most important message from this work, however, is that extreme care needs to be taken in controlling conditions during experiments on magnetoreception.

Early experiments on magnetoreception, using high field intensities, must have been equivalent to testing for star orientation while shining a bright light into a bird's eyes. For all we know, that light may still be shining, albeit from other factors, in most experiments on the magnetic sense.

8.3.2. The search for other critical factors

One reason why the identification of critical factors, and thus all progress in the study of magnetoreception, has been so slow is the constraint of having to work on birds showing migratory restlessness. This limits the times of year over which experiments can be carried out, enforces long time intervals between the treatment of young birds and the first opportunity to test them (Wiltschko 1982), and restricts treatment of the data to second-order analysis.

A long list of authors, their number entering well into double figures, have tried to free the study of the magnetic sense from this constraint. The favourite target has been the development of a method based on some form of conditioned response. If such a method could be found, it would speed up considerably the identification of those factors that interfere with magnetoreception and hence also understanding of the way the magnetic sense works.

Unfortunately, everybody reported negative results. Even Kreithen, who has had such success in extending our knowledge of the sensory world of the pigeon by his use of the technique of cardiac conditioning, reported failure (Kreithen and Keeton 1974c). Everybody, that is, until finally Bookman (1978) reported success.

Bookman trained pigeons to discriminate between the presence or absence of an Earth-strength, vertical field generated by Helmholtz coils in a flight tunnel inside a chamber shielded with mu-metal that thus excluded the Earth's magnetic field. The birds usually walked through the tunnel, flying only rarely. Successful discrimination between the presence or absence of the artificial field, however, was associated with hovering, jumping or turning by the bird. Such activity was absent in all previous experiments in which the birds were usually strapped down or otherwise restricted. The temptation is to conclude that birds can only detect the magnetic field if they are showing a minimum degree of movement.

Encouraging though Bookman's results appear, his experimental design is not particularly simple and does not bring the opportunity to study magnetoreception within the range of many laboratories. Unless there is a major advance soon in finding an experimental technique for studying the magnetic compass of birds, a technique that is simple to carry out, uses a convenient laboratory species, such as the pigeon, and can be carried out at any time of year, my own feeling is that progress in understanding magnetoreception will come from a different source entirely; from the study of man himself.

Non-visual compass orientation by humans can be studied very simply by placing a blindfolded, ear-muffed subject on a wooden chair on a turntable, rotating him, and from time to time stopping and asking him to estimate the compass direction in which he is facing. Such 'chair' experiments are a pleasingly simply way of testing for non-visual compass orientation by humans, but are not simply a question of placing someone on a chair and turning. The pattern of turning is critical. Too little rotation and the subject may follow the rotation by means of inner ear mechanisms. Too much rotation and the ability to judge compass direction disappears altogether. This and other evidence suggests that magnetic receptors dislike being spun as much as inertial receptors. Given just the right amount of turning, however, chair experiments seem to test only a magnetic compass

106 Bird navigation: the solution of a mystery?

sense (Fig. 8.5) and thus provide a simple method for testing magnetoreception.

As for migrant birds in orientation cages, compass orientation in chair experiments is only just discernible against the background noise if experiments are carried out on a randomly selected group of people with no constraints on their actions and on test conditions. The experimental technique is so simple, however, that it has been possible to identify quickly some of the factors that are critical for successful magnetoreception. Some of these factors are unsurprising. Others are totally unexpected but in retrospect could be important indicators to the way magnetoreception occurs. We may guess that a similar range of factors await discovery for birds.

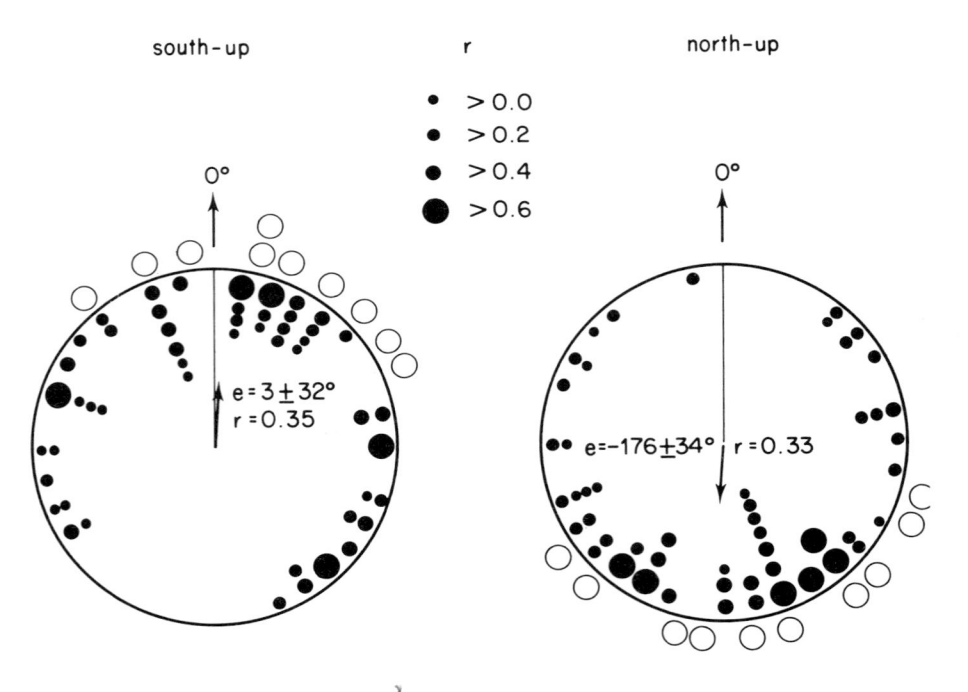

Fig. 8.5 Evidence for a magnetic compass in humans

Data were obtained from chair experiments (see text). Each black dot shows the mean vector for the 8 errors in estimating compass direction while blindfolded during a single test-run. Ten subjects were each tested 5 times (i.e. 40 estimates of compass direction per person) while wearing a bar magnet in north-up alignment on the right temple and 5 times while wearing a bar magnet in south-up alignment. Open dots show the mean error for these 40 estimates for each person. At all levels of analysis, reversal of the orientation of the bar magnet produces a significant rotation in the estimates of compass direction.

(From Baker 1984a. Photo by Les Lockey)

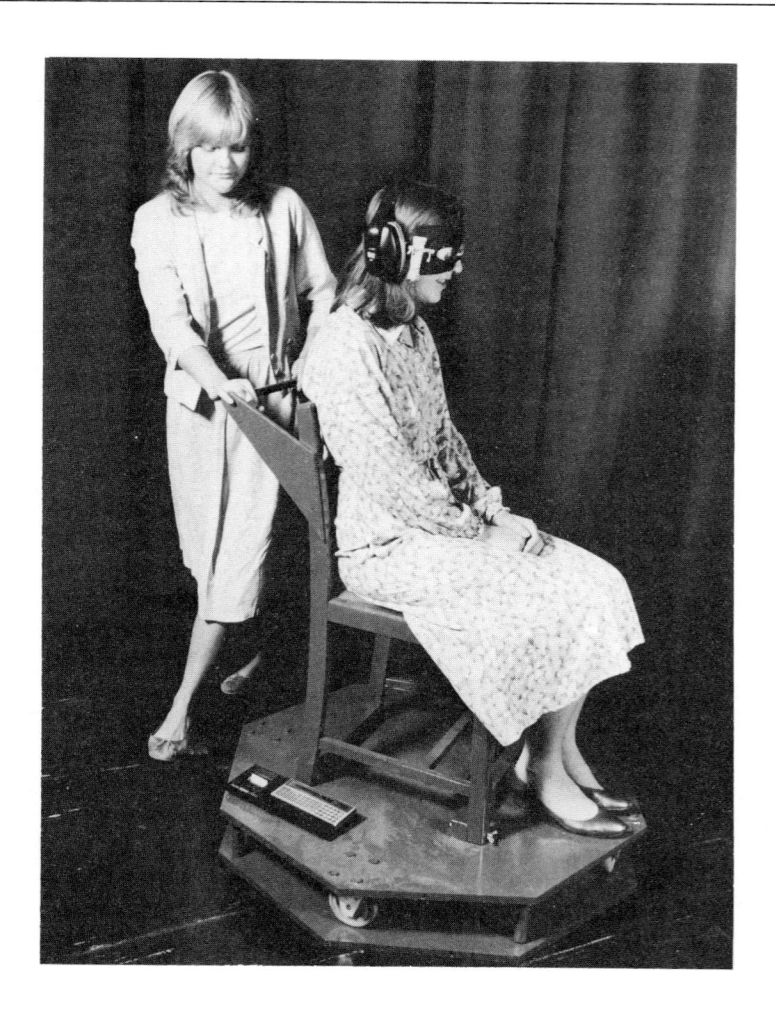

8.3.3. Electric fields

Some authors have explored the possibility for birds that magnetoreception is based on electromagnetic induction, the magnetic sense organ thereby detecting electric currents rather than the magnetic field directly (Wallraff 1978a; Rosenblum and Jungerman 1981). Support for such a mechanism in insects could be gleaned from the early observation by Schneider (1961) that the resting orientation of beetles only relates to the lines of magnetic force in the presence of appropriate electrostatic fields. Magnetoreception seems unlikely to be influenced by such electric fields unless it involves detection of electric currents.

As many of us know to our discomfort, many modern clothing fabrics generate powerful electrostatic fields. It was possible, therefore, that

108 *Bird navigation: the solution of a mystery?*

magnetoreception by humans might be influenced by these fields. Two identical robes were made, one of cotton, the other of polyester, and a group of volunteers were tested in the chair for their ability to orientate. Compass orientation while wearing the polyester robe was significantly worse than orientation while wearing cotton (Fig. 8.6).

The environment in which the human magnetic sense evolved its own particular characteristics will have been one permeated electrostatically by only the Earth's own field, distorted in places and at times by anomalies

(a)

e = 2°
r = 0.40

none

(b)

e = 7°
r = 0.37

cotton

(c)

e = −3°
r = 0.07

polyester

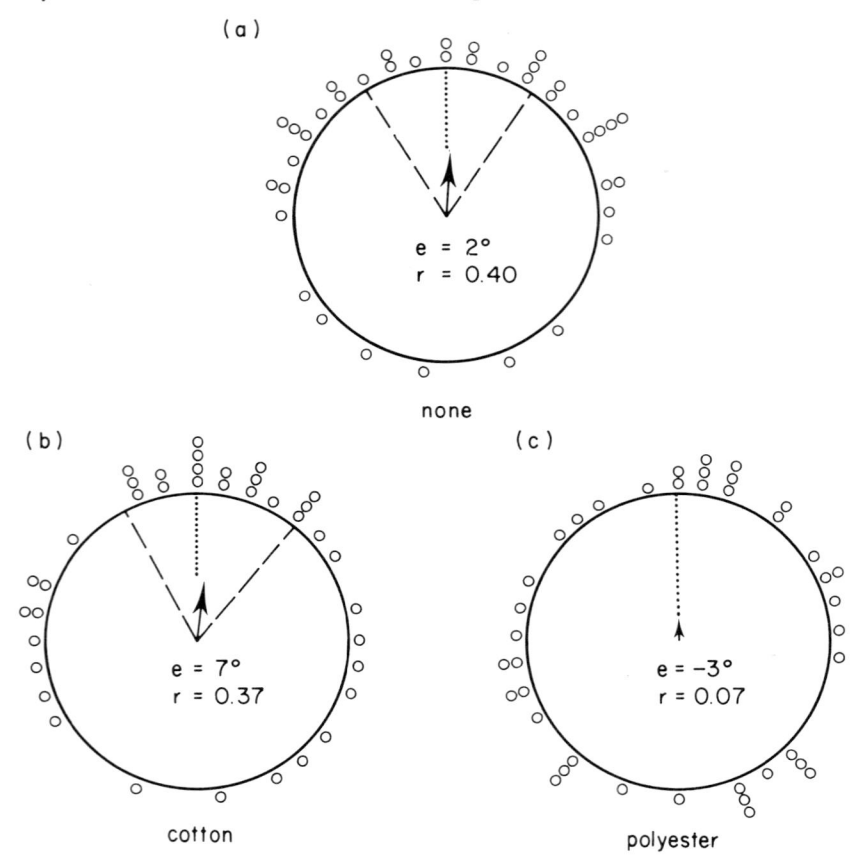

Fig. 8.6 Influence of clothing on non-visual compass orientation by humans in chair experiments

Each of 39 subjects were tested twice wearing nothing, once wearing only a cotton robe and once wearing a polyester robe. The experimenter wore nothing during all tests. Subjects had a brass bar (which they were told might be a magnet) on the right temple in all tests. Each dot shows the mean error for the 8 compass estimates made during a single test. Good compass orientation was found only if subjects wore nothing or cotton. With polyester clothing, orientation was random. The h-values (Fig. 3.4) for tests wearing polyester and cotton were significantly different.

(Data from Anon. personal communication)

such as waterfalls and events such as electric storms. Polyester clothing is a recent event and if it does interfere with magnetoreception, there has not yet been time or perhaps even selection, for man to evolve a magnetic compass that is immune to such electrostatic interference.

While thinking in evolutionary time scales, the thought may have occurred to the reader, as it did to me and my research team, that through most of the few million years that natural selection has been acting on human magnetoreception, not only did man not wear polyester clothes, he wore no clothes at all. How do we know, therefore, that human magnetoreception does not suffer further interference simply by the wearing of clothes?

It so happens that we do know. A member of my research team, who prefers to remain anonymous, volunteered to carry out a series of chair experiments (the series that gave the polyester/cotton comparison) at a naturist club near Manchester. To everybodies' relief, orientation was no better in the total absence of clothes than when wearing a cotton robe (Fig. 8.6).

It should be stressed that our experiments on humans do not so far demonstrate a direct relationship between electrostatic fields and the efficiency of magnetoreception. It is difficult to imagine what other characteristic of polyester clothing could be critical apart from the electric fields that are its most conspicuous feature. We have not yet, however, tried the critical experiment of manipulating the electric field around a subject wearing cotton. Moreover, we cannot yet say that the influence of polyester clothes is directly on magnetoreception. On the evidence so far, the interference could come anywhere on the pathway of nerves from the magnetoreceptor, where the magnetic fields are detected, to the central nervous system, where the input is processed to produce an estimate of direction. For the present, however, the important point is that there does seem to be interference.

Wallraff and Gelderloos (1978) were concerned about the 50 Hz interference that permeates modern buildings in case it disrupted magnetoreception by their caged birds. It now seems that perhaps an even more potent source of disruption may be those stray electrostatic fields that are such a feature of the modern, man-made environment. Birds subjected to laboratory and other experimental conditions are being asked to detect magnetic fields in an electric environment as distorted from that in which their magnetic sense evolved to function as are humans wearing polyester clothing.

8.3.4. Orientation during rest?

The most bizarre discovery made while testing human magnetoreception may well turn out to be the most significant. Indeed, it seemed so strange on first discovery that we found it difficult to believe. To date, however, the results have been replicated seven times using different groups of people. Each time the same effect emerges: efficiency of magnetoreception is influenced by alignment of the bed over the previous few nights (Baker 1984b).

The most accurate compass orientation in chair experiments is shown by people who sleep on a bed aligned within 45 degrees of the north–south axis (as measured by a compass on the pillow to take into account any local magnetic fields due to the bed, radiators, etc.). People who sleep with their head to the south and feet to the north are more accurate than those who sleep the other way round.

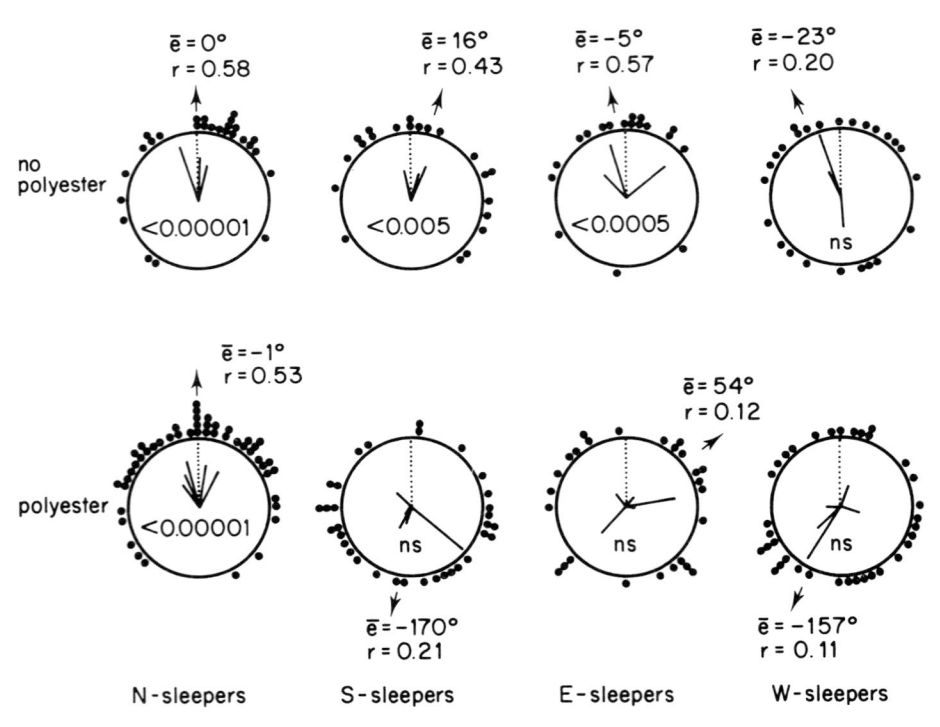

Fig. 8.7 Influence of bed orientation on non-visual compass orientation by humans in chair experiments

A north-sleeper is a person whose bed is so-aligned that when the person sits up in bed, he faces north $\pm 45°$; a south-sleeper, south $\pm 45°$; and so on. Eight series of chair experiments, testing various aspects of compass orientation, have been carried out with bed alignment of the subjects recorded as a matter of routine. Mean vectors for the mean errors (one mean error per person using all estimates) are shown for each of the four categories of bed alignment.

(From Baker 1984b)

At first, we could find no evidence at all that people who slept on beds aligned along the east–west axis could detect the magnetic field. Since testing people dressed only in cotton (or nothing), however, significant orientation by groups of east–west sleepers has been obtained. There has also been an improvement in the accuracy of compass orientation by south sleepers, people who sleep with their feet pointing south. Even under quiet electrostatic conditions, however, north–south sleepers are marginally superior to east–west sleepers.

The fact that the overnight orientation of the body influences compass orientation could imply that the magnetoreceptor can be affected in some long-term, but reversible, way by external forces, perhaps the magnetic field of the Earth. Several other lines of evidence also point to this conclusion. Firstly, the worst performance of all in chair experiments is given by people who have slept on beds aligned in different directions on the two nights preceding the test. Secondly, when a group of four east–west sleepers turned their beds to sleep with their feet pointing north, within a week an ability for magnetoreception had appeared where none could be detected before.

As yet, there is no evidence whether birds show any preference for aligning their head along the north–south axis during rest or sleep. Studies on a single cat show that, in this individual at least, there is such a preference.

The favoured interpretation at present is that the bed alignment effect is restorative, a mechanism evolved to repair any damage sustained by the magnetoreceptor during the course of the day (see Section 8.3.7). It is proposed (Baker 1984a, b) that repair is effected by aligning the head, and turning it from side to side during the night, within the Earth's magnetic field.

Evolutionary maintenance of such a repair mechanism renders the magnetoreceptor liable to disruption from exposure to the strong magnetic sources that abound in our modern environment. Birds and other animals, as well as humans, are subjected to these strong fields, particularly in laboratory environments.

8.3.5. Exposure to strong magnetic fields?

Evidence is accumulating for humans that modern magnetic artefacts such as stereo headphones, subway or other electrified trains, aeroplanes, electromagnetic coils in laboratories, etc., all disrupt magnetoreception for hours after exposure to the source. If the same is true for birds, we may wonder how many of the experiments on magnetoreception failed, or produced weak results, simply because the animals used had been inadvertently exposed to some strong magnetic field that blinded their magnetic sense before it was tested.

If a person wears a bar magnet with N-pole uppermost and a pole strength of about 200 G on their right temple for about ten minutes, then removes the bar magnet and is tested in the chair while wearing a brass bar (neither subject nor experimenter knowing which bar is a magnet and which made of brass) the ability to detect compass direction disappears (Fig. 8.8). We have not yet plotted rate of recovery from this disruption, but our preliminary results suggest that this after-effect of wearing a bar magnet lasts for the remainder of the day of the test, compass orientation not reappearing until after a night's sleep.

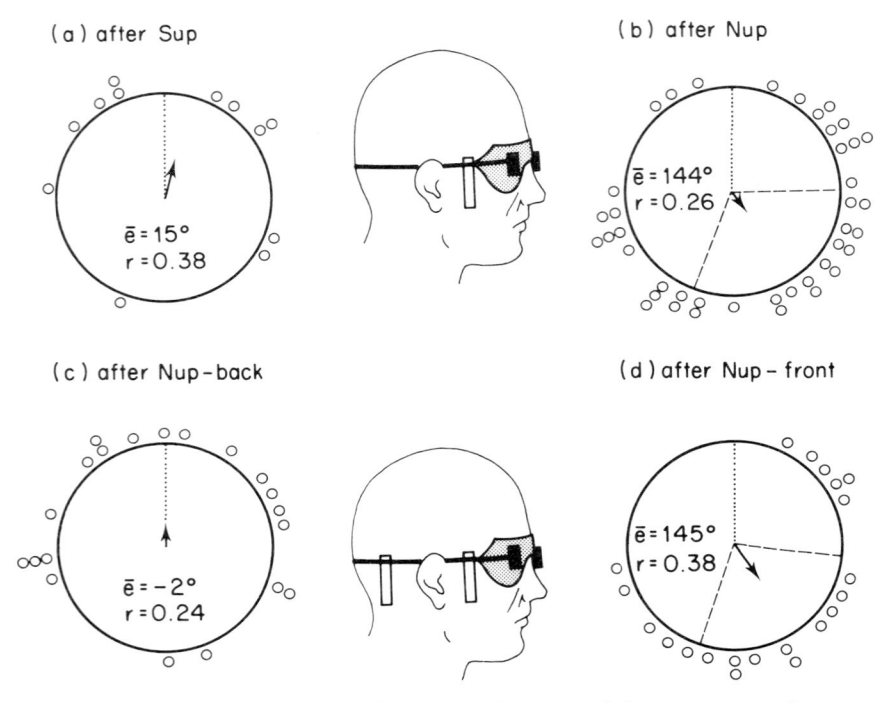

(a) after Sup

$\bar{e} = 15°$
$r = 0.38$

(b) after Nup

$\bar{e} = 144°$
$r = 0.26$

(c) after Nup-back

$\bar{e} = -2°$
$r = 0.24$

(d) after Nup-front

$\bar{e} = 145°$
$r = 0.38$

Fig. 8.8 The after effect: influence of polarity and position of a bar magnet on subsequent ability to orient to the geomagnetic field in chair experiments

Ear muffs not illustrated for clarity. All conventions as in Fig. 8.5. In (a) and (b), subjects wore a bar magnet for ten minutes as shown. The magnet was then replaced by a brass bar and the subject tested. In (c) and (d) subjects wore two bars (1 brass, 1 magnet) as shown. These were then replaced by 2 brass bars for testing. Nup or Sup describes uppermost pole of magnet

Walcott and Gould (in Walcott 1982) have tried exposing homing pigeons to strong magnetic fields and then testing them in homing experiments. Some pigeons were exposed to an alternating magnetic field produced by Helmholtz coils. Field intensity began at a level three thousand times stronger than the Earth's field and was then slowly reduced. This was done three times, with the pigeon's head in different directions relative to the field each time. When such birds were released under overcast skies, they were at first disoriented and showed worse homing performance than controls taken to the release site directly from the loft. When released under sunny skies, there was no difference between the two groups.

A second group of pigeons were exposed to a strong but steady magnetic field by placing their heads between the pole pieces of a magnet in a field more than two thousand times stronger than the Earth's field. Again this had no influence on birds released under sunny skies, but under overcast skies the treated birds seemed to be better oriented and homed faster than controls.

Evidently pigeons, like humans, are influenced in some medium- to long-term way by exposure to unusual magnetic fields, though the influence may aid, instead of disrupt, magnetoreception (see Baker 1984a). For the moment, however, let us concentrate on disruption and follow further the thesis that the bed alignment effect reflects a repair mechanism. We may wonder what natural features of the environment might have disrupted the magnetoreceptor of humans and other animals before the advent of the iron age to an extent sufficient to maintain selection for a repair mechanism such as the bed alignment effect.

Magnetic rocks occur naturally and any human or bird that chanced to come into close contact with such a rock may have been subjected to an influence similar to bringing a bar magnet near to the head. Much larger magnetic deposits in the Earth's crust give rise to anomalies in the Earth's magnetic field that can produce changes in intensity of the order of a few thousand nano Teslas over a distance of less than a kilometre. A bird flying through such an anomaly might conceivably receive minor damage to the magnetoreceptor of the type we are considering here. Another possible source of such damage is found in magnetic storms.

8.3.6. Magnetic storms?

Magnetic storms should not be confused with meteorological events such as thunderstorms (which are 'electric' storms). Unlike thunderstorms, magnetic storms are not generated within the Earth's atmosphere, but millions of kilometres away on the surface of the sun. Protons, electrons and other particles stream away from magnetic disturbances, such as sun spots, in the sun's outer layers to be carried across space by the solar wind. As these particles pass by the Earth, they generate electric currents in the upper atmosphere. These currents in turn induce rapid fluctuations in the Earth's magnetic field as measured on the Earth's surface.

Two important points have to be appreciated about fluctuations in the Earth's magnetic field caused by magnetic storms. The first is that the changes in intensity are very small relative to the intensity of the magnetic field due to the Earth itself. In mid-temperate latitudes, for example, field intensity due to the Earth's core is about 50 000 nT. Within this field, most storms cause fluctuations of only 100 nT or so, perhaps 1000 nT in the most extreme storms. The second point is that such storms can scarcely be detected from the movements of a compass needle. The direction of the geomagnetic lines of force are rarely deflected even by so much as one degree.

Fluctuations in intensity due to magnetic storms are usually of the same order as the change in intensity with time of day (Fig. 4.11). Indeed, one of the most marked influences of magnetic storms is that they may mask the more regular fluctuation with time of day.

It seems inconceivable that such tiny fluctuations in the magnetic field could be of any consequence to a magnetoreceptor, to magnetoreception, or to any aspect of physiology. It seems particularly unlikely that they could influence compass orientation because of the minute effects that the storms have on the direction of geomagnetic lines of force. Yet, as we shall see below and again in Chapter 10, all of these elements seem to be sensitive to magnetic storm activity.

A variety of indices have been developed by geophysicists to measure magnetic storm activity. The only index that has been used by students of bird orientation, however, is the K-index. This index ranges from 0 to 9, where $K = 0-3$ indicates little or no disturbance, $K = 4$ indicates weak storm activity, $K = 5$ indicates moderate disturbance, and $K \geqslant 6$ indicates moderately severe storms. K-values are measured for eight 3-hour periods each day and formally reflect the difference in nanoTeslas between the lowest and highest intensity readings during the particular 3-hour period (when 'normal' fluctuations due to time-of-day and other factors are taken into account). K-indices show daily, lunar and seasonal cycles as well as longer term variations due to cycles of, for example, sun-spot activity. In general, however, the most common K-index is between 1 and 2.

Evidence that homing pigeons are influenced by magnetic storms is presented in Chapter 10. Here we are concerned only with orientation that we can be reasonably certain is magnetic compass orientation.

Southern (1969, 1971, 1978) has collected data which suggest that orientation to the magnetic field by chicks of the ring-billed gull (*Larus delawarensis*) from a colony in the north-eastern United States is disrupted by even relatively weak magnetic storms. Southern's chicks could orientate under overcast skies only at K-values of 1 or 0. Under clear skies they were well oriented at all K-values less than or equal to 4, but not above. These results, gathered over a 15-year period, although variable and inconsistent, do rather suggest a sensitivity on the part of the birds to changes in the magnetic field of the order of perhaps only 20 nT.

As will become apparent in Chapter 10, the nature of the orientation being shown by Southern's gulls becomes a matter of some importance. When tested in an orientation cage, or rather an arena, chicks as young as 2 days of age show a southeastward orientation that is unrelated to the direction of their home colony from the test site. The only reasonable explanation for this orientation is that it is a precursor of the direction they will adopt when they set off on autumn migration. Evidence that they eventually fly in this direction is provided by banding returns of birds from the colony (Southern 1971). If this interpretation is correct, the magnetic storms are disrupting straightforward compass orientation. Moveover, it seems to be magnetic compass orientation (Southern 1972).

There have been several indications from chair experiments that the human magnetic compass may also be disrupted by magnetic storms (Baker and Mather 1982b). In some series of experiments, compass orientation during the day seemed to be influenced by magnetic storm activity during the previous night. These trends, however, where once again orientation disappeared at mean K-values greater than about 3, have not always appeared in all later series of chair experiments. More data are needed.

Over the past few years a number of studies have suggested that a wide spectrum of human behaviour and physiology may be affected adversely by magnetic storms. The incidence of aircraft accidents attributable to pilot error (Srivastava and Saxena 1980) and of road accidents (Verma et al. 1977), both of which could just conceivably involve magnetic orientation, have been claimed to increase with an increase in magnetic storm activity. Admissions to coronary intensive care units may also show such a correlation (Verma et al. 1978; Knox et al. 1979).

If such apparently unconnected aspects of human physiology can all be influenced adversely by magnetic storms, it begins to seem more feasible that both compass orientation and even the coherence of the magneto-receptor might also be adversely affected. In which case, perhaps magnetic storms are, after all, one of the reasons for maintaining a mechanism for overnight repair.

8.3.7. Time of day?

Compass orientation by humans in chair experiments varies with time of day. The best performance is around midday but from about 15.00 h GMT orientation deteriorates. At first we thought that, as seems to be the case for bees, the variation in performance mirrored the daily changes in intensity of the geomagnetic field (Baker and Mather 1982a). With more data, however, this possibility is fading. In particular, what now emerges is that the way a person's compass sense is influenced by time of day depends on the orientation of their bed during sleep (Fig. 8.9). North sleepers as a group have the most robust sense and do not lose their orientation ability until well after midnight. South sleepers begin the day with an orientation ability as good as north sleepers but lose it much more quickly, producing random results from about 15.00 h. Interestingly, after midnight, south sleepers show what can only be described as a reversal of their magnetic compass. Consistently, and significantly, they begin to misidentify north as south, east as west, etc. East–West sleepers produce random results throughout the day, but even they come nearer to producing significant results earlier in the day than later.

116 Bird navigation: the solution of a mystery?

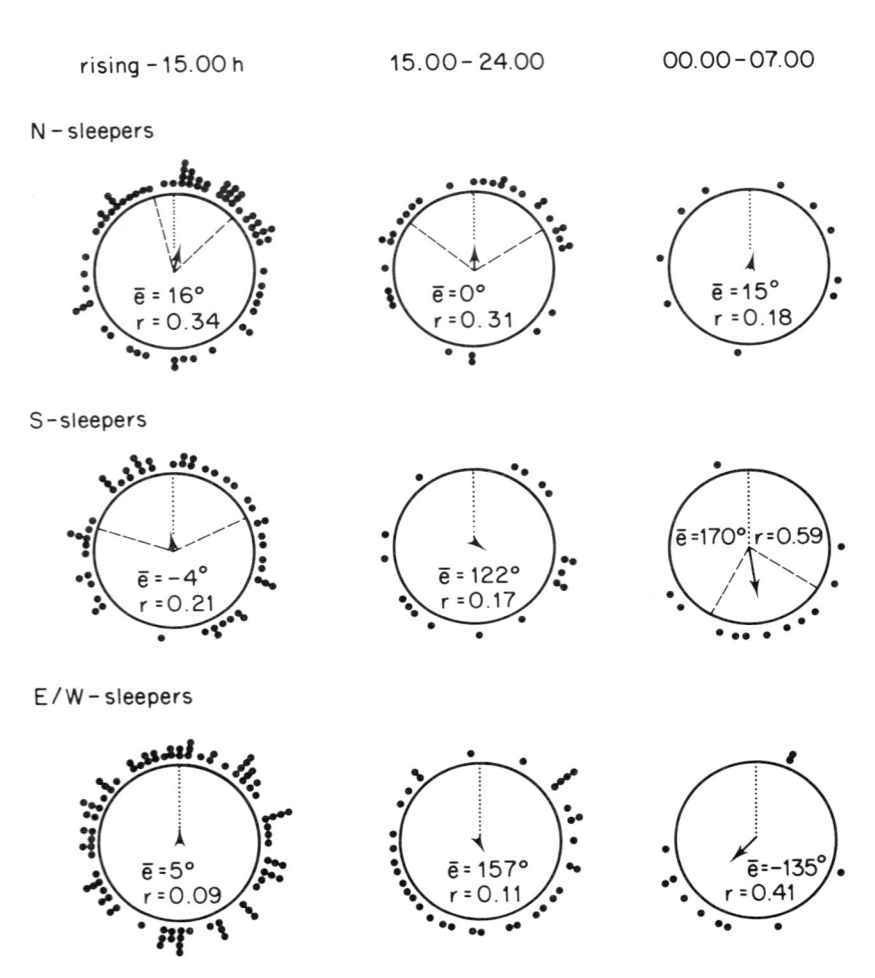

rising – 15.00 h   15.00 – 24.00   00.00 – 07.00

N – sleepers

ē = 16°, r = 0.34   ē = 0°, r = 0.31   ē = 15°, r = 0.18

S – sleepers

ē = –4°, r = 0.21   ē = 122°, r = 0.17   ē = 170°, r = 0.59

E/W – sleepers

ē = 5°, r = 0.09   ē = 157°, r = 0.11   ē = –135°, r = 0.41

Fig. 8.9 Deterioration in compass orientation by people with time of day according to bed alignment the previous night

This pattern of results is just what we should expect if the influence of bed orientation were due to the alignment of magnetic particles under the influence of the geomagnetic field during sleep followed by their gradual disruption or realignment during the day while the person is active (Baker 1984a, b). If birds are similarly influenced by the geomagnetic field during sleep and activity, the success of experiments on magnetic compass orientation may depend on the time of day at which they are carried out and how long it is since the bird last slept.

8.4. The search for the magnetic receptor

I have referred frequently and confidently over the past few pages to a magnetoreceptor, both in birds and humans. The fact is, of course, that as yet we have no idea for any animal what a magnetoreceptor looks like, how it works, or, indeed, even whether magnetoreception requires a specialized receptor at all.

This gap in our knowledge is challenging, but is not so dramatic a caveat as many people might like to think. After all, the mechanism of detection of ultraviolet light is also unknown (Kreithen 1979). So, too, is the mechanism of detection of polarized light. Birds probably detect polarized light with their eyes (Kreithen and Keeton 1974b; Delius *et al.* 1976), but in salamanders at least, polarized light detection is not by the eyes but by a receptor system located under the bones of the skull (Alder and Taylor 1973).

Early theories of magnetoreception expected birds to be able to detect and measure in a direct way the components of the geomagnetic field (Viguier 1882). For a while, however, theories of direct sensitivity were replaced by others based on indirect sensitivity.

8.4.1. Induction

Yeagley (1947) suggested that the magnetic field could be detected by a flying bird acting as a linear conductor moving through the lines of force. Theoretically, this would set up a small potential difference between the two ends of the conductor. To be useful, however, the induced voltage might need to be measured to within a millionth of a volt. Moreover, such minute voltages would have to be measured against a background of the far more powerful electrostatic field of the Earth (about one volt) and of the fluctuating effect of charged clouds (Slepian 1948).

Stewart (1957) suggested that air friction on a bird's feathers sets up electrostatic forces which react to the Earth's field whereas Talkington (1967) argued that the lymph tubes in the pleats of the pecten (a heavily pigmented projection into the eye from the optic disc) could act as conductors in which an electromotive force is generated. Danilov *et al.* (1970) also suggest a role for the pecten in magnetic sensitivity, but in a detailed search of the structure using an electron microscope and other, histological, techniques, Southern *et al.* (1982) could find nothing that looked like a magnetoreceptor.

Wilkinson (1949) suggested that a more satisfactory form of indirect detection would be a conducting loop (such as a semicircular canal) oscillated in the geomagnetic field. An alternating current would be generated by the dynamo principle and this might be easier to measure than a potential difference. Even so, measurement of the generated current would still have to be made against powerful background currents, such as those generated by other physiological processes.

Some authors still prefer indirect hypotheses. Wallraff (1978a) envisages the bird acting as a form of magnetic probe which measures magnetic intensities. The principle of measurement is based on electromagnetic induction. Important parts of the probe are suggested to be located in the wings and normal function would depend on flapping flight. Rosenblum and Jungerman (1981) have predicted the size and shape of a magnetoreceptor based on Wilkinson's conducting loop principle. They conclude that a structure millimetres in size is required to sense the Earth's field by such induction and point out that the labyrinth of the inner ear has many of the required characteristics.

Induction mechanisms for magnetoreception are still the favoured models for marine fish (Kalmijn 1978), largely because they live in a medium (salt water) that has a low-resistance return path ideal for such a mechanism but also because sharks and rays are known to have electric organs, the ampullae of Lorenzini, capable of detecting electric fields as low as 0.01 $\mu V/cm$, well within the necessary range. For terrestrial animals, however, such as birds and insects, the general concensus has moved away from such indirect methods to more direct mechanisms.

In part, the move toward models of magnetoreception based on direct perception of the magnetic field has been motivated by the observation that the magnetic field can be detected by almost stationary birds in cages. Such sedentary magnetoreception would not be predicted by theories based on induction. As Wallraff (1978a) has pointed out, however, caged birds seem only just able to sense the magnetic field and, as we have already seen, conditioning experiments in which birds were unable to move all failed. Human subjects in chair experiments are stationary when making their estimate. However, before each estimate they are moving through the magnetic field, albeit about a stationary axis, and it could be that magnetoreception occurs during the final part of the turning sequence while the subject is still moving.

Whether these small amounts of movement by birds and humans are enough for mechanisms based on electromagnetic induction to be feasible is a moot point. The influence of polyester clothes on magnetoreception by humans could certainly be used as evidence in support of such models. The fact remains, however, that most experimental and theoretical effort at present is aimed at the production and evaluation of hypotheses based on more direct methods of magnetoreception.

8.4.2. Optical resonance

Leask (1977) suggested that detection of the magnetic field takes place in the eye, in the molecules of the retina, as a by-product of the normal visual process.

Specifically, Leask proposed an optical or radio frequency double-resonance process involving the lowest excited molecular triplet state of a particular molecule, e.g. rhodopsin. The triplet states of such a molecule have a magnetic moment and their energy variations with the magnetic field are anisotropic, depending on field magnitude as well as on field direction.

Such molecular response is axial not polar. Only the field component parallel to the molecular axis matters, not its polarity. This is consistent with the conclusion that birds have only an inclination compass. Similarly, as the molecular response to the ambient field is axial and as the retina is a directionally ordered structure, for a particular eye orientation the resonance condition would be satisfied only on particular areas of the retina. Altering the ambient field would shift the resonance condition to some other region of the retina. This could confuse the bird initially, whereas a day or so later it could have learned to respond to the altered cue. Leask suggests that this could explain the adaptation period to altered field intensities found to be necessary in European robins (Section 8.3.1).

As originally formulated, Leask's model predicts that some light is necessary for magnetoreception to occur. A review of the literature and personal communications led Schmidt-Koenig (1979) to conclude that in most successful experiments on magnetoreception by birds, some residual light was present. A direct test of the influence of light has since been carried out by Wiltschko and Wiltschko (1981).

As described in Chapter 11, when young pigeons are transported to the release site in a distorted magnetic field, there is often an increase in scatter of their vanishing bearings, sometimes to the point of randomness. The Wiltschkos tested Leask's hypothesis by transporting pigeons to the release site in total darkness, while at the same time controlling for detection of familiar smells, visual isolation from outside, and psychological disturbance due to novel transport in total darkness. They found that birds displaced in total darkness were as disoriented as birds transported in a distorted magnetic field. Their results were thus consistent with the major prediction of the optical resonance hypothesis. As the Wiltschkos point out, however, their results are not conclusive. Had their results been positive, the evidence would have been against Leask's hypothesis. Negative results, however, although consistent with the hypothesis, do not prove it. Disorientation could have been due to some other factor, irrelevant to the mechanism of magnetoreception, such as birds being more prone to sleep in total darkness.

In chair experiments on humans, experiments at night have been carried out in a windowless hut with the only light, sufficient for the experimenter to see directions, coming from a miner's lamp worn on the experimenter's forehead. The subject wears two blindfolds (the inner blindfold made from several layers of cloth, the outer from a pair of swimming goggles lined with plasticine) and is in what seems to be total darkness. Yet significant orientation does occur under such conditions. From my own experiences as a subject, I should say that darkness was total. There has to be some possibility, however, that a subliminal amount of light from the miner's lamp leaked through or round the double blindfold, without this being perceptible to the subject. Although the data seem to run counter to Leask's model, therefore, not everybody may regard them as absolutely critical.

8.4.3. Magnetite

Although Leask's model aroused a great deal of interest when it was first published, most effort in recent years has been directed to the possibility that magnetoreception might be based in some way on deposits of magnetic material within biological tissue. The material in question is magnetite, a ferric/ferrous oxide (FeO . Fe$_2$O$_3$) that many people know better as lodestone.

The magnetite story in fact begins as long ago as 1962 when Lowenstam (1962) discovered that the radular teeth of chitons (Polyplacophora) were capped with the material. At the time of his discovery, Lowenstam was ridiculed for suggesting that magnetite might be manufacturered by an animal, a cold-blooded animal at that, for, as every physicist knew, enormous temperatures and pressures were necessary for its synthesis. Now, in the early 1980s, magnetite is well on its way to becoming recognized as the most widespread inorganic material synthesized by living organisms.

Although Lowenstam first suggested that magnetite might be involved in magnetoreception by animals, the main impetus came with the discovery that some species of mud-living bacteria could orient and swim relative to the ambient magnetic field (Blakemore 1975), that they contained conspicuous inorganic particles (Kalmijn and Blakemore 1978), and that these particles were in fact magnetite (Frankel et al. 1979). Now the search for biogenic magnetite is really under way. At a conference held at San Francisco in December 1981, no fewer than eleven separate research projects aimed at finding magnetite and testing its role in magnetoreception were described (abstracts in *EOS*, 1981, **62** pp. 849–850 and 1982, **63** p. 156). Alongside this practical progress, theoretical models are being developed (Kirschvink and Gould 1981; Yorke 1981) for possible ways in which magnetite could impart to organisms the observed level of sensitivity and those few other characteristics of magnetoreception that have so far been identified.

Apart from chitons and magnetotactic bacteria, magnetite, or at least magnetic material, has been found in honey bees (Gould et al. 1978), mice (Mather and Baker 1981, Mather et al. 1982), dolphins (Zoeger and Fuller 1981), green turtles (Perry et al. 1981), tuna fish (Walker and Dizon 1981), monarch butterflies (Jones and Macfadden 1981), and others. As far as birds are concerned, magnetite came to the fore when Walcott et al. (1979) announced the discovery of magnetite associated with the dura mater in the heads of homing pigeons.

At the time, many people felt that the mystery of magnetoreception by birds had been all but solved. The first dampening of the euphoria over magnetite came when Presti and Pettigrew (1980) reported magnetite in the neck muscles of homing pigeons but not in the dura as in the original discovery. The second came from a more recent report by one of the authors of the original paper.

The first stage in the original discovery was to test tissue from a pigeon's head in a magnetometer after the tissue had been exposed to a strong magnetic field, several thousand times stronger than that of the Earth. If magnetic material is present in the tissue, the magnetometer should indicate the presence of magnetic remanence. Walcott and Walcott (1982) now report that they have examined over 80 birds since the original sample and in no case could they find the induced magnetic remanence described in the original report (Walcott et al. 1979). Nor could they find magnetic material in the neck tissue as reported by Presti and Pettigrew. Histological examination at the original site on the dura mater also now failed to find iron.

Disillusioned by their magnetometric search, Walcott and Walcott have turned instead to a slow and laborious but systematic histological search for tissue containing iron. They are producing serial sections of the pigeon head from the beak backwards and searching for tissues that consistently stain for iron in a number of individuals. So far, they report three major sites that seem consistently to contain iron. One site is the Harderian gland, a lachrymal gland that lies medially to the eye within each orbit. A second site is in the base of the beak just posterior (toward the head) to the cere. This site is unpaired, usually on the left side, and lies in a lateral fold of the dorsal surface of the oral cavity. The third site is also unpaired but more central. The cells are in close association with a bony ledge just ventral to the olfactory nerves and olfactory lobes of the brain.

The one vital piece of evidence that is missing for birds and all other animals except humans is experimental evidence for the most likely position of the magnetoreceptor. Early experiments on magnetoreception by humans, using electromagnetic helmets, found an association between the direction of the lines of force in one part of the head and estimates of direction in 'homing' experiments (Chapter 10) (Baker 1981a). The region was therefore suggested to be the most likely site for magnetoreception to occur. This predicted site was on the mid-line, between but slightly below the eyes, about 3–4 cm in from the front of the head (Fig. 8.10).

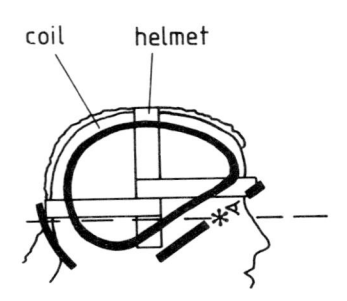

Fig. 8.10 Asterisk shows the predicted site of a magnetoreceptor in the human head, based on the results of early homing experiments

(Simplified from Baker (1981a))

More recently, experiments have been carried out in which a bar magnet is placed on the head and aligned vertically with its north-pole uppermost. A magnet so aligned and placed on the right temple, was known to rotate estimates of compass direction in chair experiments by 180 degrees (Fig. 8.5). It was also known to produce a striking after-effect if the subject was then tested in a normal magnetic field. These more recent experiments have now shown that the influence of the magnet differs according to whether it is placed at the front of the head or toward the back; an after-effect being discernible only if the magnet has been worn forward of, as opposed to behind, the ear (Fig. 8.8). These results are consistent with the magnetoreceptor being in the front half of the head.

A magnetometric search of the heads of human cadavers revealed that the bones forming the walls of the sphenoid and perhaps ethmoid sinuses, just in the region predicted, were magnetic whereas other bones were not (Baker et al. 1983). Moreover, histological examination showed that these bones possess a layer of a compound containing ferric iron running just beneath their surface (Fig. 8.11). We are awaiting the results of tests to determine whether any of this material is magnetite.

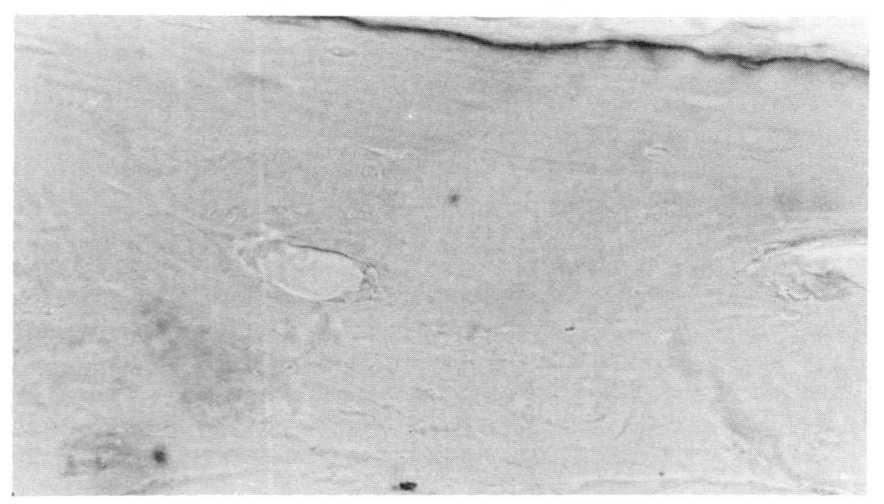

Fig. 8.11 Section through the thin bone forming the wall of the sphenoid sinus of a 57-yr-old woman showing a continuous layer of a ferric iron material approximately 5 μm below the surface.

Stained using the Perl reaction.

[From Baker et al. 1983]

Woodmice (*Apodemus sylvaticus*) have a magnetic sense and also show greatest magnetic remanence in the anterior dorsal part of the head in the olfactory region (Mather and Baker 1981). A histological search reveals that the bones of the ethmoturbinal region show an iron-staining layer visually identical to that seen in humans (Mather et al. 1982). The woodmouse head also contains magnetite (Kirschvink, personal communication), though whether this is due to the layer in the bone is not yet known. Walker and Dizon (1981) report that the region of the bones of the ethmoid sinus complex of tuna fish is also magnetic and that this remanence is due to magnetite. Finally, to a lesser or greater degree, all three of the sites that contained iron in the pigeon head, as reported by Walcott and Walcott (1982), were near the olfactory region. Moreover, these authors report that bones in this region also stained for ferric iron.

At one stage, therefore, early in 1982, the results looked extremely promising that perhaps for vertebrates generally, including birds, the magnetoreceptor might be closely associated with the olfactory region and be based in some way on magnetite deposits in or on the bones of that region. The first complication came when Mather and Kennaugh (pers. comm.) found that in marmosets (*Callithrix*) the layer of iron beneath the surface of the bone extended throughout all the bones of the head, albeit more dense in some bones than others. The same pattern was later confirmed for woodmice, all bones, not just the ethmoturbinals, containing the iron-staining layer. Nor is the layer confined to the bones of the head, having also been found in the femur.

We may assume that the same widespread occurrence of an iron layer beneath the surface of bones will be found for humans and perhaps birds and that its primary function may therefore be for something other than magnetoreception, perhaps bone repair (Baker et al. 1983). Magnetite is already known to occur in the adrenal glands of humans (Kirschvink 1981).

These findings suggest that either: (1) the iron layer in the bone is not the source of the magnetic remanence in the sphenoid sinus bones of humans and the anterior dorsal region of woodmice; (2) the magnetic characteristics of the layer may differ in different regions, perhaps due to variation in particle size (Kirschvink and Gould 1981); or perhaps even (3) in other bones, the iron deposits may be the precursors of magnetite, not becoming magnetic until breakage, whereas in the sinus bones perhaps magnetite is normally formed.

The widespread occurrence of ferric compounds, indeed of magnetite, throughout the vertebrate body in organs that seem unlikely to be involved in magnetoreception does not preclude the possibility that magnetite is an important element in magnetoreception. Nor does it negate the experimental evidence for humans that the magnetoreceptor may be in the front of the head in the region of the sinuses, or preclude the indications of a common association for vertebrates between olfactory and magnetic senses. It does mean, however, that we still have not established a clear link between magnetoreception, tissue with magnetic remanence (usually due to the presence of magnetite) and histological identification of iron. In the absence of evidence for such a link, the magnetite theory of magnetoreception is still no further validated than the optical resonance theory of Leask or the induction theories of other authors.

Wherever the magnetoreceptor may be located and however it may function, the nerve impulses generated by it may well travel along a pathway taking them through the pineal body in both birds and mammals. Recent work by Semm and his colleagues (e.g. Semm *et al.* 1982) has shown that nerve cells in the pineal body of pigeons (and guinea pigs) respond to changes in an Earth-strength magnetic field by increasing their rate of firing. Moreover, each cell seems to fire in response to relatively specific changes in the ambient field. Some, for example, respond to changes in intensity, others to changes in direction, and so on. The authors also point out that the pineal gland of birds is known to be a light-sensitive, time-keeping organ, and would, therefore, be an ideal site for the integration of combined magnetic and time-compensated sun compasses.

If the optical resonance or induction theories are correct or even if some nerves respond directly to the magnetic field, there is no specific organ of magnetoreception to be seen. If it is based on magnetite, there may well be such a separate and discrete organ. As those of us engaged in the search for the magnetic sense organ pore through the samples of tissue and sections, we may already have seen the sense organ for which we are searching. On the other hand, there may be nothing for us to see. The fact of magnetoreception by birds and other animals now seems firmly established. The search for the magnetoreceptor and an understanding of the characteristics of magnetoreception is as frustrating and exciting as ever.

9 Bird compasses: relationships

If any aspect of orientation and navigation is clear, it is that birds need never be without a compass. Sun and polarisation patterns by day, stars and Moon by night, and the magnetic field continuously (except at the equator and perhaps at magnetic anomalies or during magnetic storms) should combine to give them a compass at all times. In fact, on a good day or night there is more information available than a bird really needs, for any one of these compasses on its own would seem to be adequate.

9.1. The compass hierarchy and redundancy

One of the biggest conceptual advances in the study of bird orientation in recent years was made by Keeton (1972). In retrospect, it seems so simple, but at the time was very important. Keeton stressed that bird orientation and navigation was not based on just one factor but was characterised by considerable redundancy of information. At any one time birds used only a fraction of the information available to them.

It was this first clear statement of the principle of redundancy that brought home the importance of the distinction between information that normally is used for orientation and navigation and information that is essential. We have already used this distinction in discussing whether visual landmarks and smells have a place in the avian landscape (Chapter 5). The first use of the principle, however, was in the initial demonstration that pigeons used a magnetic compass.

Early authors had assumed that if birds were able to use a magnetic compass then interfering with that compass would prevent orientation. Thus a number of experiments were carried out in which birds had magnets glued to their body and wings before being released on homing experiments. Most of these experiments took place on sunny days, for only on such days are pigeons easily motivated to home. The magnets had no effect and it was concluded that pigeons could not detect the geomagnetic field (Matthews 1968).

Keeton (1971, 1972) showed that magnets indeed have no effect if the pigeons are released only on sunny days. Train a group of birds to home on overcast days, however, and then attach magnets to them, and an effect is found. It is not, therefore, that birds cannot use magnetic compass information in homing, only that they prefer to use the sun if it is available. On sunny days, the magnetic compass can appear redundant.

Keeton (1972) stressed the hierarchical arrangement of compasses, taking as his starting point the fact that pigeons prefer the sun to the magnetic field. Later, I formalized this idea into a 'least navigation' model. The suggestion was that in solving any particular problem of navigation or orientation, birds use only that subset of available information which gives the best trade-off between accuracy and effort (Baker 1978). Under different conditions, different subsets are optimum.

Following Keeton, most authors have conceded that, for some unknown reason, the magnetic compass is toward the bottom of the hierarchy of information used by birds. Such a conclusion was inevitable in view of the difficulty of detecting a magnetic compass in orientation experiments on caged birds and the weak and variable effects of interfering with magnetoreception by homing pigeons (Chapter 11). Nevertheless, for many, the conclusion came as a big disappointment. Throughout the first half of this century it had been more or less anticipated that if magnetoreception by birds could be demonstrated, the entire mystery of bird navigation would be solved at a stroke. Magnetoreception was demonstrated, but the mystery failed to disappear.

Without doubt, an ability to detect the Earth's magnetic field is part, but only part, of the answer to the puzzle of bird navigation. It is possible, however, that at present we underestimate its level in the hierarchy of information. Suppose that magnetoreception by birds and other animals appears so weak and relatively unimportant because, as discussed in Chapter 8, we house and test birds under conditions that destroy their magnetoreceptive abilities. If this were so, it is possible that in the environment in which birds evolved, uncluttered by those magnetic and electric human artefacts now so abundant everywhere, magnetoreception played a much more important role. Asking a bird to orientate by the magnetic field nowadays may be equivalent to asking a human to use star orientation in a brightly lit modern city, just about possible but difficult. Indeed, many city dwellers never learn to recognize star patterns, to a large extent because, through the barrier of haze, artificial lights and tall buildings, they can rarely see the stars clearly enough to become familiar with them.

The possibility that, while being observed, birds may never have the opportunity to use their magnetic compass as efficiently as they could should be borne in mind in the sections that follow.

9.2. Evidence for an inborn compass

A great deal of interest in recent years has centred on trying to discover if birds are born with one of their compasses already programmed within them.

For many birds, of course, an inborn compass is not essential. At first sight, birds that do not perform seasonal migrations have no use for such a compass. While in the nest during the days or weeks after crawling out of the egg or, perhaps more likely, when they first begin to explore after fledging, young birds could observe that not all directions are the same. Different directions feel different, magnetically. Moreover, the daily arcs of the sun, Moon and polarization patterns traverse magnetic directions in a regular way, and a particular region of the night sky never moves in relation to the magnetic field. Within a short time, all the various compass cues could be integrated with each other and with the local landmarks on their rapidly expanding familiar landscape.

On such a system, nothing is inborn, everything is learned. The situation has to be different, however, for seasonal migrants, for there is no doubt that the young of such species are born with a preferred compass direction already programmed within them. This fact is most strikingly illustrated by those species of which the young regularly have to carry out their first autumn's migration without any contact with their parents.

The European cuckoo (*Cuculus canorus*) is one of the clearest examples. It is a parasitic species that lays its eggs in the nests of a variety of passerines. Not only does the young cuckoo never associate with its genetic parents, or indeed with any adult cuckoo, during its first few weeks of life but also it is not fit for long-distance flight until most, if not all, such adults are well on their way to the wintering grounds. These grounds are in Africa, a few thousand kilometres away. Nevertheless, the young cuckoo, when its time comes, sets off in the appropriate compass direction and unites over winter with the rest of its species. It is difficult to escape the conclusion that the

Fig. 9.1 Seasonal migration of the cuckoo (**Cuculus canorus**)

Solid outline, breeding range; dotted outline, non-breeding range. Arrows show the suspected migration route of the population breeding in Britain.

[From Baker (1978), after Moreau and Mead]

young cuckoo has programmed within it a preference to fly in the direction traditional for its race or species. In the 1940s and 50s the results of a number of experiments showed that this conclusion could be applied equally well to a number of seasonal migrants.

Rowan (1946) showed that when young American prairie crows (*Corvus brachyrhynchos*) were held back in their natal area until all members of their species had left on autumn migration and then released, they were nevertheless recovered in the standard direction. Schüz (1949, 1950) overcame the objection that Rowan's birds may have been forced to take up their final direction by local topography and winds when he carried out similar experiments in Europe using white storks (*Ciconia ciconia*).

Conveniently, white storks show a migratory 'divide' (Fig. 9.2), birds from the west setting off for Africa to the southwest, birds from the east setting off to the south-southeast. Schüz took young white storks from the Baltic, in the east of their range, to Essen, Germany, in the west, where they were reared and released. Their behaviour seemed to depend on when they were released. Young storks released while stork migration through the area was still in progress produced recoveries to the southwest, characteristic of the local population. Young storks released after storks had left the area produced recoveries and sightings to the south-southeast characteristic of the parental population. Evidently, the storks had an inherent preference for the south-southeast but this preference could be overriden by the example of other birds.

130   *Bird navigation: the solution of a mystery?*

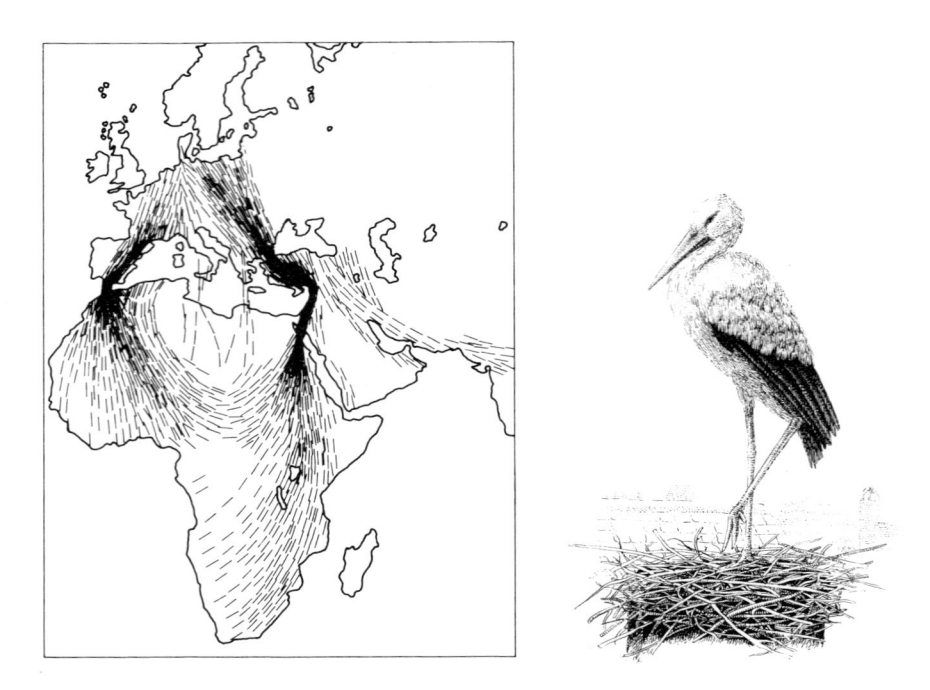

Fig. 9.2 Autumn migration of the white stork (**Ciconia ciconia**) showing the 'migratory divide' in western Europe

[From Baker (1978), after Verheyen and Rüppell. Photo (p. 129) by Arthur Christiansen, courtesy of Frank W. Lane]

The most famous experiment of this type and, despite recent criticisms (Gerrard 1981) justly so, if only for the industry and effort involved, was carried out by Perdeck (1958). Over 11 000 young starlings (*Sturnus vulgaris*) were captured in The Netherlands and displaced 600 km or so south-southeast to Switzerland. The young birds were netted while on autumn migration and controls, simply caught, ringed and released in the Netherlands, continued west-southwest to traditional wintering grounds in Britain and France. Displaced birds also travelled a route to the west-southwest and were recovered in wintering grounds primarily on the Iberian peninsula (see Fig. 2.2).

The inborn preference of young birds to travel in a particular compass direction during their first migration cycle is not confined to the autumn phase of their journey. Hooded crows (*Corvus corone cornix*), captured during spring migration and displaced from Rybachiy, on the eastern coast of the Baltic Sea, to Flensburg, northern West Germany, 750 km to the west, showed a similar tendency to fly in the traditional compass direction rather than to return to their original breeding range (Fig. 9.3).

It was already clear from field data, therefore, that the young of various species of seasonal migrants were born with an inbuilt preference for particular compass directions during their first autumn and spring migrations. Further confirmation through studies of birds in orientation cages has recently been obtained for garden warblers (*Sylvia borin*) (Wiltschko and Gwinner 1974; Wiltschko *et al.* 1980; Gwinner and Wiltschko 1980), savannah sparrows (*Passerculus sandwichensis*) (Bingman 1981) and pied flycatchers (*Ficedula hypoleuca*) (Beck and Wiltschko 1982).

Garden warblers, for example (see Fig. 8.4), normally migrate on a broad front. Birds from central and western European populations fly first

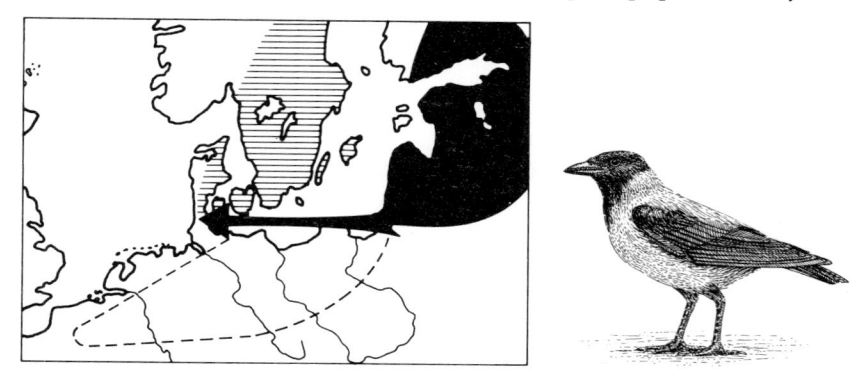

Fig. 9.3 Shift in breeding area of young hooded crows (**Corvus corone cornix**) displaced during their first spring migration

Crows were captured at Rybachiy and displaced to Flensburg, 750 km to the west. The normal breeding and wintering areas of crows that migrate through Rybachiy are shown, respectively, by solid black and dashed outlines. The subsequent breeding area of displaced crows is shown by hatching. [From Baker (1978), after Rüppell]

to the southwest or south-southwest, arriving at the southern coast of Iberia at about the end of September. Here they seem to alter their track to a more southerly or southeasterly direction to reach their winter quarters in central Africa. Less is known about the spring migration but again the warblers seem to move on a broad front northwards with perhaps no changes in direction comparable to the one during the autumn migration. Hand-raised garden warblers, tested in orientation cages, orient to the southwest in August and September, to the south in October and November, and to the northeast in spring (Fig. 9.4).

The advantage of such internal programs of preferred directions is obvious to species such as the cuckoo which do not associate with experienced birds until after their first migration. Even to birds that do so associate, however, the advantage is probably equally real if there is a significant risk of having to perform parts of the migration solitarily (Baker 1978; Wiltschko and Wiltschko 1978). The advantage to birds that do not perform seasonally directed movements is less obvious. Nevertheless, I have argued (Baker 1978, 1982) that even these birds might evolve inborn directional preferences.

132 *Bird navigation: the solution of a mystery?*

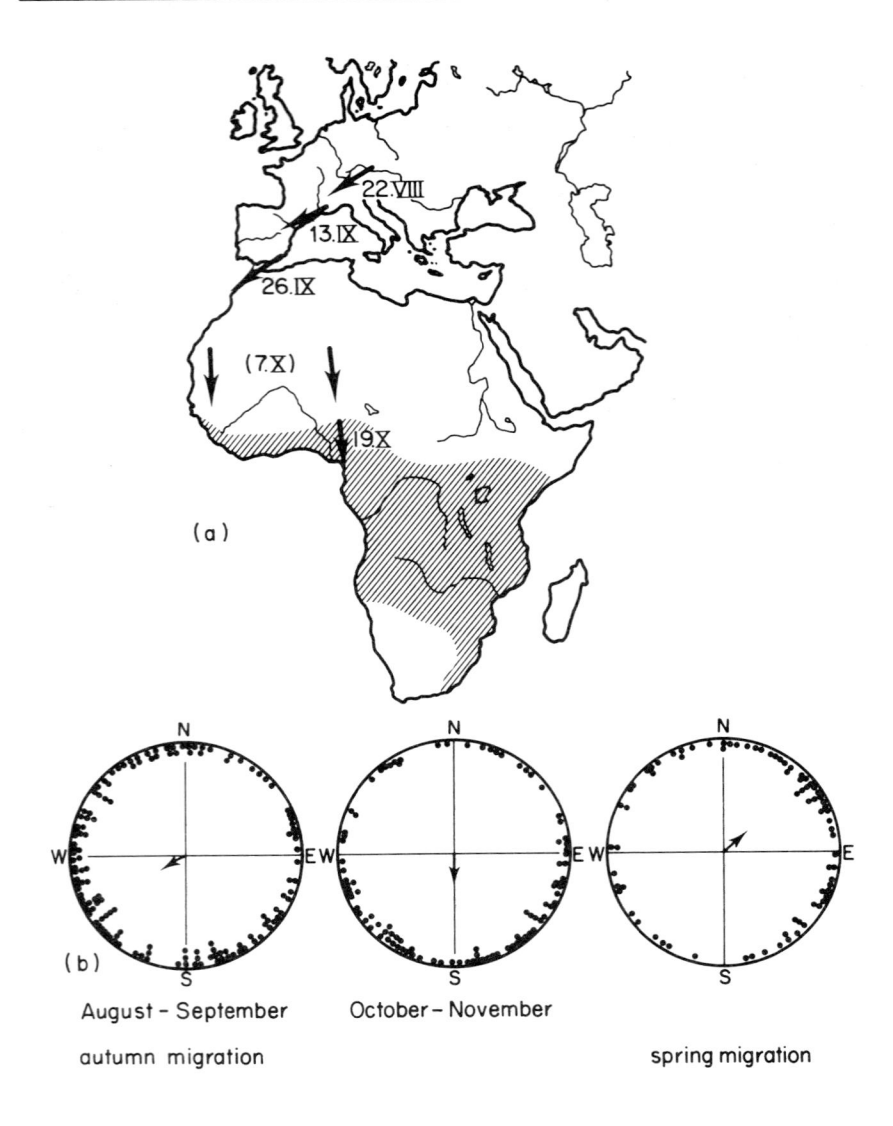

Fig. 9.4 Seasonal change in inborn preferred compass orientation of garden warblers (**Sylvia borin**) relative to the geomagnetic field

Map shows the African winter range and autumn migration route. Circles show the orientation of young birds hand-raised in the normal geomagnetic field without ever seeing the sun and stars. Tests were carried out in orientation cages under the same conditions.

[Compiled from Gwinner and Wiltschko (1978, 1980) and Wiltschko (1982), after Wiltschko, Gwinner and Wiltschko. Photo by Walther Rohdick, courtesy of Frank W. Lane]

Unlike seasonal migrants, which are characterised by a very narrow spread of directional preferences within the population, the spread of preferred directions should be much wider in these 'non-migrants', even, perhaps, with all compass directions being equally represented. At the individual level, however, the inborn preference for a particular direction may be no less real. Briefly, such directional preferences within the individual are suggested to evolve through combined selection (1) on the individual to straighten out its track during post-fledging exploration while it searches for suitable moulting, feeding and particularly breeding sites; and (2) on each parental generation to produce the optimum dispersal of their offspring (Baker 1978, 1982).

This model suggests, therefore, that all individual birds are born with directional preferences but whereas all individuals of seasonally migrant species beyond a certain date in autumn show a preference for more or less the same direction, different individuals of 'non-migrant' species prefer different directions. There are two particular consequences of this model.

Academically, it explains how inborn directional preferences came to be present in bird populations in the first place. These individual preferences then provided the raw genetic material on which selection acts to produce the highly biased preferences found in populations of seasonal migrants. Practically, it means that an inborn compass can meaningfully be sought in 'non-migrants', such as the homing pigeon, as well as in seasonal migrants, such as cuckoos and garden warblers.

As it happens, most work on 'non-migrants' has tried to determine whether the sun or magnetism forms the innate compass of day-flying birds, whereas most work on seasonal migrants has tried to determine whether the stars or magnetism form the innate compass of night-migrating birds. In both cases, however, it is expected that each bird is born with an instinctive preference to orient such that at least one of its compasses 'feels' one way rather than another (e.g. towards the sun at midday, away from magnetic north, away from that part of the night sky that rotates least).

9.3. Compass development from birth

9.3.1. Day-time compasses

Most experiments on day-time compasses have been carried out on starlings and homing pigeons. We already know from Chapter 6 that these birds are not born with an ability to compensate for the sun's movement across the sky during the day but have to learn how to do so. If the sun is the primary compass, therefore, the one with which the bird is born, such learning would have to take place by observing the sun's movement over the landscape. The available evidence, however, suggests that the sun compass is calibrated in the first place against a magnetic compass.

When experienced homing pigeons are released on sunny days, magnets attached to their backs have no effect on their initial orientation at distances from home of 27–50 km, though they may have some effect at longer distances (Keeton 1971). Young birds, however, about 3 months old, that had been given daily exercise flights at the loft but had never before been displaced in experiments, were influenced by magnets when released under sun, vanishing points being random. Since such first-flight birds had previously been shown also to need the sun if they were to depart in the home direction (Keeton and Gobert 1970), it seemed that both solar and magnetic information was necessary for homeward orientation on a bird's first displacement from the home site.

In these early experiments of Keeton's, pigeons were raised in normal aviaries. In later collaborative experiments with the Wiltschkos, however, Keeton found (Wiltschko et al. 1976) that pigeons raised without ever seeing the sun were able to orient properly when released under overcast skies. It seems that if deprived of a sun compass from birth, pigeons quickly

Bird compasses: relationships 135

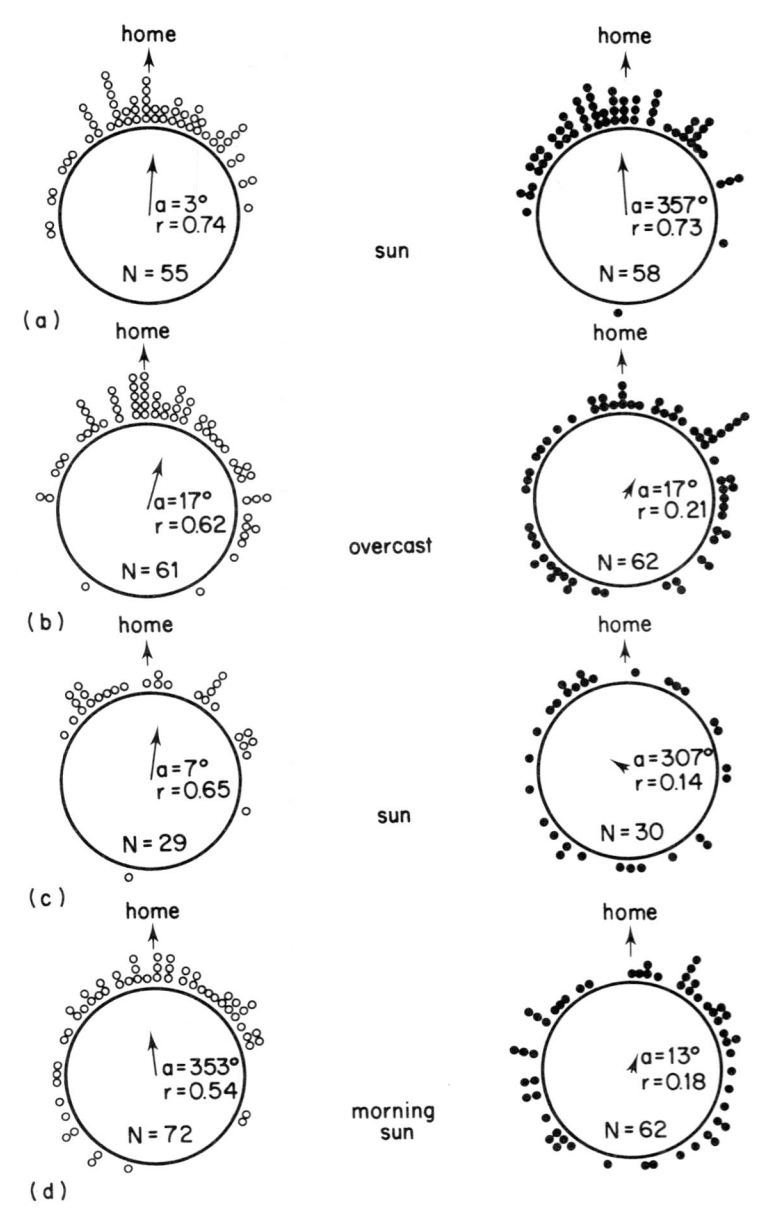

Fig. 9.5 Influence of sun and magnets on the initial orientation of homing pigeons

Open symbols, controls wearing brass bars; solid symbols, experimental birds wearing bar magnets.
(a) Experienced birds under sun. (b) experienced birds under overcast. (c) Inexperienced first-flight birds under sun. (d) Releases under the morning sun of birds that were experienced but only under the afternoon sun.

((a)–(c) compiled from Keeton (1972), 1974a). (d) re-drawn from R. Wiltschko et al. (1981))

learn to make use of a magnetic compass on its own. If raised in such a way that they are able to learn both the sun and the magnetic field, it is the relationship between the two compasses that is learned. If young birds are then released under conditions (e.g. overcast or wearing magnets) where this relationship is unfamiliar, the pigeons are confused. With greater experience at orientation and homing experiments, however, it seems that the birds become able to use either compass system independently of the other.

Two further experiments have a bearing on the initial integration of sun and magnetic compasses by young birds.

The first of these also involved the Wiltschkos (Wiltschko et al. 1981) and made use of their earlier discovery that if young pigeons that had observed the sun only in the afternoon were tested in their (pigeon) morning, they were unaffected by clock shifts when released with a view of the sun and were still able to set off in the home direction (Chapter 6). Evidently, such birds do not use a sun compass in the morning. Nevertheless, they are homeward oriented. When released wearing magnets, however, they are disoriented, suggesting that their orientation is the result of using a magnetic compass.

The second experiment was an early, and still problematic, study by Keeton (1971). He placed either a magnet or a brass bar on a large number of young birds when they first began to fly, and left these on throughout the birds' early training at distances of up to 16 km from the loft. Birds with magnets and brass bars were trained together in mixed flocks so as to reduce losses during this early period. The birds were then tested at sites 50 km from home. Just before release, however, the magnet or brass bar was removed. Three test releases were made, all in overcast conditions. In all three, both control and experimental birds were homeward oriented. However, also in all three test releases, the birds formerly wearing magnets took longer to disappear from the release site, though this was not reflected in slower homing speeds.

Although the wearing of magnets during training has some influence, therefore, on subsequent homing under overcast without magnets, pigeons are still able to orient. Unfortunately, we have no way of knowing whether the 'magnet' birds in this experiment were using a magnetic compass in determining their orientation. If not, and given the absence of a sun compass because of overcast, the birds may well on this occasion have homed using solely the landscape, thus accounting for their longer delay in deciding on home direction. At perhaps only 24 km beyond the edge of their familiar area, the birds could easily have been within the limits of their familiar area map (Chapter 5).

The current state of our understanding of the development of compasses from birth in homing pigeons, therefore, is as follows. If young pigeons are not allowed to see the sun, or even if they see it for only part of the day, they are still able to use a magnetic compass. On the other hand, they either do not develop a sun compass at all or develop one only for the time of day that they have first-hand experience of the sun's movements. Even if they are allowed to see the sun and thus to develop a sun compass, young pigeons are disoriented by wearing magnets when released in sunny conditions. So far, therefore, the results suggest that pigeons set off in life with a magnetic compass and only slowly learn to use a sun compass.

Of these two day-time compasses, the magnetic compass is the only one that may still prove to be inborn. As yet, however, critical data are not available. The results so far could just as easily be taken to show that the magnetic field is simply the first one that the birds notice. After all, even an embryo could perceive the magnetic field and learn to associate different directions with different magnetic feelings.

What is certain is that pigeons can have a magnetic compass without having a sun compass. We do not yet know if they can also have a sun compass without having a magnetic compass. We need an experiment in which a bird is deprived of all magnetic information from the time when it is an embryo to the time of first testing, yet is allowed to see the sun. Would such a bird prove to have developed a sun compass but not a magnetic compass?

The only information we have on the development of day-time compasses in seasonal migrants comes from Hoffmann's study of the development of a sun compass by starlings (Chapter 6) and from Southern's work on the compass orientation of ring-billed gull chicks (Chapter 8). The latter study shows that gull chicks may orient to the southeast within two days of hatching from the egg, that they do so in the first instance almost certainly by using a magnetic compass (Southern 1978), and that, even under clear skies, they are influenced by magnetic storms. Apparently, therefore, the pattern of compass development could be similar to that for homing pigeons but with the additional inference that the, presumably inborn, preference to orient to the southeast is expressed first in relation to the magnetic field. We find better evidence that the inborn preference of seasonal migrants can be encoded to the magnetic field, however, when we examine night-time compasses.

9.3.2. *Night-time compasses*

Many of the conclusions reached about the development of night-time compasses are similar to those reached for pigeons. More species have been studied, however, and it is beginning to look as though the details are not always the same for all species.

Experiments were described in Section 9.2 in which garden warblers (Fig. 9.4), pied flycatchers and savannah sparrows were hand-raised and then tested in an orientation cage for an inborn preference to migrate in the appropriate compass direction. Each of these experiments also showed that birds, hand-raised without a view of sun or stars, could still orient in seasonally appropriate directions and that they did so by means of a magnetic compass.

Evidently, in these birds, as in pigeons, a magnetic compass can develop without the help of celestial cues. The experiments also show, as was inferred for ring-billed gulls, that the birds are born with a preference to migrate at a particular angle to the magnetic field. Finally, the fact that garden warblers changed their preferred directions from southwest to south to northeast at the appropriate times of year, despite being cut off from outside day-length and temperature fluctuations, suggests that these inborn preferences are part of some long-term internal program.

This program, which we now know runs on a more-or-less year-long or 'circannual' clock, is described further in Chapter 12. For the moment it is enough to point out that the program is independent of at least obvious phase setters, such as day-length and temperature, and that it organises not only the sequence of preferred directions for migration but also the sequence of moulting, fat deposition, gonadal development and migratory restlessness.

There is one other feature of their development of magnetic and celestial compasses in which garden warblers and homing pigeons are similar. Whereas the magnetic compass of the garden warbler can mature without the bird seeing the sun or stars, if these are present the use of the magnetic compass is influenced.

Groups of garden warblers were hand-raised in an open aviary with a view of the natural sky until the time for autumn migration. They were then tested in an orientation cage in a normal geomagnetic field but in the absence of sun or stars. They failed to orient. As for homing pigeons (Keeton and Gobert 1970), it seemed that once young warblers had developed a celestial compass they were then unable, or unwilling, to orient in its absence.

This influence of a celestial compass on magnetic orientation, however, may not be as general as it at first appeared. Pied flycatchers and savannah sparrows, also hand-raised with a view of the stars, still showed orientation when tested in their absence (Beck and Bingman in Wiltschko 1982).

Although the maturation or use of a magnetic compass in young birds is influenced by whether or not they develop a celestial compass, it remains true that nocturnal migrants are born with a preference to migrate at particular angles to the magnetic field at particular times of year. This does not mean, of course, that they are not also born with a preference to migrate at particular angles relative to a star or Moon compass. Indeed Emlen's experiments on the development of the star compass in indigo buntings (Chapter 7) could well be taken to imply an inborn preference to migrate away from the axis of rotation in autumn. As the magnetic field was not manipulated in these experiments, however, and as it is now known that indigo buntings do have a magnetic compass (Emlen et al. 1976), it is still possible that an inborn preference relative to the magnetic field had already, by the time of testing, been translated into a preference relative to the axis of rotation.

Fergenbauer (in Wiltschko 1982) hand-raised young garden warblers and as soon as they became self-sufficient placed them under artificial but rotating night skies with one revolution per day. They were also exposed to the natural geomagnetic field, but magnetic north and the axis of rotation did not coincide. Moreover, the relationship was different for different

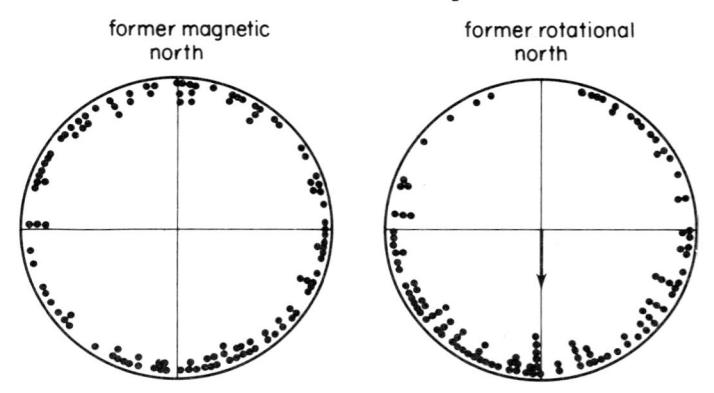

Fig. 9.6 Possible evidence for an inborn preference for orientation to the axis of rotation of the night sky in garden warblers (**Sylvia borin**)

Warblers were hand-raised and then placed under a rotating, artificial night sky in the natural geomagnetic field. Geomagnetic north and the axis of rotation of the night sky did not coincide. When tested in orientation cages during autumn migration under a stationary sky and in the absence of meaningful magnetic information, the birds oriented to the former rotational north rather than to the former magnetic north.

[Modified from Wiltschko (1982), after Fergenbauer et al.]

groups of birds. During migratory restlessness in autumn, the birds were tested under a similar but now stationary sky in the absence of meaningful magnetic information. Preliminary results show that the birds are clearly oriented away from the axis of rotation of their different night skies rather than toward south as should have been indicated by the geomagnetic field. Wiltschko (1982) warns, however, that not all factors, such as asymmetrical distribution of light, were controlled and that the results may not yet be critical. Doubts are raised particularly because the change in migratory direction that would have been expected as August and September passed into October and November (Fig. 9.4) did not occur. Moreover, results were not random relative to the magnetic field, but bimodal, albeit along a SSE–NNW axis rather than along the more appropriate SW–NE axis. In a second series of experiments the warblers were exposed to rotating skies and the geomagnetic field for 6 weeks, instead of 4, before testing. In these experiments, southward orientation relative to what had been magnetic north was found.

Although at face value these results suggest that garden warblers may have inborn preferences relative to both the magnetic field and the axis of rotation, the question is still not settled.

9.4. The use of compasses by experienced birds

When we consider the compass sense of older birds, more experienced in finding their way around, we discover that they soon learn to use their different compasses in a way that seems both integrated and efficient. Each compass is used in those roles for which it seems best suited, giving more than a hint of 'least-navigation' in action.

9.4.1. By day

Experiments on day-time compasses have again largely been carried out with pigeons, and little has changed since Keeton (1971) demonstrated that although experienced birds could use a magnetic compass, they preferred to use a sun compass if one was available. Unlike young birds that had developed a sun compass, however, birds with sufficient experience of overcast conditions would use their magnetic compass if released under overcast. Even under sunny conditions, one group of experienced birds at a distance of 85 km from the loft did seem to make some use of a magnetic sense (Keeton 1971).

Since these experiments by Keeton, relatively few authors have looked for a possible interaction of magnetic and sun compasses, the general feeling being that most often birds simply use their sun compass when the sun is

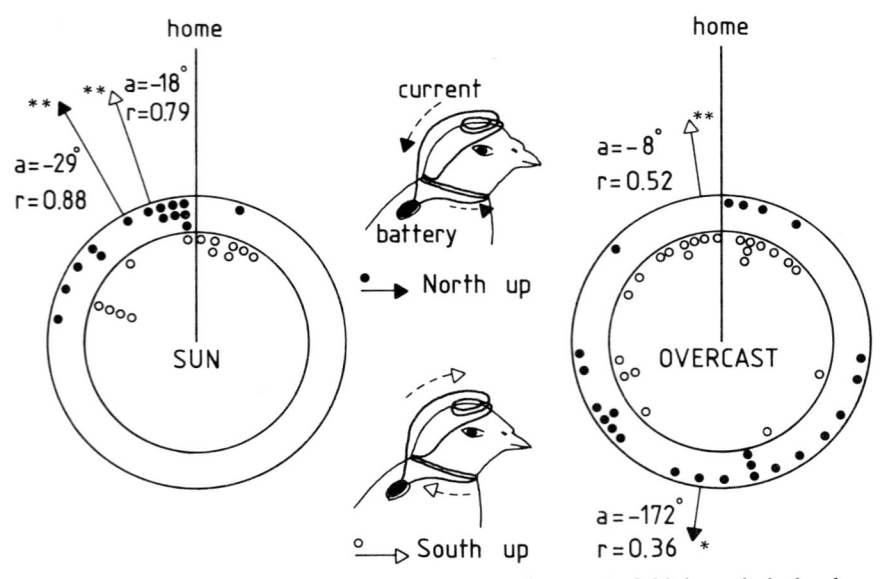

Fig. 9.7 Influence of reversed vertical component of magnetic field through the head on the initial orientation of pigeons

When birds are released for the first time under overcast, individuals with a north-up field through the head disappear from view in a direction away from home.

[From Baker (1981a), after Walcott and Green]

visible but switch to their magnetic compass under overcast. This view was reinforced when Walcott and Green (1974) released pigeons with a cap-and-collar system of coils (Fig. 9.7) powered by a small battery on the bird's 'shoulder'. Depending on the direction of the current through the coils, the magnetic field through the bird's head was either north-up or south-up. Walcott and Green found that if the birds were released in homing experiments under overcast conditions, the two groups set off in more or less opposite directions. Under sunny conditions, however, both groups seemed to be equally oriented toward home.

More recent experiments have compared the performance under sunny skies of birds wearing either coils or bar magnets with the performance of suitable controls (Walcott 1977; Visalberghi and Alleva 1979). Although it was still found that birds with an artificial magnetic field through their head were, by and large, homeward oriented, they nevertheless showed greater scatter to their vanishing bearings than controls, as well as some delays in homing.

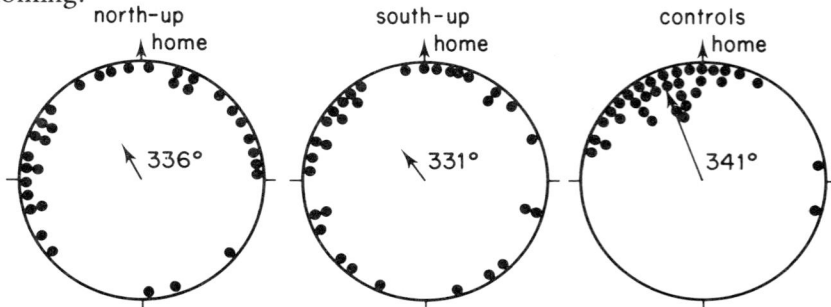

Fig. 9.8 Influence of an artificial magnetic field through the head on the scatter of vanishing bearings of homing pigeons when released under sun

Controls have a normal magnetic field through the head. North-up and south-up birds have an artificial magnetic field through the head as for Fig. 9.7. Under sun, all groups show homeward orientation, but those with an artificial field show more scatter to their vanishing bearings than those with a normal field.

[Simplified from Visalberghi and Alleva (1979)]

It seems, therefore, that when sun and magnetic compasses give conflicting information, the birds do not simply ignore the magnetic information, though it is still true that they give priority to the information provided by the sun.

Further evidence that a pigeon's preference for using a sun compass when the sun is shining does not simply reflect an abandonment of the magnetic compass under such conditions is provided by studies of clock-shifted birds (Alexander and Keeton 1974; Edrich and Keeton 1977).

As we have already seen on several occasions in earlier chapters, clock-shifts are the primary means of identifying that a time-compensated sun compass is being used. The efficiency of clock-shifting as an experimental technique, however, depends on how much and what type of activity the bird is allowed under the natural sky during the clock-shifting period. In clock-shifts of the order of 6 hours or less, there is always a period during the day when both the 'pigeon day' and the natural day are light rather than dark, and birds can be allowed to exercise and to see the natural sky during this period. When they do so, however, there is a risk that the birds will realize the sun is not in its 'correct' position according to the bird's subjective time of day (e.g. if the bird thinks it is 06.00 h and thus expects the sun to be in the east, it could notice that the sun is actually in the south if the time is really 12.00 h). It could notice the anomaly in at least three ways: (1) the sun will be in the wrong position relative to local landmarks; (2) it will also be in the wrong position relative to the Earth's magnetic field; and (3) the sun's altitude may be wrong (in the above example it would be high in the sky instead of low on the horizon as expected).

Despite the abundance of information which could indicate to the birds that something is wrong, Keeton and his colleagues found that if the bird was allowed only to sit in the aviary and observe the sun, the clock-shift technique was still effective. This suggests that the bird either does not detect the misalignment of the sun and familiar landmarks and the incorrectness of the sun's altitude or, if it does, it prefers to believe the evidence of the light/dark regime in which it is living. On the other hand, free flight around the loft, or exercise flights in large aviaries, does reduce the degree of deflection achieved by clock-shifting when eventually the birds are tested in homing experiments. The amount by which subsequent deflections are reduced, however, is greater when the birds are made to fly in aviaries aligned east−west than when they are made to fly in aviaries aligned north−south.

This series of experiments, in conjunction with the experiments by Wiltschko et al. (1976) on young pigeons, described in Chapter 8, shows fairly clearly that even once pigeons have calibrated their sun compass, they continue to check it at intervals against other factors and to recalibrate it when necessary. The evidence is circumstantial, but Keeton's experiments hint that recalibration, perhaps as with the initial calibration when young, is carried out relative to the bird's magnetic compass.

9.4.2. By night

Experienced birds continue to use their magnetic and celestial compasses at night in an interactive way. Wiltschko and Wiltschko (1975a) tested wild-caught garden warblers in situations where directional information from the magnetic field and stars was contradictory. During migratory restlessness in spring, the birds oriented always according to magnetic north, despite the conflicting evidence from the natural stars. When the magnetic field was removed, however, the birds continued to orient in the direction relative to the stars that had been magnetic north in the earlier part of the experiment. This suggested that the star compass had been recalibrated according to the artificial magnetic field.

Further evidence for such recalibration was obtained with European robins, caught in Spain during northward migration in spring (Wiltschko and Wiltschko 1975b, 1976a). At first, however, the robins failed to respond when the magnetic field was rotated through about 120 degrees relative to the stars. The birds instead continued to orient toward stellar north. Two nights after the rotation, however, the bird's orientation changed to align with magnetic north. When the magnetic field was partially compensated, so that an inclination compass could no longer be used, birds previously used as controls continued to orient to stellar north. Those used earlier as experimental birds initially behaved the same but then orientated to the east-southeast.

In later experiments, wild-caught robins were placed under an artificial sky with an unnatural star pattern and at the same time a reduced magnetic field. The birds were disoriented. A magnetic field of normal intensity but rotated clockwise by 80 degrees was then introduced, whereupon the robins oriented to magnetic north. Finally, the original conditions of reduced magnetic field and artificial stars were reinstated. Now, however, the birds continued to orient by the stars in the direction that had been magnetic north.

Fig. 9.9 Recalibration of the star compass by the magnetic compass in the European robin **(Erithacus rubecula)**

Robins captured during spring migration in Spain were exposed in orientation cages to reduced magnetic fields and the artificial star pattern shown left. Both patterns were new to the birds and they were disoriented (a). The magnetic fiield was then reinstated but rotated by 80°. The birds adopted the traditional migration direction, toward magnetic north (b). Evidently they also recalibrated their star compass, adopting the artificial pattern. When the conditions shown in (a) were reinstated, the birds could now continue to orient in the appropriate direction (c).

[From Baker (1981a), after Wiltschko and Wiltschko]

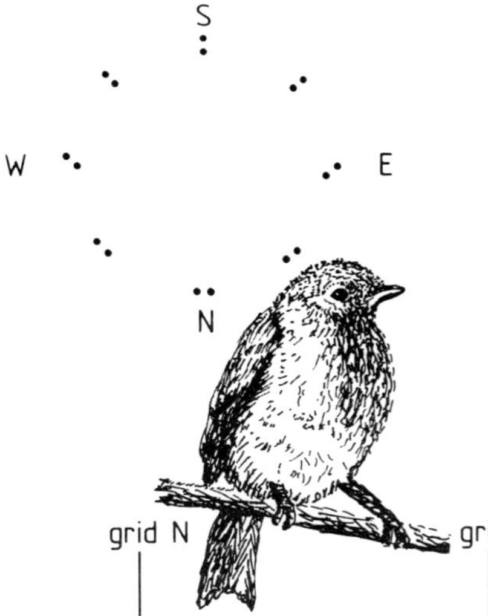

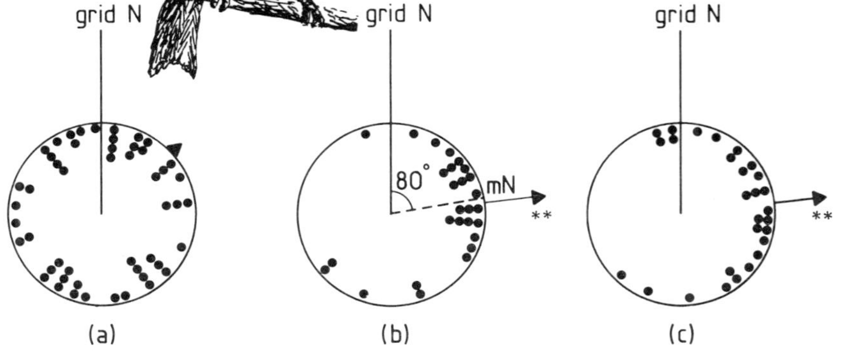

The general impression, therefore, is that both garden warblers and robins from time to time check their star compass against a magnetic compass. The robin, however, seems to do so less frequently than the garden warbler. The available evidence suggests that as garden warblers migrate through temperate latitudes, they frequently recalibrate the night sky against their magnetic compass. Being such a long-distance migrant (Fig. 8.4), of course, the pattern of stars will change considerably as the migration progresses.

We can only assume that, once the garden warbler reaches equatorial latitudes and its magnetic compass disappears, it switches entirely to orientation by its star compass. When it emerges on the southern side of the equator at a latitude where its magnetic compass is once again functional (assuming it reaches that far south), something interesting must happen. Now its migration track takes it poleward, not toward the equator. This requires a reversal of its preferred orientation relative to an inclination compass.

There is no indication for hand-raised garden warblers in cages that, late on in the autumn at about the time they would normally be crossing the magnetic equator, they have any inborn preference to migrate poleward (Fig. 9.4). Evidently, such a reversal is not part of the inbuilt circannual program. We can only assume, therefore, that as they cross over into the Southern Hemisphere, they recalibrate their re-emerging magnetic compass by means of the star compass that took them so successfully through the equatorial region. Similarly, in spring the apparently inborn preference to migrate poleward shown by caged birds, must be overriden in natural migrants setting off from the Southern Hemisphere, again perhaps by giving priority to their star compass. The evidence suggests (Wiltschko and Wiltschko 1975a), however, that by the time they reach as far north as southern Europe during their spring migration, they are once again giving priority to their magnetic compass.

For the robin, the situation is different. Being a short-distance migrant, the starry sky changes little during migration, recalibration of the star compass with such frequency as in the garden warbler may be less important, and there are no problems due to crossing equatorial regions.

One other feature of the way experienced birds use their star and magnetic compasses was revealed in these studies. It was observed for both species that the consistency of orientation within the orientation cage was far greater if stars were visible than if only the geomagnetic field were available (Wiltschko and Wiltschko 1978a; Wiltschko 1982). It seemed that even when the magnetic compass was used to select the direction in which to migrate on any particular night, both species preferred to use their star compass to maintain that direction.

This preference for using stars to maintain direction may, of course, reflect simply the difficulty that caged birds seem to have in reading their magnetic compass. There are indications, however, for both moths (Baker and Mather 1982a) and humans (Baker 1981a) that a magnetic compass is again not favoured for maintaining direction. It may well yet prove to be a feature of the magnetic compass sense in animals that its primary use is for the selection or identification of direction, rather than the maintenance of direction while moving.

9.5. Links

So far, we have concentrated on the interactive use of compasses during either the day or the night. Some evidence is available that birds also have mechanisms to provide links between compasses used at different times of day. This evidence brings with it, however, some unexpected problems.

9.5.1. *The setting sun*

It has been suspected for some time that a view of the setting sun or early-evening stars may be involved in some way with the selection of the migration direction for the coming night. Some of Kramer's earliest orientation studies were on red-backed shrikes (*Lanius collurio*), blackcaps (*Sylvia atricapilla*) and whitethroats, (*S. communis*), all of which are nocturnal migrants (Kramer 1949). Orientation in cages under the night sky was observed but only when the birds were exposed to it at the test site at or before sunset. Moreover, there were indications that they mistook the sky-glow over a nearby city for the sunset and oriented accordingly in inappropriate directions.

When Saint Paul (1953) then showed that the red-backed shrike and another nocturnal migrant, the barred warbler (*Sylvia nisoria*), had a well developed time-compensated sun compass, and the list was extended still further by Shumakov (1965), it was suggested that night migrants might determine their direction of flight around sunset. This direction would then have to be maintained in some rough-and-ready way through the hours of darkness. Vleugel (1953, 1959, 1962), for example, proposed that, having fixed their flight direction from the position of sunset, nocturnal migrants could then maintain that direction after dark by flying at a fixed angle to the wind.

Recent studies have shown that at least part of this behaviour may indeed often be shown. F. Moore (1978, 1980) has shown that the orientation of savannah sparrows (*Passerculus sandwichensis*) in orientation cages at night is significantly better if they are allowed to see the sun set. This has since been confirmed in the white-throated sparrow (*Zonotrichia albicollis*), another short-distance migrant (Bingman and Able 1979). When allowed to see the sun near the time of sunset, or to see the sunset glow and the stars, the birds were able, later in the night, to orient in the appropriate migratory direction. When allowed to view only the stars after dark, orientation was much poorer and, in fact, many individuals failed to orient.

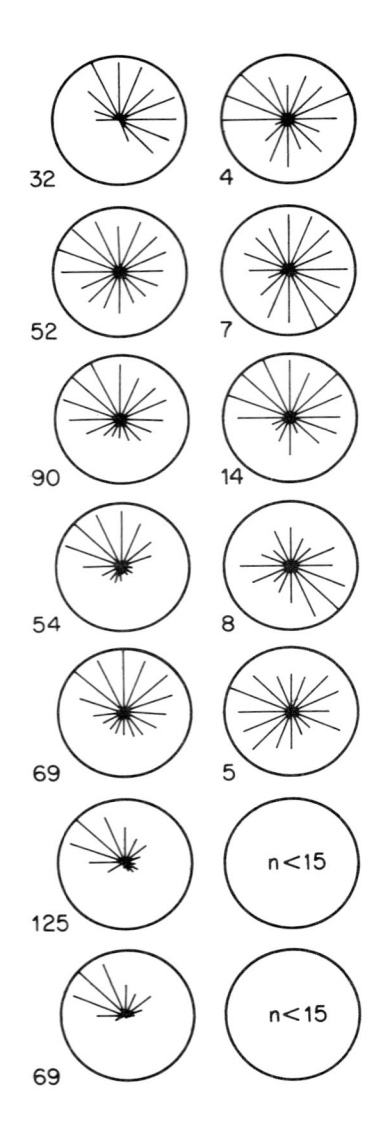

Fig. 9.10 Influence of sunset on the orientation of caged savannah sparrows (**Passerculus sandwichensis**) in spring

Each pair of vector diagrams shows the orientation of one individual. In both cases the birds could see the stars but in the right-hand diagram the birds were not placed in the cages until all traces of sunset had disappeared. The numbers at the lower left of each circle give the number of units of activity that define the longest vector. When denied a view of sunset the birds' activity was not oriented and was also much reduced.

[Simplified from Moore (1978). Photo by Verne Bingman]

The importance to selection of the migration direction of being able to obtain celestial information during the sunset period, as deduced from studies of caged migrants, has been confirmed for free-flying migrants using tracking radar. When migrants had seen the sun and/or the stars near the time of sunset, they were found to orient during flight in appropriate migration directions (Able 1978, 1982a, b). When thick overcast set in several hours before sunset and continued until after dark, migrating birds were often disoriented or headed in seasonally inappropriate directions.

When solid overcast began near dusk, allowing the birds a view of the sun near sunset but preventing them from seeing stars, the headings were indistinguishable from those under clear skies.

Whether the importance of the dusk period is due to the migration direction being set by reference to a compass based on sun, stars or polarization patterns is undecided. Viehmann's results for blackcaps using an artificial sunset, as described below, and Able's evidence that polarization patterns during twilight are in some way important to the white-throated sparrow (Chapter 6), suggest that it may well be polarization patterns that are the critical factor rather than the position of the sun below the horizon. Either way, however, the obvious point of interest in these findings on the importance of sunset glow concerns the apparent absence of a role for the magnetic compass. Viehmann (1982) has provided fascinating but perplexing data on this question.

Blackcaps and robins were caught locally around the University of Frankfurt. After exposure to particular environmental conditions, they were tested in orientation cages at three different times of day: in the afternoon; at night; and in the morning.

First, Viehmann confirmed that when caged and exposed to the normal geomagnetic field, both species would orient in appropriate migration directions even in the absence of celestial cues. Then he changed conditions at particular times of day. If the magnetic field was normal during the day, but then at night reduced in intensity (from the local 46 000 nT down to 5000 nT) and the lines of force made horizontal and therefore useless for magnetic orientation, blackcaps nevertheless continued to orient-correctly. Viehmann concluded that the birds were selecting their orientation direction before nightfall and then maintaining it at night, perhaps by reference to features within the cage or test room. If the horizontal component of the geomagnetic field was reversed during the day, but then restored to normal at night, the birds were disoriented. On the other hand, if the magnetic field was reduced during the day, but then restored to normal at night, orientation still occurred, suggesting that the selection of migration direction can also occur at night.

Taken together, these three results suggest to me that reversal of the magnetic field influences the birds' ability to orient to the magnetic field, whereas simply changing intensity (between reduced and normal) does not. I can see no consistent pattern suggesting that these birds determine their nocturnal migration direction by reference to the geomagnetic field in the afternoon.

Having established that reversing the magnetic field sometime during the day produced disorientation at night, even if a natural field was given at night, Viehmann used this technique to see if any other cues would permit orientation under such conditions. Providing the bird with an artificial sunset (but probably without appropriate polarization patterns) failed to produce orientation later on in the night. On the other hand, if the birds were given an artificial night sky with a pattern of just 16 star-like points of light, significant orientation in appropriate directions was obtained.

Unfortunately, it is impossible, from the way the experiments are described, to determine the sequence of cues to which each individual bird was exposed. We cannot decide, therefore, whether the birds had the opportunity to learn their artificial star compass in an unchanging magnetic field or whether they were able to do so even in a field that changed from day to night.

At present, data for the day- and night-time use of compasses by birds seems to conflict with data for the transition between these two sets of compasses. On the one hand, there is clear evidence of the importance of the magnetic compass in the recalibration of sun and star compasses and in the selection, though perhaps not the maintenance, of an appropriate direction for migration. On the other hand, there is the equally clear evidence that if celestial information is not available around sunset to link day- and night-time compasses, birds seem unable to select their migration direction. We even know for one of the species for which this is true, the savannah sparrow, that the bird does possess a perfectly good magnetic compass (Bingman 1981). For some reason, the magnetic compass, which is the obvious means of 'linking' day- and night-time compasses, seems inadequate for this purpose, whereas earlier in the day and later in the night it assumes a significant role.

The solution to this dilemma may have been found in recent studies on the savannah sparrow (Bingman, personal communication). These suggest that only adults make use of the setting sun in determining their migratory direction overnight. When the birds are young, it seems that the geomagnetic field is still the primary cue and that the setting sun alone is not enough.

9.5.2. Wind

The final element in the array of compass information available to birds is the wind.

Wind direction cannot, of course, be used as a compass in its own right except in those parts of the Earth where its direction is consistent for long periods of the year. Such areas are few and are largely those that experience the trade winds of the tropics or the Roaring Forties of the Southern Ocean. Even here, variation in direction does sometimes occur as meteorological disturbances move through the otherwise uniform wind fields. Elsewhere on the Earth, wind direction is so variable from day to day, hour to hour, or even minute to minute, that it is useless as a compass cue in its own right. As described by Able (1980), however, the direction of wind aloft is often quite predictable on the basis of information available on the ground.

If birds are as good at meteorology as evidence suggests, they could obtain crude compass information even in the absence of more reliable systems. They can, for example, detect changes in barometric pressure (Kreithen and Keeton 1974a). If they are able to obtain directional information from infrasounds (Chapter 5), they may also hear the approach of storms or frontal systems from distances of hundreds or thousands of kilometres (Kreithen 1978). Frontal systems, in particular, often bring with them winds that blow in roughly predictable directions. In the northeastern United States, for example, the passage of a cold front in autumn is usually signalled by falling temperature and relative humidity, clearing skies, and rising barometric pressure. A bird, on detecting the approach of such a front, and then perceiving its passage overhead, could gamble, with a high chance of being correct, that the winds aloft will be blowing from the northwest and therefore favourable for southward migration. By selecting the appropriate night for migration, and then flying simply downwind, the bird could have a high probability of flying in the intended direction, even in the absence of all other cues. The most likely use of wind direction, however, is not as a compass cue in its own right, but as a link between one compass system and another.

Most recent information on the orientation of birds with respect to wind has come from tracking radar studies of free-flying nocturnal migrants (e.g. Able 1982a) or from birds taken aloft in boxes by balloon, released, and then again tracked by radar (Emlen and Demong 1978; Demong and Emlen 1978; Able 1982b). The results show that birds do indeed often orient downwind, on occasion even in seasonally inappropriate directions. Discussion is best left, however, until Chapter 12 when we consider navigation in action.

In some ways, the study of the compasses used by birds has come a long way since the pioneering experiments of Kramer, Hoffmann, Schmidt-Koenig, Matthews and others in the 1940s and 50s. In other ways, there is still a long way to go before we can feel that we have a firm grasp of how these compasses are used. In particular, we seem some distance from a full appreciation of the way the different compasses are developed and integrated and why specific compasses are sometimes used and sometimes not. The one undisputed fact, however, is that birds live in a world that is strongly polarized into compass directions.

This certainty stands in stark contrast to our frame of mind over the question of whether birds can refer to any form of universal grid that will extend their 'map' beyond the limits of that familiar landscape we have called their familiar area map. The final set of evidence that we have to examine, therefore, before we look at orientation and navigation in action, concerns the possible detection by birds of universal coordinates, such as latitude and longitude. We find in the next chapter that the data are still weak and equivocal. We still have to ask whether avian grid coordinates are fact or fiction.

10 Grid maps: fact or fiction?

Everybody accepts that, at one level or another, birds are able to learn and use landscapes. Similarly, few people see anything surprising in the idea that birds are able to detect compass direction, at least by the sun and stars. Neither of these abilities causes the phenomenon of bird navigation to transcend the ordinary. It is instead the possibility that birds might be able to read some form of absolute grid map, stretching far beyond the limits of their familiar area, that for many people raises the study of bird navigation from the level of the humdrum to the mysterious.

As I write, it is just a century since Viguier (1882) first proposed such a map, based on the geomagnetic field. Yet at times we seem little nearer to knowing whether grid maps are indeed a fact or, although a good idea, simply fiction.

Most people agree that for a grid map to work it has to be based on at least two gradients. Moreover, the nearer the angle at which these gradients intersect each other approaches 90 degrees the better, at least in terms of the ease with which the map could be 'read'. The major possibilities were outlined in Chapter 4 and in this chapter discussion is restricted to the available experimental evidence relevant to the different possible gradients.

It is convenient to discuss the evidence in two parts: data relevant to gradients giving roughly 'latitudinal' information; and data relevant to roughly 'longitudinal' information. It should be stressed, however, that except for gradients generated by the Earth's rotation, much of this information gives only approximate north—south and east—west gradients and therefore does not lead to neat grids, with isolines (i.e. lines of equal value) crossing each other at 90 degrees.

10.1. 'Latitudinal' gradients

10.1.1. Coriolis force

The first serious attempt to obtain experimental evidence for a grid map was made by Yeagley (1947, 1951), who favoured a grid based on the geomagnetic field and Coriolis force. Both of these are in fact 'latitudinal' gradients, but Coriolis force is based on the Earth's rotation, and thus the geographic poles, whereas the geomagnetic field is based on the geomagnetic poles. The two poles are separated by only 2000 km or so and thus produce isolines crossing each other at a a fairly narrow angle (Fig. 4.12). Moreover, not all positions on the Earth's surface are unique. Every intersection of two isolines has a second 'conjugate' point, where the same two isolines cross somewhere else on the Earth's surface. This was the phenomenon exploited by Yeagley in an ambitious experiment that received much publicity.

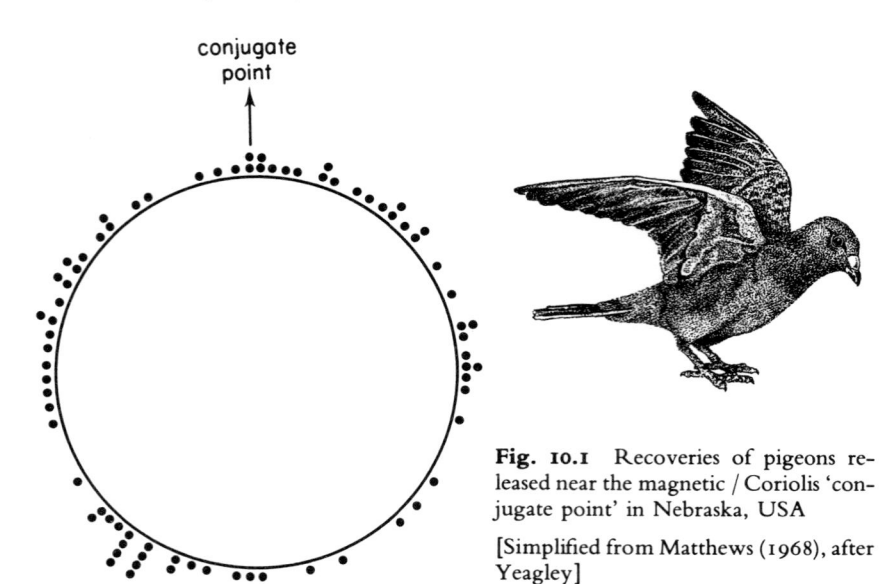

Fig. 10.1 Recoveries of pigeons released near the magnetic / Coriolis 'conjugate point' in Nebraska, USA

[Simplified from Matthews (1968), after Yeagley]

The idea was to train pigeons to home to one position and then to release them near to the conjugate point of that position. They should then mistake the conjugate point for the original home and respond accordingly.

Yeagley began by training pigeons to home to a conspicuous mobile loft in Pennsylvania, USA. To get them used to returning to the loft in different surroundings, the whole set-up was moved at intervals during training. The birds were then taken about 2300 km to the conjugate point in Nebraska, USA.

Releases were made between 40 and 130 km from the conjugate point, most birds being released at distances greater than 80 km. Of 459 birds released, only 8 regained a loft. Moreover, later analyses of the results (Matthews 1968) showed that the directions in which birds were recovered from the release point were essentially random. Finally, whereas the average distance of release from the conjugate point was about 110 km, the average distance of recovery was about 170 km. Birds actually ended up further away from the conjugate point than they were when released.

Although Yeagley saw some support for his grid model in his results few other people could, and no experiment to test for the possible involvement of Coriolis force in a grid map has been carried out since. In large part this was because attention was diverted almost immediately to the much more attractive proposition that birds might have a grid map based on gradients derived from the sun's arc (Matthews 1951, 1953).

10.1.2. Altitude of the sun

In Matthews' formulation of his sun-arc model he suggested that birds extrapolated observed movement of the sun over a small part of its arc to the position it would reach at noon, when it is due south. This position, the highest point on the day's arc, is known as the 'culmination point'. Pennycuick (1960a, b; 1961) slightly modified Matthews' thesis by suggesting instead that birds might compare the instantaneous altitude of the sun above the horizon relative to its memorized altitude at the same time at home. Finally, Nazarchuk et al. (1969) proposed that the angle of slope of the sun's arc would be a more usable measure. Whichever measure of the sun's arc is considered to be important, however, in essence all variations of Matthews' model suggest that birds obtain a north–south gradient by measuring the height of the sun above the horizon.

Matthews tested the possibility that pigeons may be using the sun's altitude to determine latitude in the following way. He reasoned that, upon release, pigeons may compare the altitude of the sun's arc at the release point with the altitude of the last remembered sun's arc at home. At the autumn equinox in late September, the sun's altitude at noon decreases by about 23' each day, equivalent to a shift of about 40 km of latitude. Matthews (1953) kept birds at the home site without a view of the sun for 6–9 days. As long as such birds are released less than 200 km to the south of home, the sun's altitude will still be lower than the last remembered altitude at home and the birds should fly south, even though home is really to the north. In Matthews' experiment that is precisely how the birds behaved (Fig. 10.2(a)).

When the experiment was repeated by Rawson and Rawson (1955) and Kramer (1957) both controls and experimentals flew toward home to the

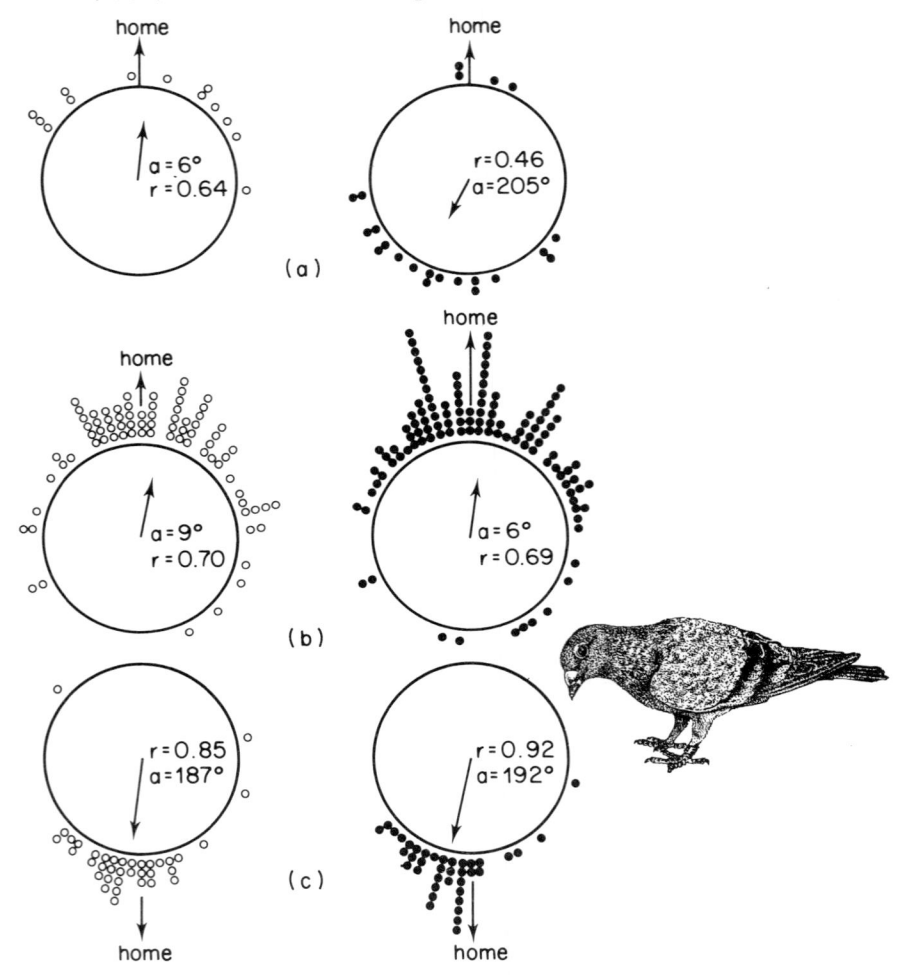

Fig. 10.2 Equinoxial tests of the hypothesis that pigeons determine the latitude of their release site by comparing the sun's altitude with its last observed altitude at the home site

Open circles, control birds allowed to see the sun throughout the equinoxial period. Solid symbols, birds prevented from seeing the sun during the equinoxial period. Release sites were selected so that if the hypothesis were true, experimentals should head away from home. (a) Matthews' original 'successful' experiment. (b) and (c) Summarised 'unsuccessful' replicates by other authors during autumn (b) and spring (c) equinoxes.

[Modified from Schmidt-Koenig (1979), after Matthews, Rawson and Rawson, and Kramer, Hoffmann and Keeton]

north. The birds used in these later experiments were from the loft in Wilhelmshaven, West Germany, and, as we have seen (Chapter 3), have a tendency to fly north upon release, irrespective of home direction. Hoffmann (1958) used Matthews' stock and lofts in Cambridge but again could find no difference between experimentals and controls; nor could Keeton (1970, 1974a, b), using Cornell birds at Ithaca, New York, in tests at the autumn and spring equinoxes.

Walcott and Michener (1971) tried the different approach of using mirrors at the loft to alter the sun's altitude at noon by 31−70′, appropriate to positions 58−130 km north or south of the loft depending on whether the altitude was raised or lowered. By choosing suitable distances for release (e.g. 40 km north of home when the arc had been lowered) it is possible to predict on the basis of Matthews' hypothesis that controls and experimentals should depart from the release site in opposite directions. Once again, however, all groups departed toward the real home.

In this last experiment, the birds were deprived of a view of the real sun-arc for 10−14 days and in the equinoxial experiments the deprivation was for a similar period. None of the experiments would have been expected to work, therefore, if the original premise were wrong and birds did not relate the sun's arc at the release site to the last remembered (and unmanipulated) arc at the home site. Perhaps the pigeons involved were capable of extrapolating the seasonal change in sun's altitude for long enough to bridge the period of experimental treatment. It is not impossible that birds will have evolved such an ability. Otherwise (on Matthews' model) periods of overcast of several day's duration, particularly during the equinoxial periods, when many birds other than pigeons are migrating, would play havoc with their navigational ability.

So far, I have referred fairly casually to birds measuring the sun's altitude, in other words its height above 'the' horizon. Which horizon, however, is a matter of some importance. The level of the Earth's horizon depends on the height at which the bird is flying and the height of topographical features such as hills and mountains. The only usable horizon would seem to be some form of internal 'subjective' horizon based on the bird's horizontal axis relative to gravity. In an elegant experiment, Benvenuti (1976) fitted pigeons with prismatic goggles that changed the altitude of the sun at the release site by about 3 degrees. Such birds could still orient towards home. However, the goggles used would also have shifted the visible horizon by the same amount. The experiment seems, therefore, to argue against some form of 'internal' horizon, which is the only realistic horizon if Matthews' model were to be viable.

Since about 1970, a succession of authors have rigorously reviewed and repeated those of Matthews' experiments that seemed to support his model. They have more or less unanimously concluded that the weight of the evidence is against him. In my view, however, less rigour has been shown toward the experiments that run counter to the sun-arc model. For example, in none of the experiments quoted above that produced no difference between experimentals and controls was there a control for the possibility that the birds might follow their outward journey by some form of route-based navigation (Chapters 4 and 11). In which case both controls and experimentals may already have decided on home direction even before they saw the sun at the release site. It is possible that in Matthews' original and successful experiment some quirk of the magnetic or olfactory conditions during transport prevented route-based navigation, thus forcing the birds to use the sun's altitude and in turn producing the observed difference between experimentals and controls. In later experiments, conditions may have been favourable for route-based navigation, thus allowing both controls and experimentals to use the same mechanism.

We are back to the principle of redundancy. The data certainly show that accurate perception of the sun's altitude is not essential to navigation. They do not show that under normal conditions, birds fail to refer to the sun's altitude. If we are being rigorous, therefore, I feel we have to conclude that critical data on the possible use of the sun's altitude by pigeons do not yet exist.

Operant conditioning experiments (see Chapter 6) carried out by Whiten (1978) showed that, with a fixed and artificial sun, a pigeon could learn to adjust the sun's height relative to a learned altitude. The level of accuracy achieved would, in the real world, have allowed the bird to detect displacements of about 33 km, but this calculation makes no allowance for inaccuracies that would derive from the sun's movement with time of day, the seasons, and the bird's own movements when normally active. Attempts to obtain data for more normal sun movements were less successful. On the other hand, in field experiments, using the same operant techniques, Whiten showed that his pigeons could learn to distinguish between three sites around Bristol solely on the basis of the sun's altitude. Some authors have objected (e.g. Schmidt-Koenig 1979) that this study does not show that the birds detect home direction from the sun's height, only that they associate keys in particular compass directions with differences in the sun's altitude. The fact remains, however, that the birds could learn to distinguish sites on the basis of the sun's height at those sites.

Under some circumstances, therefore, pigeons can learn to use the sun's altitude to select a particular compass direction. The relevance of this to free-living birds may be marginal, but in the absence of critical experiments against the use by birds of the sun's altitude, combined with the positive results in Matthews' original equinoxial experiment, we have at least to consider that a gradient based on the height of the sun is still a possibility.

Having made this point, I should stress that there have been experiments in which birds seem to have neglected apparently ideal opportunities to use the height of the sun. If ever the sun's altitude were to be useful, it should be after long-distance displacement along a north–south axis.

Recently, for reasons unconnected with the possible use of the sun,

Wallraff and his colleagues (Wallraff 1981a, Wallraff et al. 1981) have begun subjecting pigeons to intensive training programmes along a north–south axis, leading eventually to training releases as far as 170 km N and 150 km S of the home loft in Florence, Italy. If pigeons were ever going to learn the significance of the height of the sun as a means of working out north–south displacement, they should have done so by the end of such training. The birds were then displaced beyond the Alps to release sites in Northern Bavaria, 700 km to the north of the home loft. Before displacement, the pigeons were divided into control and experimental groups, the latter having their nostrils plugged throughout the journey. The plugs were then removed before release.

Both groups of birds disappeared to the west upon release, and later recoveries suggested that the experimentals had continued in that direction until giving up. The controls, however, seemed later to turn to the south and to head towards home. The results suggest that it was information collected during the outward journey that was important to the controls. The obvious lowering of the height of the sun at the release site, relative to its height at home, should have been as useful a cue to the experimentals as to the controls. It is possible, of course, that the trauma of having had plugged nostrils could have reduced the motivation of the experimentals to fly home, but if this were the explanation we should have expected their behaviour to differ from the controls immediately upon release rather than later.

10.1.3. Night sky

The only authors to favour the possibility that birds might use the night sky to detect position on a north–south gradient were the Sauers. Even their own work, however, offered little by way of convincing experimental support.

Sauer (1957) claimed that when he tilted the axis of his planetarium projector to produce night skies appropriate to lower latitudes, one of his birds, a lesser whitethroat (*Sylvia curruca*), showed a shift in preferred orientation in an orientation cage. The change was from south-southeast to south as if it were migrating from Germany, through the Levant and then down through Africa, Moreover, this bird had been born in captivity. Its behaviour implied, therefore, that its apparent appreciation of the changing pattern of stars with latitude was innate.

Sauer offered no statistical support for his interpretation of the data. Wallraff (1960) showed that the proposed shift in direction, which was by no means gross, could just as easily be interpreted as showing the same orientation throughout. Moreover, Sauer and Sauer (1960) found no such effect with garden warblers (*Sylvia borin*) or blackcaps (*S. atricapilla*).

In an ambitious experiment, the Sauers shipped warblers to South West Africa at the beginning of the autumn migration and tested them in orientation cages under the natural night sky. Only one bird, a whitethroat

158   *Bird navigation: the solution of a mystery?*

(*S. communis*) was really active yet still headed to the south despite being at the southern limit of its wintering range. What little activity was shown by three garden warblers and one lesser whitethroat was also to the south, even though the latter was some 3000 km to the south of its normal wintering range.

The Sauers still managed to find some encouragement in their results, but the possibility that birds might determine their latitude from the stars has not been resurrected by later workers.

10.1.4. *Geomagnetic field*

Most research activity into the possibility of a grid map in recent years has been generated by modern versions (e.g. Gould 1980, 1982; B. Moore 1980; Walcott 1980b) of Viguier's magnetic map hypothesis. As far as magnetic 'latitude' is concerned, a north–south gradient could be determined by the angle of inclination, or the total field intensity, or the intensity of vertical or horizontal components and none of the data below are more relevant to one measure than another. It is convenient in discussion to refer primarily to total field intensity.

In order to use the magnetic field to obtain map information, birds would need a seemingly impossible level of sensitivity to variation in the magnetic field of about 1 in 2000. Experiments by Schmidt-Koenig and Walcott (1978), using pigeons fitted with frosted-glass contact lenses, showed that of 74 such birds released at distances of up to 20 km, 19 returned to within 0.5 – 5 km of the loft. If this performance is interpreted

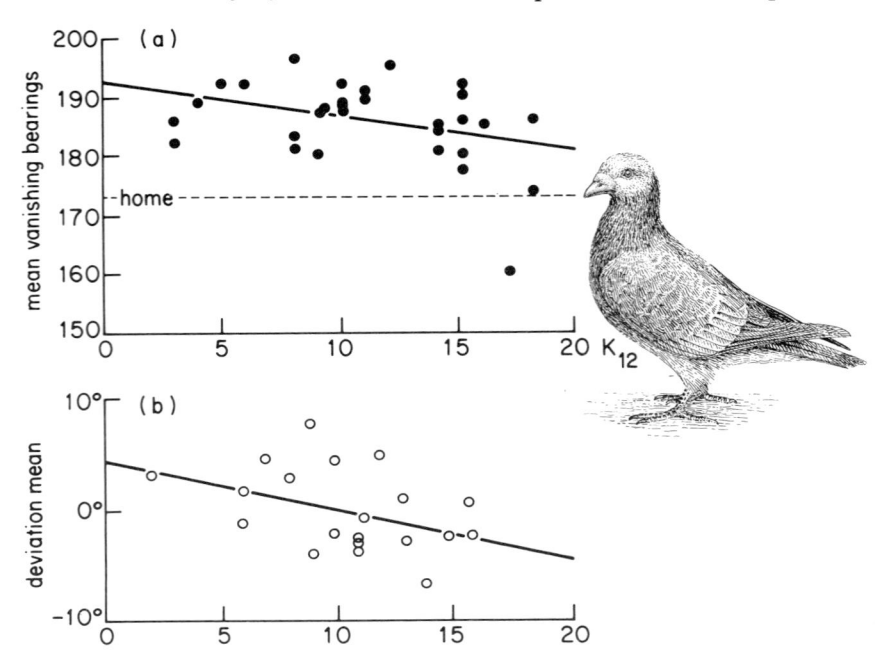

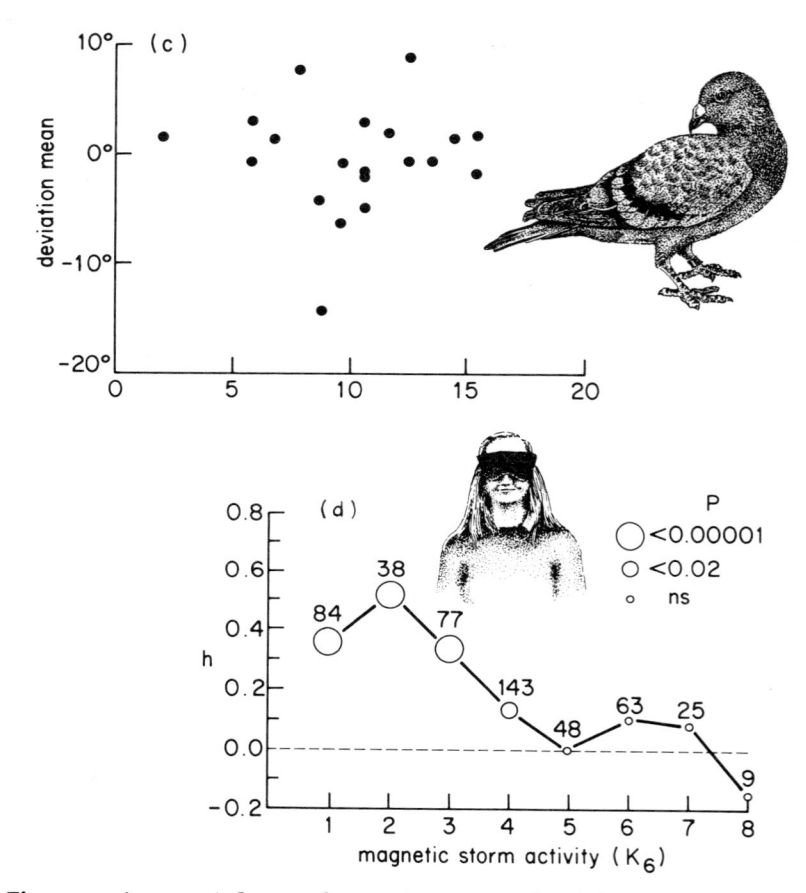

Fig. 10.3 Apparent influence of magnetic storms on: (a–c) the mean vanishing bearings of pigeons from Ithaca, New York; and (d) the strength of homeward orientation by people.

Birds were released at Weedsport, New York. The index used for magnetic storm activity is the sum of the four K-values for the 12-hour period ending at the time the release was completed.
(a) Mean vanishing bearings for 33 releases expressed relative to home bearing. (b) and (c) The bearing for each pigeon is expressed relative to its individual mean over many releases before calculating the mean deviation for the group. (a) and (b) normal magnetic field; (c) birds wearing bar magnets.
(d) Correlation between homeward orientation and magnetic storm activity in 'bus' experiments

Numbers above dots, number of individuals tested at each value for K_6. A single mean error / individual was calculated for all estimates of the compass direction of home following displacement at a given K_6-value. Individual means were then subjected to higher-order analysis to calculate h, the homeward component (where perfect homeward orientation is given by h = 1, uniform orientation or a mean error of $90°$ gives h / 0). The significance of h (v-test) is shown by dot size.

[Compiled from Keeton et al. (1974), Larkin and Keeton (1976) and Baker 1984b] solely in terms of a magnetic map, it suggests a sensitivity to within 10–30 nT, given that, in the northeastern United States, intensity increases at the rate of about 5 nT per kilometre travelled to the north (Gould 1982). Such sensitivity indeed seems impossible, yet evidence has slowly accumulated that birds may well be as sensitive as this.

Yeagley (1951) provided the first indication of such apparent sensitivity when he showed that the speed at which winning pigeons return home in pigeon races correlates with sun-spot activity the day before the race. Schreiber and Rossi (1976, 1978) found similar evidence in an analysis of 18 years of pigeon racing in Italy.

As we have seen, sun-spots are the major cause of magnetic storms on Earth, and in the early 1970s Southern had shown that ring-billed gull chicks were influenced in their orientation by such storms (Chapter 8). Some radar studies (Richardson 1974; Moore 1977) but not others (e.g. Able 1982a) reported that the scatter of directions of migrant birds correlated with the K-index of magnetic storm activity.

In a more direct study, Keeton et al. (1974) showed that the mean vanishing direction of homing pigeons from two sites, Weedsport and Campbell, in New York State were rotated counter-clockwise by an amount that correlated with magnetic storm activity. In this case, it was the sum of the K-values for the four three-hour blocks preceding the end of the experiment that provided the best correlation. In other words, magnetic storm activity for the 12 hours leading up to the experiment was the important factor. Later, Larkin and Keeton (1976) found that birds released at Weedsport wearing magnets glued to their backs showed no correlation with K-values whereas control birds wearing brass bars again showed the correlation.

In bus and walkabout experiments on people (Chapter 11), the ability to describe the compass direction of home also correlates ($r_s = -0.905$, n = 8, P < 0.02) with magnetic storm activity (Fig. 10.3(d)). The results differ from those for pigeons in two ways: (a) magnetic storms are associated, as for migrant birds, with a scattering, not a rotation, of estimates; (b) the best (but not the only) correlation is with storm activity for the six hours before the experiment. As for pigeons, there is no correlation for subjects with an artificial magnetic field through the head during the experiment ($r_s = -0.310$, n = 8, P > 0.10).

Gould (1980, 1982) calculates that the magnetic storm effect implies a sensitivity in birds of about 10–30 nT, which fits nicely with calculations based on the frosted-lens experiments.

Further evidence for a sensitivity to small changes in the magnetic field of the order required to detect location on the north–south gradient of the geomagnetic field comes from the behaviour of birds when released at the site of magnetic anomalies. At such sites, however, the change in magnetic field can be much greater than during magnetic storms. Whereas even a maximum K-rating of 9 indicates fluctuations in magnetic intensity of only 500 nT, many magnetic anomalies, caused by magnetic rocks buried in the

Earth's crust, can generate much larger changes of intensity over short distances. The Iron Mine anomaly used by Walcott (1978, 1980a, 1982), for example, involves changes in intensity of 3500 nT over 1 km.

The first indications that magnetic anomalies may influence the vanishing directions of homing pigeons were obtained by Graue (1965) and Talkington (1967) and then later by Wagner (1976), Frei and Wagner (1976), Frei (1982) and Kiepenheuer (1982a). Walcott (1978, 1980a, 1982) found a correlation between the degree of scatter of vanishing bearings and the strength of the anomaly at which the pigeons were released.

Gould (1980, 1982), using some of Walcott's data, has plotted the tracks

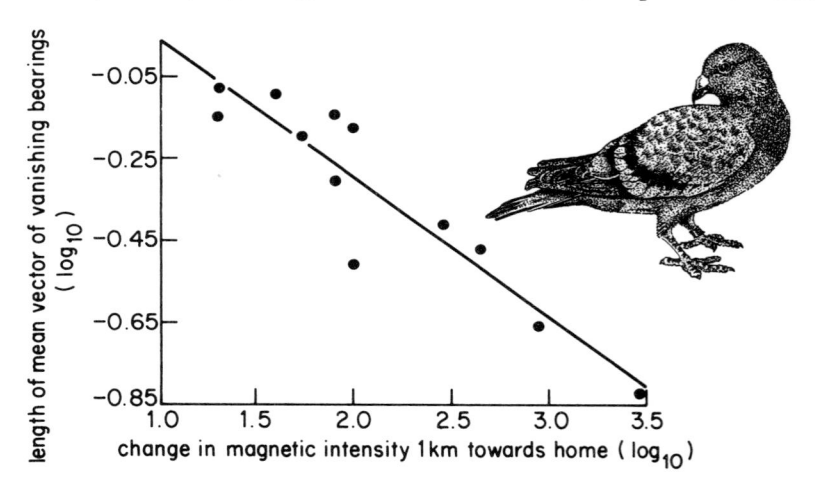

Fig. 10.4 Relationship between the amount of scatter in the vanishing bearings of homing pigeons and the intensity of the magnetic anomaly at which they are released

Scatter is quantified by r, the length of the mean vector (see Fig. 3.4). Intensity of the magnetic anomaly is given by the change in field intensity along a 1 km line towards the home loft.

[Modified from Walcott (1980b)]

of some pigeons against magnetic topography (fig. 5.11). One of the trends that emerged was that the birds seemed to fly down magnetic gradients. Such a pattern had also been demonstrated previously by studies at anomalies in western Switzerland (Wagner 1976; Frei and Wagner 1976). Here, pigeons were found to fly down magnetic 'slopes' with gradients of about 30 nT/km, regardless of home direction. As stressed by Gould (1982), such down-slope flight could be adaptive. Any gradient greater than about 5 nT/km should indicate an anomaly, and almost all anomalies are elevations in intensity, depressions being extremely rare. By flying down steep magnetic gradients, therefore, the bird eventually returns to the magnetic plains where reliable map information can again be obtained. This could be the explanation for the peculiar behaviour of pigeons from Ithaca when released in a magnetic basin at the Jersey Hill fire tower in

162 Bird navigation: the solution of a mystery?

western New York State. Tracked by plane the birds are found to wander around within a few kilometres of the release site.

Even once birds escape from an anomaly, it takes some time for homeward orientation to appear. Some form of after-effect is evident. On the other hand, birds released elsewhere fly straight over magnetic anomalies on their way home. On the basis of these observations, Gould (1982) suggested that the magnetoreceptor needs a relatively long time to interpret the magnetic field. Alternatively, it could simply mean that cues used for selecting direction upon release are different from those used for the maintenance of direction while flying home.

In the Iron Mine experiments, Walcott found that his birds showed no improvement upon successive releases at the same anomaly. Kiepenheuer (1982a), however, working in Germany, found that by their second release, his birds were as well oriented as controls released nearby but outside of the anomaly.

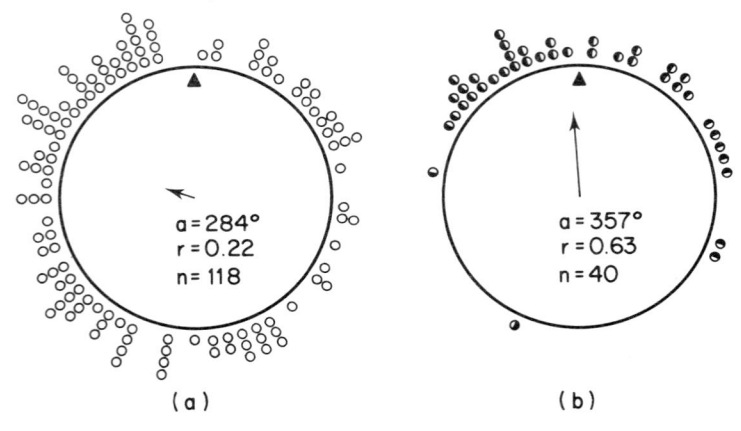

(a) $a = 284°$
$r = 0.22$
$n = 118$

(b) $a = 357°$
$r = 0.63$
$n = 40$

Fig. 10.5 Influence of previous experience of a magnetic anomaly on vanishing bearings of homing pigeons

(a) Pigeons released at an anomaly for the first time.
(b) Pigeons released for the second time at the anomaly

[Simplified from Kiepenheuer (1982a)]

Taken as a package, the evidence looks fairly convincing that birds are influenced in some way by relatively small changes in the magnetic field. The level of sensitivity that seems to emerge even appears to be of about the right order for them to be able to locate their position on the global north–south gradient. Thus far, the evidence, although entirely circumstantial, looks encouraging. Can we really conclude, therefore, that birds are able to read a magnetic coordinate as a contribution to a grid map? It seems we cannot, or at least, not yet.

Many of the problems still to be resolved have been outlined by Lednor (1982). In particular, he points out that the global change in magnetic field with latitude that is suggested to offer birds a north—south gradient is perturbed, not only by anomalies a few kilometres across, as exploited in the experiments just described, but also by larger anomalies thousands of kilometres across. These can destroy the global gradients over large areas. Lednor has shown that one such anomaly has a marked effect on the magnetic gradients around Ithaca, where Keeton and his colleagues studied the influence of magnetic storms. Here, along a NNW—SSE axis, field intensity increases in both directions from Ithaca over just the range of distances that most homing experiments are carried out. It is thus useless from the point of view of a magnetic grid. Yet, in this region, Keeton and his colleagues found that homeward orientation by their pigeons was disturbed by magnetic storms.

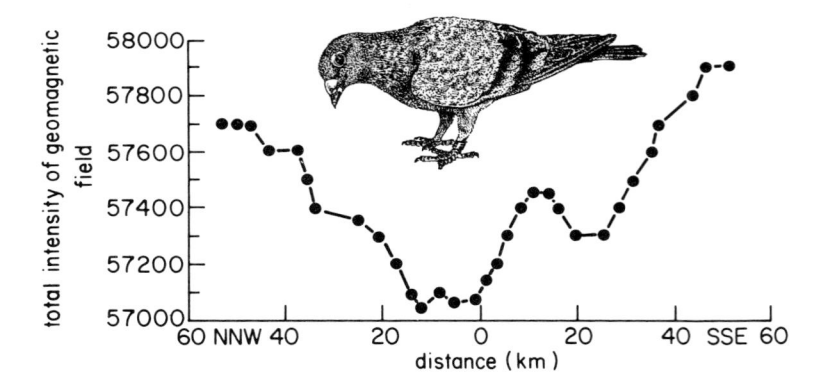

Fig. 10.6 Total intensity of geomagnetic field along the axis of maximum rate of change as plotted against distance from Ithaca, New York

[Redrawn from Lednor (1982)]

Lednor also points out that bar magnets do not destroy the homeward orientation of pigeons when released under sunny skies (Matthews 1955b; Keeton 1971; Walcott and Green 1974; Wiltschko and Wiltschko 1976b; Walcott 1977; Visalberghi and Alleva 1979), even if the magnets are on the wings and therefore produce a field with rapid changes in intensity. Under such circumstances it would seem impossible for the bird to detect the small changes in natural intensity necessary to read a magnetic map.

Similarly, Walcott and Gould (in Walcott 1982) found that apparently they could destroy the magnetic sense of pigeons, as indicated by failure to orient towards home under overcast, by placing each bird's head in an alternating and powerful, but decreasing, magnetic field. When the birds were released under sunny skies, however, they could still work out the compass direction of home. It seems that the magnets had damaged the birds' magnetic sense, but had not influenced their ability to work out home direction.

We always have to beware, of course, of the principle of redundancy. It is possible that in both of these experiments the pigeons, unable to read their magnetic coordinates, used some other mechanism, not involving magnetism. One technique would be for them to determine the compass direction of home from the release site by picking up familiar smells (either during the journey or at the release site; Chapters 5 and 11) and then to set off for home using their sun compass. On the other hand, we have to remember that the influence both of magnetic storms (Keeton *et al.* 1974) and of magnetic anomalies (e.g. Walcott 1978) was observed under sunny skies. Why, then, did the birds not also switch to alternative methods on these occasions?

Neither of these last two experiments, nor the fact of regional difficulties in finding useful magnetic gradients, such as around Ithaca, prove that birds do not use the geomagnetic field to fix their position on the north–south axis. On the other hand, they do raise fairly major doubts. If birds do not have a magnetic map, however, we are left with the question of why they should be influenced by magnetic storms and anomalies. Is it possible, perhaps, that these events and locations influence the bird's ability to read, not their magnetic map, but their magnetic compass?

Walcott (1980b, 1982) and Gould (1980, 1982) argue that, since even a 1000 nT storm could not rotate a compass needle by as much as 2 degrees, the magnetic storm effect cannot be due to disruption of the bird's magnetic compass and, at first sight, this is difficult to contest. Wiltschko (1972, 1978) has shown, after all, that variation in magnetic intensity from 38 000 to 57 000 nT does not disrupt compass orientation in robins. Yet here we are asking whether changes as small as a few tens of nanoTesslas may be disruptive. Nevertheless, several of the examples of orientation being disrupted by magnetic storms could have been based on a compass, rather than a map. Thus racing pigeons and birds in the act of migration are both highly likely to select and maintain their direction by means of a compass. In particular, the orientation of ring-billed gull chicks studied by Southern (e.g. 1971) cannot seriously be considered as anything but compass orientation.

There is no way, at present that we can resolve the question of whether birds can determine their position on a north–south gradient by checking with the magnetic field. On the one hand, the data seem to indicate quite convincingly that birds have the sensitivity necessary to detect tiny changes in the magnetic field. On the other hand, regional variations in the global gradient suggest that in at least one of the regions used to study pigeon homing a useful map is not even there to be read. Also, some experiments in which it should have been impossible for the birds to read the magnetic grid produced perfectly good orientation. Equally perplexing is the fact that disruption of a bird's magnetic compass by magnetic storms and most magnetic anomalies seems impossible, yet there is more than a hint that this may be just what happens. If it is the magnetic compass that is influenced by these small changes in magnetic field, then all support for the presence of a magnetic axis on the grid map evaporates.

10.2. 'Longitudinal' gradients

10.2.1. Geomagnetic field

Far fewer suggestions have been made about the way in which birds could detect coordinates along an east—west axis. Viguier's original suggestion was that the best feature of the magnetic field to provide 'longitudinal' information was declination, the angle between geomagnetic and geographic north. Declination isolines show a complex pattern over the Earth's surface (Fig. 4.13), but would serve as a crude analogue of longitude.

While not supporting Viguier's suggestion, Matthews (1968) accepted that, if birds could detect geomagnetic north, as we now know is possible (Chapter 8), they should also be able to measure declination because they could obtain true north from the sun and stars. Gould (1982), neglecting the principle of redundancy, dismisses the possibility that declination might be used because birds can detect home direction even under overcast skies when geographical north cannot be detected. Keeton (1981) proposed a solution to this objection by suggesting that gravity gradients could be used to detect geographical north on overcast days.

This suggestion derived from results obtained by Larkin and Keeton (1978) for releases of birds at the same site over a period of four years. They found that the mean vanishing bearing varied with the day in the lunar month. Sometimes the bearings were in phase with the New Moon to New Moon cycle, at other times with the Full Moon to Full Moon cycle. Either way they showed only one cycle per lunar month whereas gravity, the factor to which they attributed the effect, shows two cycles per lunar month. Keeton then pointed out that gravity varies along a north—south axis because the Earth is not a perfect sphere. This gravity gradient could, therefore, serve to provide birds with a north—south coordinate. It could also allow them to identify geographic north and thus to use declination to form the longitudinal coordinate of a magnetic grid map. Keeton (1981) also states that preliminary results obtained by Larkin and himself suggest that declination is the parameter most influenced by magnetic storms.

Gould (1982) argues that, without declination as a cue, birds would have to compare horizontal intensity with vertical or total intensity if they were to have a purely magnetic grid. Even so, this comparison produces a poor grid, with isolines from the two gradients crossing each other in the northeastern United States at an angle of only 20 degrees. Moreover, it would predict that birds should identify home direction better from the east-northeast or west-southwest. Although Gould feels there is some trend towards this in navigation data, it is by no means striking.

Both Gould (1982) and Lednor (1982) conclude that there is no satisfactory magnetic parameter by which to obtain a coordinate along the east-west axis.

10.2.2. Night sky

In addition to suggesting that nocturnal migrants might be able to use star patterns to provide a latitudinal coordinate, Sauer suggested that they might also be able to use the night sky to obtain longitudinal information (Sauer 1961, Sauer and Sauer 1960, 1962).

Sauer rotated the projector in his planetarium to produce a night sky appropriate to a location 15-90 degrees of longitude to the east of the true location but on the same latitude and at the same time of night. One lesser whitethroat (*Sylvia curruca*) and one blackcap (*S. atricapilla*) were reported to compensate for the shift by orienting to the west in their orientation cage. Wallraff (1960) pointed out that at least part of the response was disorientation, rather than compensation. Moreover, the Sauers found that their birds did not orient to the east when shown a sky appropriate to a location to the west of home.

No later workers have supported the Sauers' thesis and data on an ability of birds to determine longitude from the night sky. On the contrary, Emlen (1967a,b) found that indigo buntings (*Passerina cyanea*) were unaffected by rotation of the sky to produce star patterns appropriate to sites separated in longitude from the home site by 45, 90 or even 180 degrees.

10.2.3. Local time from the sun's arc

Most of the experimental effort in the search for a factor which might provide birds with an east-west axis on their grid map has been directed at the sun's arc, following the theory proposed by Matthews and modified by Pennycuick and Nazarchuk *et al.* (Chapter 4).

No matter how the sun's arc is interpreted, the measurement of longitude ultimately reduces to judging the difference in local time between the release site and home. Moreover, the measure of time has to be accurate. In Britain, for example, a shift of 1 degree in longitude, about 75 km, is characterized by only a 4-minute difference in the times the sun reaches the highest (culmination) point of its day's arc in each place.

If birds really are using the sun's arc and local time to measure longitude, they should interpret clock-shifts as longitudinal displacements. For example, if a bird's clock is set 6 hours early, so that at local noon its subjective clock reads 18.00 h, it should interpret the change to mean it has been displaced a quarter of the way round the world to the west. If released on a homing experiment, such a bird should, therefore, fly east. In all such experiments (e.g. Schmidt-Koenig 1961, 1972; Walcott and Michener 1971), however, clock-shifted birds behave as if only their sun compass,

Grid maps: fact or fiction? 167

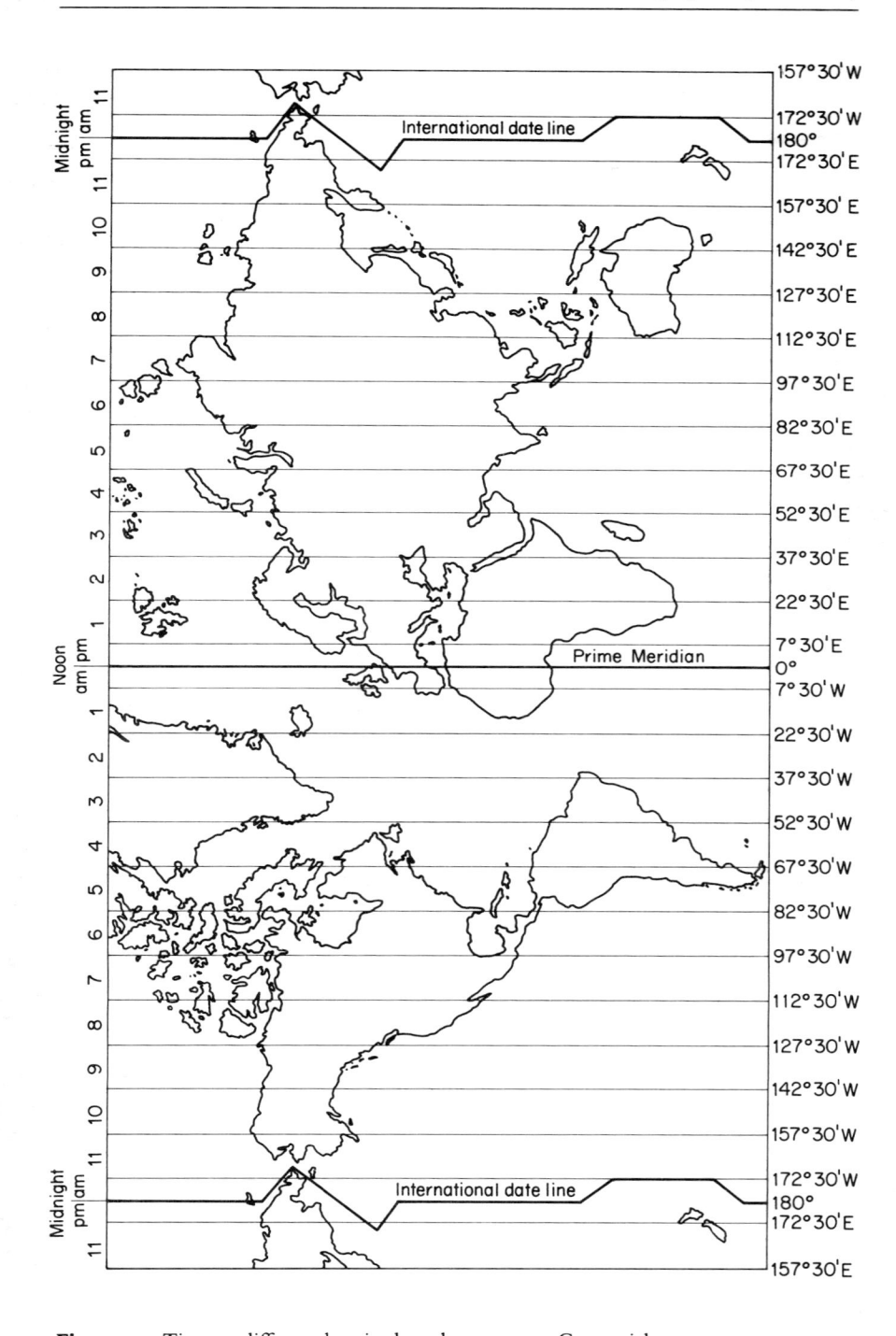

Fig. 10.7 Time at different longitudes when noon at Greenwich

not their interpretation of longitude, has been affected. In the above example, for instance, birds displaced to the west would fly north.

Evidently, clock-shifting experiments only influence that part of the bird's biological clock which controls its sun compass. However, we concluded in Chapter 4 that, in order to use the sun's arc to judge longitude, a bird would need two clocks: one set to the local light/dark cycle, which would be used in sun-compass orientation; the other, let us call it a 'chronometer' to distinguish the two, set irrevocably to home time. Perhaps the clock-shifting experiments just quoted changed only the compass clock and not the chronometer.

Matthews (1955a) tried to disrupt the chronometer by first exposing his birds to an irregular light/dark cycle and then clock-shifting them by 3 hours each day. When released in an appropriate direction (e.g. in the

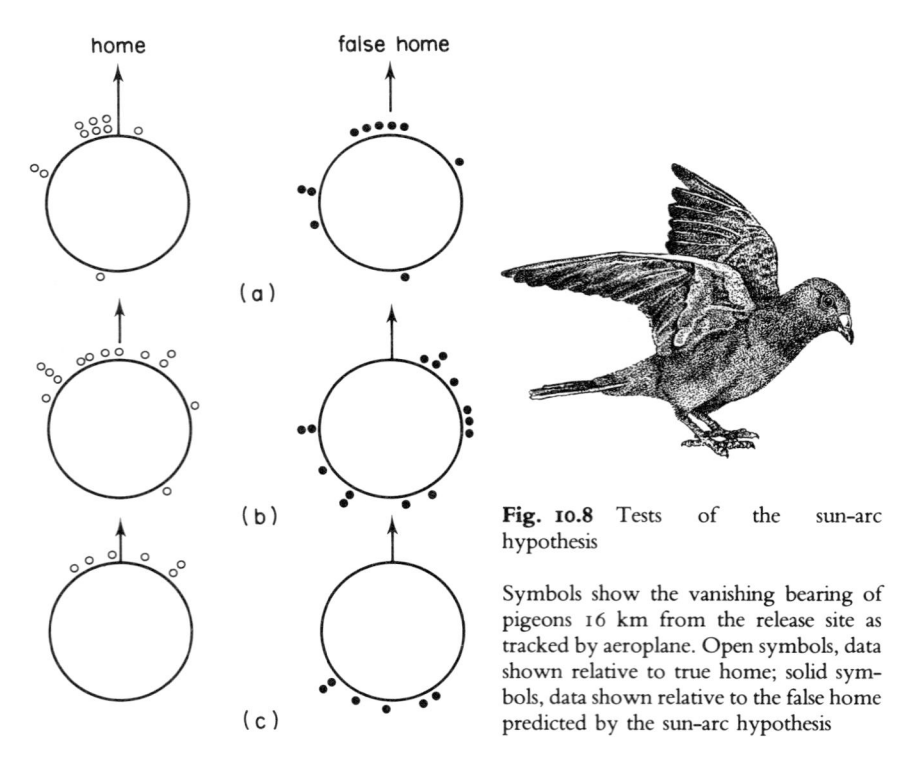

Fig. 10.8 Tests of the sun-arc hypothesis

Symbols show the vanishing bearing of pigeons 16 km from the release site as tracked by aeroplane. Open symbols, data shown relative to true home; solid symbols, data shown relative to the false home predicted by the sun-arc hypothesis

(a) Altitude of the sun as perceived at the home loft manipulated by mirrors. (b) Birds subjected to clock shifts of ten minutes or less, which according to the sun-arc hypothesis should influence the birds' estimate of the longitude of the release site relative to home. (c) Birds fed 'heavy water' and exposed to irregular photoperiods in an attempt to disrupt their chronometer and then entrained to the light−dark regime of another site, far to the west of the home loft. In all cases the birds oriented toward home rather than the false home.

[Compiled from Walcott and Michener (1971)]

above example, when released to the east of home so that controls should fly west whereas experimentals should fly east), the birds did set off in roughly the expected directions. However, the data showed a great deal of scatter and, even more disquieting, the experimentals actually homed almost as fast as the controls. They must therefore have rapidly realized their mistake, but how? After all, as pointed out by Kramer (1957), they could not have reset their chronometer to home time until they actually got home.

Walcott and Michener (1971) also tried to disrupt the chronometer by using irregular photoperiods and by the ingenious technique of giving the birds 'heavy' water to drink, which has been shown to slow their free-running rhythm by 45–75 min/day. After 9 days, the pigeons were taken by plane 400 km to the west on the same latitude and then allowed 6–9 days to reset their chronometers to local time. They were then released at a site 84 km west of their original loft along with birds displaced directly from the loft. If the experimental manipulations had been successful, and if Matthews' theory were right, the controls should have flown to the east whereas the experimentals should have flown west. In fact, there was no difference in the orientation of the two groups.

It could always be argued, of course, that the chronometer is so rigid that none of the experimental treatments (except perhaps that of Matthews) succeeded in shifting it, or perhaps wherever negative results were obtained they were due to the birds solving the particular navigation problem by means of route-based navigation. As this should give the answer in terms of a compass direction, then an influence of clock-shifting on the sun compass is all that could be expected.

At a more anecdotal, but nevertheless forceful level, Keeton (1974a) points out that when birds are settled at a new loft they may nevertheless from time to time revisit the old loft. To do this, using any time-based system of longitude, they would need more than one chronometer; one, in fact, for each 'home' site to which they may need to navigate, plus one biological clock for the sun compass. This thought applies with even greater force to any seasonal migrant with a migration path taking it across lines of longitude, since such birds need to navigate to a succession of different sites (Chapter 12).

Finally, in his operant conditioning experiments, Whiten (1978) could find no evidence in support of any of the possible parameters of the sun's arc being used in a way that would allow the birds to judge longitude.

10.3. Fact or fiction?

We began this chapter with the hope that when the evidence was reviewed some firm indication of whether or not birds possess a form of absolute grid map would emerge. We end it with the impression that, despite the wealth of effort which has gone into trying to resolve the situation, the critical data are still not available. Even if we stand back from the details, little emerges that is clear.

Of the range of possibilities for a latitudinal (north—south) gradient, there is no support for Coriolis force or for star patterns. Most other authors can see little indication, either, that the sun's altitude provides such a gradient. Although I am inclined to agree, I feel that the positive results by Matthews and Whiten have not yet been countered by critical negative evidence. Finally, the data in support of the idea that the apparently exquisite sensitivity of birds to small changes in the magnetic field is in some way relevant to the presence of a magnetic gradient on a grid map are by no means convincing.

Factors that could provide longitudinal (east—west) axes are no better substantiated. Any suggestion based on time differences lacks the support of evidence that any animal can possess several (or even one) chronometers that are rigidly and highly accurately locked onto local times at each of the places to which navigation may need to occur. The only factor that does not involve time-keeping, the declination of the Earth's magnetic field, if it is to be used under cloud, requires the possession of some other, yet to be discovered, mechanism, such as gravity detection, by which to identify geographical north in the absence of celestial cues.

My own feeling, from all the evidence and argument, is that none of the global gradients of potential use to birds in establishing a grid map, has sufficient support from critical evidence to be considered viable. There are still no firm grounds for giving the grid map the status of 'fact'.

If grid maps are not a fact, it means we have to try to account for all of the characteristics of bird orientation and navigation during shorter-distance homing and longer-distance migrations solely in terms of landscapes and compasses.

The next two chapters examine navigation in action, first in relation to the data from short-distance homing experiments and then later in relation to data for long-distance migrants. In both, we address the question: are landscapes and compasses enough?

11 Navigation in action: short-distance homing

We now know that pigeons live within a vivid landscape of sights, smells and perhaps sounds and that this landscape is polarized on their mental map according to a variety of compasses.

The question we now have to ask is: is this enough? Is this combination of a polarized landscape and an integrated batch of compasses sufficient to take a bird through the travels and explorations of its lifetime? Will such a picture of bird navigation explain the confusing blend of facts, half-facts, hints and inconsistencies that have emerged from the many experiments carried out over the past few decades? This chapter is an attempt to answer these questions.

As we survey the data, it is no longer always important to ask what cues a bird might be using. In this chapter we can concern ourselves with generalities about mechanisms instead of details about senses. It is enough in most cases to decide that a bird is using a compass to do a particular job, without being sidetracked by which compass is being used. I shall also refer to 'the map', meaning either a mosaic map or a map of learned gradients (e.g. intensity of a particular smell; topographical slopes) within the familiar area map, without always specifying the type of topography (e.g. visual, olfactory, acoustic).

11.1. Map and compass?

There is little doubt that, when a pigeon arrives at an unfamiliar site, through either its own exploration or experimental displacement, and asks itself the question 'where am I?', the answer at which it arrives is invariably in the form of a compass direction (e.g. 'I am north of home, so where is south?'). We know this, both from clock-shifting experiments (Chapter 4) and from reversing the magnetic field through the bird's head at the release site just before the pigeon sets off for home (Fig. 9.7). Whether the bird ever comes up with an answer relative to its own body, as required by an inertial mechanism (Barlow 1964) and some forms of route-integration (Mittlestaedt and Mittlestaedt 1982) (see Chapter 4) is unknown, but it seems that humans may do so over short distances.

One advantage of working with people is that they can be asked to estimate home direction in two ways: (1) by pointing; (2) by describing its compass direction. This gives an opportunity to decide whether they answer the question 'where am I?' by first deciding that 'home is over my shoulder to my right' and then asking 'so what compass direction is that?' or by deciding 'home is southwest' and then asking 'so where is southwest?'. If they use the first method, they should point toward home more accurately than they can describe its compass direction, if they use the second method, the converse should be true.

Since 1976 I have been carrying out 'homing' experiments on humans in which groups of blindfolded subjects are transported by bus or van over tortuous routes to test sites up to 52 km from Manchester University, 'home' for the purposes of these experiments (Baker 1980c, 1981a, 1984b). At the 'release point' the people were asked to estimate home direction by both of the methods just described. The results show that, although by no means highly developed, humans do have a significant ability to indicate the direction of home under such conditions. Replications of these experiments in the United States (e.g. Gould and Able, 1981), although contentious, also provide evidence for such human homing ability (Baker 1984a, b). In addition to these bus experiments, I have carried out, in collaboration with Janice Mather, a series of what we call 'walkabout' experiments. In these, subjects walk without blindfolds through woodland and from time to time again estimate the direction of the starting point of their journey.

Figure 11.1 gives the results of five years of these bus and walkabout experiments. It can be seen that up to about 7 km from home, pointing estimates are the more accurate, whereas beyond that distance, compass estimates are the more accurate. It seems that over short distances, people

Navigation in action: short-distance homing 173

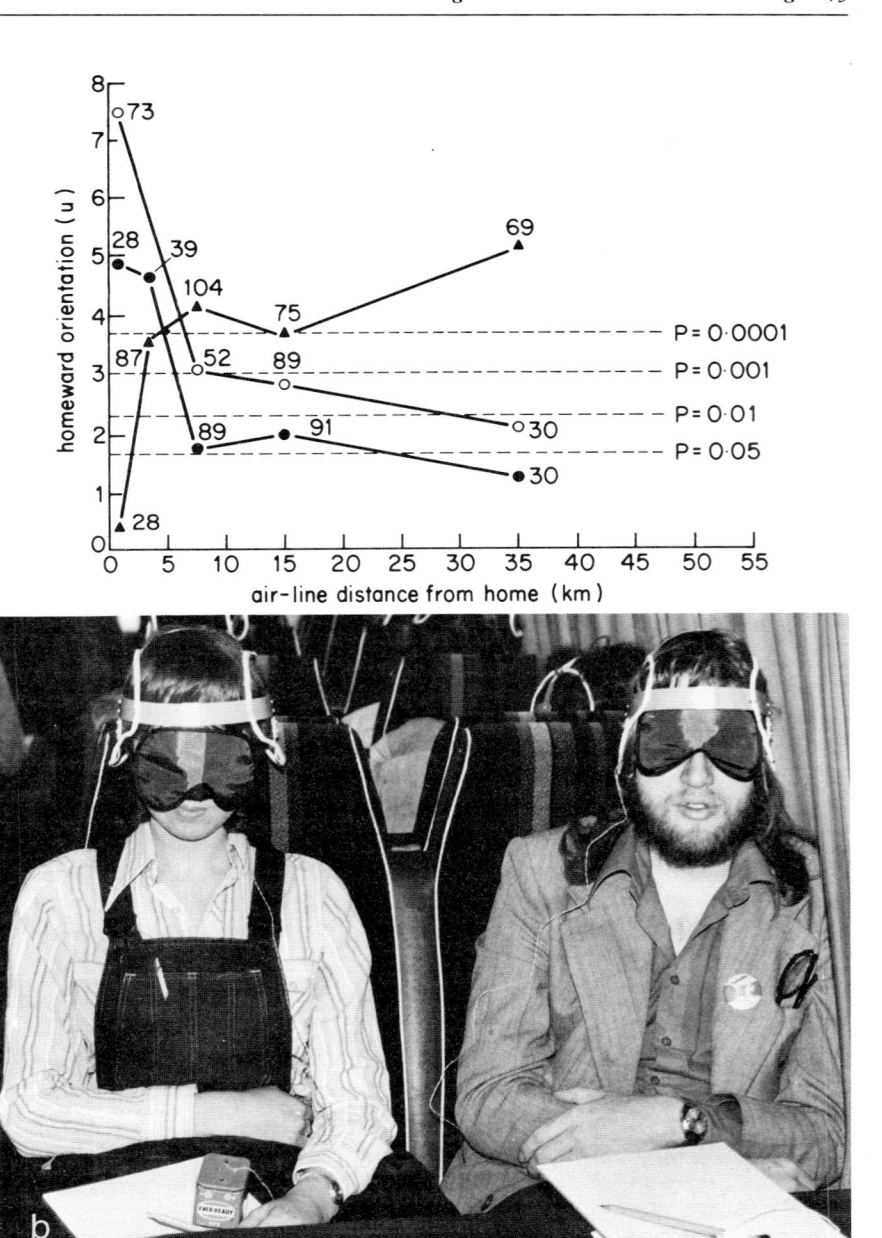

Fig. 11.1 The influence of release-distance on the ability of humans to estimate home direction after displacement

Homeward orientation is expressed as the u-value of the V-test (Chapter 3). Subjects were blindfolded during displacement. Solid symbols, estimates made while still blindfolded; open symbols, estimates made after blindfolds are removed. Circles, estimates made by pointing towards home; triangles, estimates of the compass direction of home. Numbers show the number of individuals who have made an estimate at that distance.

keep an awareness of home relative to their own body, but at longer distances they first solve the navigational problem in terms of compass direction in the same way as birds do.

Clock-shifting experiments on homing pigeons (Graue 1963) show that even at distances as short as 1 km from the loft, pigeons seem to give home direction as a compass direction. Whether this remains true over even shorter distances we do not yet know, but a technique developed by Saint Paul (1982) may soon give us the answer.

Goslings of several species travelled over short dog-leg routes and were then released to return home. Distances involved were short, no more than 1500 m, and under most conditions the birds could set off in a straight line (i.e. not retracing their outward route) to home (Fig. 11.2). Orientation was observed no matter whether the birds walked to the test site, following a human foster-mother, or were transported there in an open sided, hand-pushed trolley. Orientation also occurred under all conditions of sun and overcast and when the top of the trolley was covered. Only when the cover extended down the top third of the sides did orientation disappear. We do not yet know whether the birds were vectoring home direction by landmarks, relative to their own body, or in terms of a compass, but the technique has potential for the study of short-distance navigation.

Fig. 11.2 Departure directions of seven domestic geese after passive displacement one by one in an uncovered cage

H, home; dotted line, outward journey; arrows, departure directions. Drawing shows the view towards home from the release site, S, in a forest clearing.

[Redrawn from Saint Paul (1982)]

11.2. Following the outward journey

Dog-leg or detour experiments similar to those on goslings have been carried out on homing pigeons, but over longer distances (see early review by Matthews 1968), in an attempt to see whether pigeons take any notice of their outward journey. All produced negative results until Papi *et al.* (1973) performed such experiments in Italy. Two groups of birds were taken to the same release site but by different routes, the first leg of the journey for each group being in the opposite direction from the other group. Papi found that his pigeons then set off from the release site in a direction biased by the direction taken during the first leg of the outward journey.

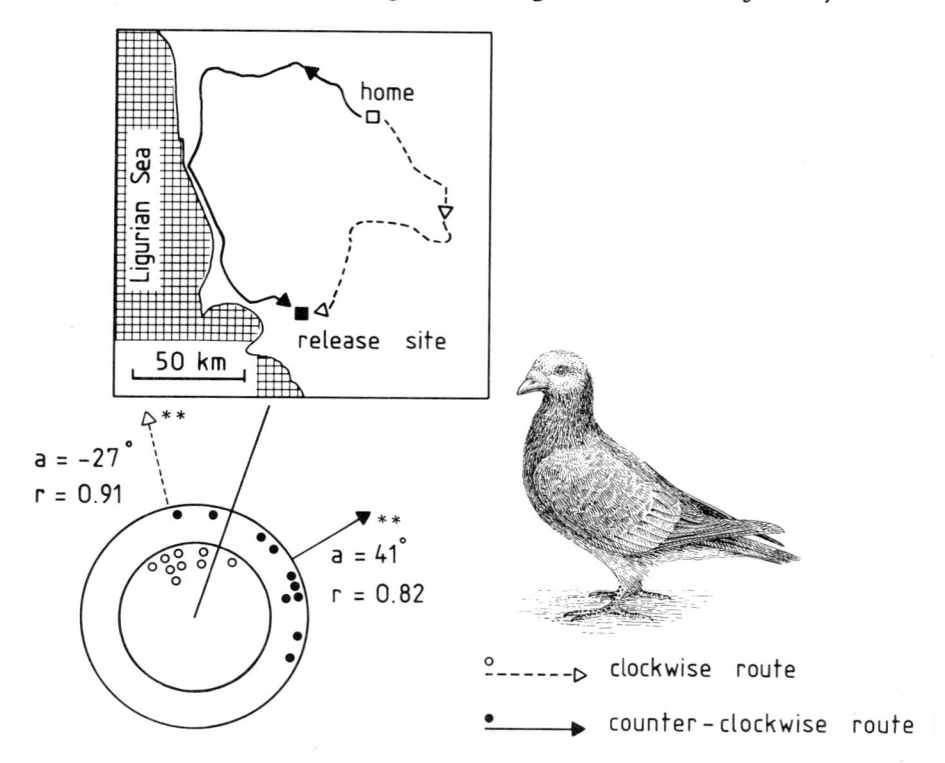

Fig. 11.3 Demonstration of route-based navigation by means of a detour experiment [From Baker (1981a), after Papi]

To obtain positive results in such an experiment, two factors have to apply: (1) the birds must for some reason take more notice of the first part of the journey; (2) their route-based estimate of home direction must not be overriden by map information obtained at the release site. If either of these two factors is not realized, such detour experiments will not give positive results. Nor will experiments in which pigeons are anaesthetized or rotated on turntables during the outward journey. It is not surprising, therefore, that earlier experiments (see Matthews 1968), and particularly attempts to replicate Papi's experiments e.g. Keeton (1974c) in the United States; Fiaschi and Wagner (1976) in Switzerland; Hartwick et al. (1978) in Germany; and Papi et al. (1978b), again in the United States) have given such variable results (reviewed by Papi et al. 1978a). In such experiments, however, one positive result is more meaningful than several negative results.

With sufficient evidence that birds are able to take notice of their outward journey, attention turned to how this enables them to estimate for home direction in terms of a compass.

11.2.1. By magnetism

In an unplanned 'experiment', the Wiltschkos noticed that birds taken to the release site in their Volkswagon squareback were sometimes more disoriented than if they were transported in some other vehicle (Wiltschko and Wiltschko 1978b). The effect was particularly noticeable if the crate containing the pigeons was transported on top of the engine in the rear of the car. This 'VW' effect was later confirmed by Walcott (in Schmidt-Koenig 1979).

The Wiltschkos suspected that the effect may be due to the magnetic fields produced by the generator. As a result they tried manipulating the magnetic field during transport in a more predictable way to see if they could influence the initial orientation of the birds upon release. They found that if they used Helmholtz coils to reverse the horizontal component of the Earth's field, thereby reversing both an inclination and a polarity compass, the birds were disoriented on release. Replications were sometimes highly successful (Wiltschko and Wiltschko 1981) and sometimes more variable (Wiltschko et al. 1978).

Kiepenheuer (1978b, c) displaced pigeons in a crate within Helmholtz coils on top of his van. When he reversed the vertical component of the Earth's field, thereby reversing only an inclination compass, he obtained a significant deflection of the experimental birds relative to the controls of about 30 degrees. When he reversed the horizontal component, thereby reversing both polarity and inclination compasses, the birds were disoriented but with a tendency to go either towards home or away from it. In addition, however, part of his treatment also influenced one of his groups of controls. In order to reverse the horizontal component throughout an entire journey, it is necessary to prevent the coils and crate from changing their relationship relative to the geomagnetic field. The coils have to remain aligned north–south if they are to continue to oppose the Earth's field. Kiepenheuer, therefore, arranged a wheel-and-belt system so that, whenever his car turned, so did the crate and coils but in the opposite sense so that they maintained their original geomagnetic alignment. As a result, two types of control birds were used. One group was also attached to a wheel and belt system, but without the reversed field. The second type of

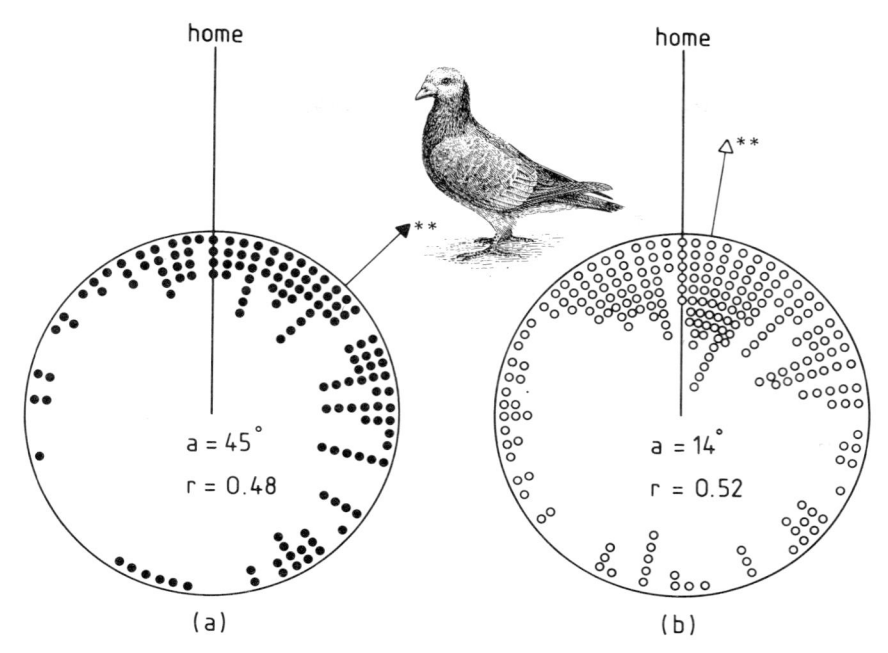

Fig. 11.4 Influence of a reversed magnetic field during displacement on the initial orientation of pigeons upon release

(a) Experimental birds exposed to reversed field
(b) Control birds exposed to normal geomagnetic field

[From Baker (1981a), after Kiepenheuer]

control bird was displaced normally, subjected to all the twists and turns of the journey. This second group oriented toward home as expected, but the first group did not, producing, as for birds with a reversed magnetic field, a bimodal distribution along the home-and-away axis. Evidently, although these birds could still work out the compass axis along which they were being displaced, the twists and turns of the journey were important to them if they were to work out in which direction along that axis they were travelling.

In the flush of experiments to identify the factors being used by pigeons to follow their outward journey, Papi et al. (1978a) displaced birds in containers made either of iron or aluminium. There was a clear indication that the birds were less well-oriented upon release after being displaced in the iron container, though interpretation was complicated by the different air supplies to which the birds were also exposed. Benvenuti et al. (1982) carried out similar experiments, without the olfactory complications, sometimes using iron containers and sometimes using Helmholtz coils.

One of the features of the earlier experiments in Germany by the Wiltschkos and Kiepenheuer was that, despite reversal of the inclination compasses, and sometimes also the polarity compasses, during the outward journey, the birds were not reversed in their orientation upon release. This is not so surprising for it could simply mean that each bird's estimate of the direction of displacement was coloured, not only by magnetic information during displacement, but also perhaps by both familiar smells during the journey and these and other types of landscape information at the release site (Baker and Mather 1982b). In which direction each bird departs will

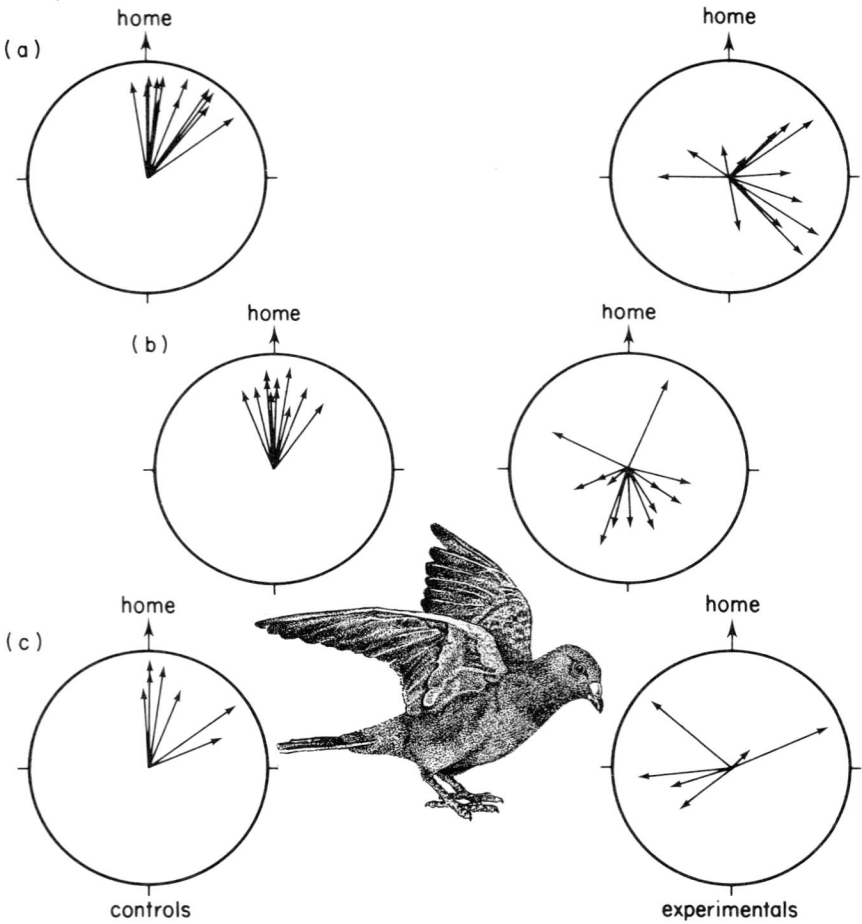

Fig. 11.5 Influence of an altered magnetic field during displacement on vanishing bearings of homing pigeons

Arrows inside circle, mean vectors (see Fig. 3.4) in single experiments. Controls were displaced in a normal magnetic field, usually in an aluminium container. Experimentals were displaced either in an iron container (a, b) or between three pairs of Helmholtz coils (c) which produced an oscillating induced field that should have offered little if any meaningful information. (a) and (c) Birds from the loft at Arnino; (b) birds from Antella.

[Compiled from Benvenuti et al. (1982)]

depend on which of the conflicting sources of information it believes most. Moreover, as we saw in Chapter 8, we cannot yet be certain in any particular situation that we are manipulating the magnetic field in a way and under circumstances in which it can be properly read by the magnetoreceptor. One of the features of the experiments by Benvenuti and his colleagues, however, is that often they did obtain reversal of orientation of their pigeons. Ironically, however, such reversal was obtained following displacement in an iron container, with its non-specific and reduced field.

In most of these experiments, the pigeons were released soon after arrival. In experiments by Wallraff *et al.* (1980), however, the pigeons were not released until the day following displacement, and the conditions during displacement had much less effect. This could suggest that the birds had obtained map information during their wait to be released.

W. and R. Wiltschko (1981) found that displacing pigeons in total darkness led to disorientation as readily as displacement in an altered magnetic field. Whether this indicates a direct influence of darkness or whether, as on Leask's optical pumping model, the darkness prevented magnetoreception during displacement cannot be decided.

Exactly how the birds use the magnetic field during the outward journey in order to arrive at an answer for home direction in terms of a compass is not known. The obvious suggestion is that the birds monitor the compass directions of the various straight stretches of the journey and then vector these, perhaps in a crude way (Chapter 4), to produce the compass direction of home. Homing would then be achieved by what Schmidt-Koenig (1979) terms vector navigation, the birds flying in the deduced compass direction for an appropriate distance, or until they strike a familiar target area. This is the way I envisaged humans were using the magnetic field on the outward journey during bus experiments (Baker 1981a) and it is the mechanism that Lednor (1982) suggests is the most likely use of the magnetic field by birds. However, until rotation of the magnetic compass during the outward journey produces a predictable rotation of orientation (e.g. reversal of the field producing a reversal of orientation) such vector navigation cannot be considered proven. For the moment, it is still possible that the birds are feeling their way across a magnetic map during displacement, unlikely as this may seem given the magnetic noise generated by vehicles on most modern roads. Less likely is the suggestion of Benvenuti *et al.* (1982) that disruption of the magnetic field during displacement causes a general and unspecific interference. As they themselves point out, if this were the case, experimental birds should simply be disoriented. Most often, however, particularly in their own experiments, such birds are well-oriented (but not towards home) and their orientation is rotated relative to controls (but not as predicted). This implies some rotation of directional information during the outward journey, but rarely by the amount predicted by the rotation of the magnetic field.

11.2.2. By smell

The same arguments can be applied to the manipulation of the smells reaching the birds during the outward journey. Early experiments (Papi *et al.* 1978a; Wallraff *et al.* 1981) showed simply that if birds were prevented from smelling natural air during transport, they were disoriented upon release. In the latest experiments (Baldaccini *et al.* 1982), nine experiments were carried out in which both controls and experimentals were displaced in air-tight containers. Experimentals were fed by pure bottled air while controls were fed by natural air sucked through the container. The control birds were well oriented toward home in seven of the nine experiments whereas the experimentals were oriented on only three of the nine occasions. Moreover, the homeward component (Section 3.5) for the controls was always greater than that for the experimentals. When the results were pooled, the experimentals showed random orientation whereas the controls were well oriented (Fig. 11.6). As in so many pigeon experiments, there was no difference in the homing performance of the two groups of birds in terms of homing speed and percentage return, but we shall return to this important point toward the end of the chapter. For the moment, we shall concentrate on what happens during the period from release to the point of vanishing.

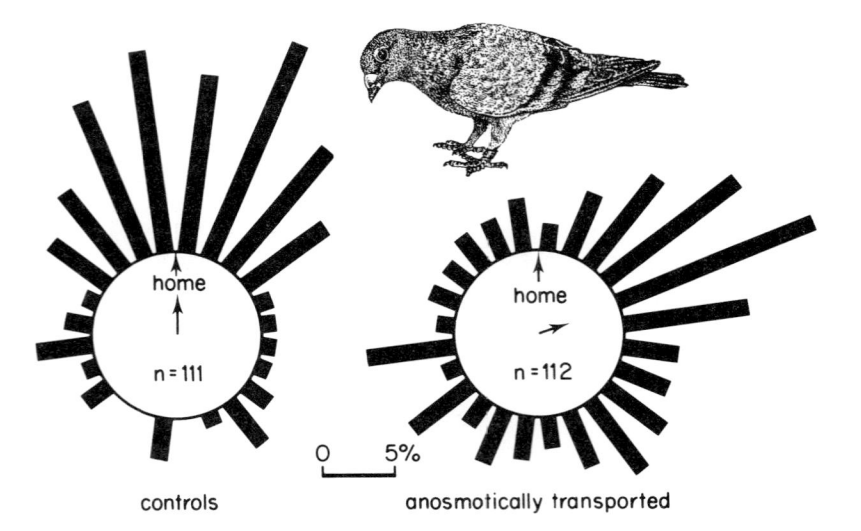

Fig. 11.6 The importance of being able to smell during displacement in route-based navigation by pigeons

Controls and experimentals were transported inside identical, air-tight containers. Pure bottled air flowed through the containers housing experimentals. Controls were provided with air sucked in from outside. The diagrams show the pooled results of nine releases. The length of each bar is proportional to the percentage of vanishing bearings in a 15-degree sector. [Modified from Baldaccini et al. (1982)]

Wallraff and Foà (1981) have shown that if the air being sucked into the containers during transport is unfiltered or filtered with fibreglass paper, which removes only solid and liquid particles, birds can still orient homewards. If, on the other hand, the air is filtered with charcoal, the birds are disoriented on release. This clearly implicates molecules being carried by the air as critical, but still does not avoid the possibility of some non-specific disorienting effect.

The first real attempt to overcome this objection has just begun in a remarkable series of experiments in Italy by Baldaccini et al. (1982). A group of pigeons is trained to two sites, east and west of the loft. They are then taken to a release site to the north and released. The morning of the experiment, teams of researchers travel east and west along the training route and collect bags of air at successive points along each of the two

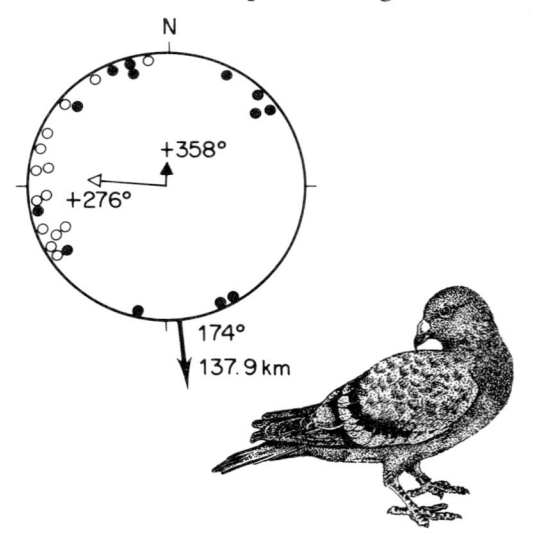

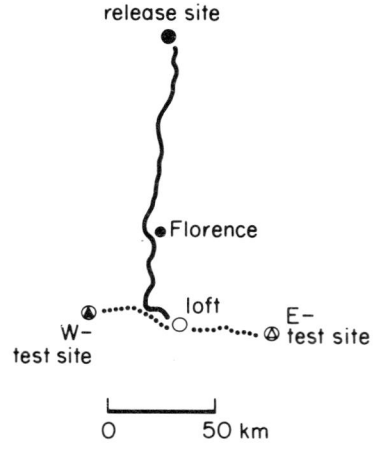

Fig. 11.7 The possible influence of the sequence of smells during the outward journey on the vanishing bearings of homing pigeons

Birds were trained to both western and eastern test-sites along the routes shown by dotted lines, then released from a site to the north. During displacement, the birds were in two groups. Air from a succession of positions along the westerly training route was passed through the containers housing one group (solid symbols), which was therefore expected to fly east upon release. The other group (open symbols) was exposed to air collected along the easterly training route and was thus expected to fly west.

[Modified from Baldaccini et al. (1982)]

routes. During displacement the birds, now separated into two groups, are fed with air from the succession of bags from one of the two training routes. From their northern release site, therefore, the prediction would be that birds given air from the westerly route should depart to the east whereas those fed air from the easterly route should fly west. Only preliminary results are so far available, but the team already seem to have established that the two groups orient differently, and that at least some of the groups orient in the predicted direction.

Once the teething problems with the experimental technique have been identified, we can look forward to intriguing information from this work. For the moment, however, we can only conclude that birds are by no means oblivious to where they are going during displacement. There is also encouraging evidence that they are in fact following their route using a combination of magnetic compass and olfactory information. If we are to be rigorous, however, we have to accept that detailed route-based navigation has not yet been proven owing to the lack of predictable rotations in the orientation of experimental birds. If such precise monitoring does occur, however, we should expect both magnetic and olfactory information to lead the bird to produce an answer for home direction in terms of a compass bearing. This is obvious for the magnetic method, but is also the case given the use of familiar smells. According to Papi's hypothesis (Chapter 4) the birds learn the compass direction from which different smells arrive at the home site. On detecting a familiar smell during displacement, therefore, the bird can work out the compass direction in which it is travelling.

11.2.3. *Age effects*

Experiments which manipulate cues during the outward journey give notoriously variable results. In large part, this is for the reasons already discussed. In part, also, it is due to the age and experience of the birds.

The Wiltschkos have shown that the disruptive influence of manipulating the magnetic information available during the outward journey is most marked when using young birds with little flight experience around the home loft (Wiltschko and Wiltschko 1982). Using birds inexperienced in displacement and homing experiments, they tested groups of different ages from 8 weeks onwards. The greatest disruption obtained was with birds less than 10-weeks old. Older birds, up to 15 weeks in age and with more flight experience around the loft (many individuals having gone off on long and spontaneous exploratory flights), were influenced far less by conditions during the outward journey. Trained birds were influenced hardly at all.

The Wiltschkos interpret this trend as indicating that only on their early flights and displacements do the young birds follow their outward journey with a magnetic compass. Instead, with increasing experience of their surroundings and a steadily increasing familiar area map, they come to rely more and more on map information, part of which, of course, could be olfactory and still perceived during the outward journey.

A second trend to emerge from this series of experiments is both intriguing and perplexing and concerns the accuracy of homeward orientation with age. Trained birds have long been known to be superior to first-flight birds in homeward orientation (Gronau and Schmidt-Koenig 1970) and individuals improve considerably with increasing experience of

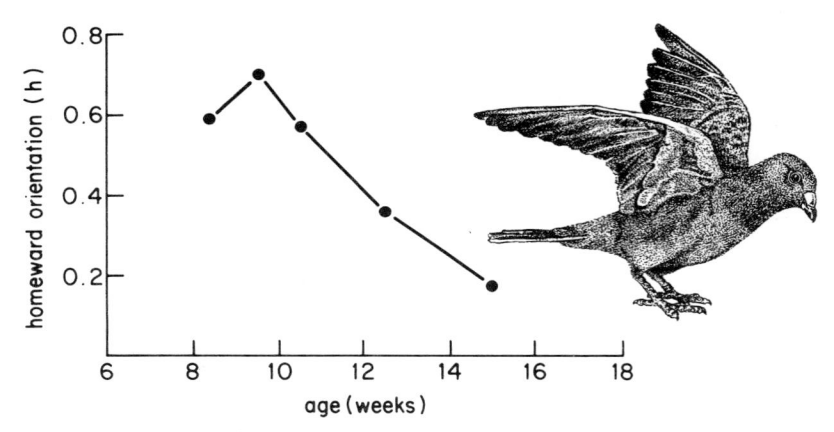

Fig. 11.8 Variation with age in the homeward orientation of inexperienced pigeons

All birds had been allowed exercise flights around the loft but were being displaced for the first time. Homeward orientation is measured by the homeward component (h) (see Fig. 3.4). [Drawn from data in Wiltschko and Wiltschko (1982)]

homing experiments. Using inexperienced birds of different ages, however, the Wiltschkos found that homeward orientation reached a peak at about 9–10 weeks of age and then gradually decreased to a relatively low level by 15 weeks. They interpreted this as reflecting a gradual switch by the birds from accurate route-based mechanisms when young to less accurate map-based mechanisms as they become older and extend their familiar area. The reduction in accuracy is suggested to be due to the fact that the birds are as yet inexperienced at reading their maps, perhaps still relying on gradients rather than on learned landmarks with known compass relationships.

I find it strange that, at this highly critical period of its life when it is beginning to build up a sizeable familiar area as a springboard for longer-distance explorations, a bird should switch from an evidently efficient mechanism that it has used with great accuracy in its early life to a less efficient mechanism leading to gross errors in estimates of home direction. My own interpretation would be as follows: (1) The birds are in a highly exploratory phase of their lives and are motivated to visit a wide range of places. They may well, therefore, treat experimental displacement as 'free' exploration and use the opportunity to visit novel sites rather than immediately to return home. (2) When young, their familiar area, which serves as their target area following exploration or displacement, is small. The range of directions that will return them to this target area is, therefore, relatively narrow. The more they explore around their loft, the larger becomes their familiar area and hence their target area. The older they get, therefore, the 'lazier' they can afford to be in the accuracy of their navigation, for even relatively crude 'homeward' orientation will cause them to strike their target area (Fig. 4.3) and thus home.

11.3. Regional effects

In the Wiltschkos' series of experiments in Germany the disruption of cues during the outward journey was found to have less and less effect as the birds gained in experience. They pointed out, however, that a similar trend did not seem to occur in Italy; there disruption of magnetic information during the outward journey seemed to have an effect on all ages of birds (Benvenuti et al. 1982).

Experiments by Wallraff and others (Wallraff 1980b; Wallraff et al. 1980) provided a similar indication of such a regional difference. In these tests, a procedure that ought to become more standard in future years was used. An attempt was made to block all means by which a bird could pick up information during the outward journey. Pigeons were rotated on a turntable, fed bottled air and subjected to a distorted magnetic field, though no attempt was made to block acoustic information. In Germany, the birds were homeward oriented upon release whereas in Italy, although homeward oriented, homing performance was reduced. However, the birds were not released until the day after displacement. This gave the experimental birds a chance to recover from the turntable rotation, but also gave them the opportunity to pick up local map cues before release.

Combined with the other evidence just quoted, these experiments suggest that in Italy birds, as they age, continue to use magnetic information during displacement whereas in Germany they do not, or at least they give it lower priority than map cues at the release site. Whether this is associated with another difference between birds in the two regions is not clear. Nevertheless, it seems to be well established that in Italy pigeons find navigation and homing to be a lot easier than they do in Germany (Kiepenheuer et al. 1979; Foà et al. 1982).

In Kiepenheuer's experiments, birds from Italian and German stocks were raised side by side to control for genetic differences. More birds failed to return home after release in Germany, though of those that did return, homing performance was faster. Homeward orientation was more accurate in Italy, though more birds settled near to the experimenters after release. In both countries, birds of domestic stock homed faster than birds from the other country.

Foà and his colleagues present results for a transect of pigeon releases from Wilhelmshaven on the northern coast of Germany down across the Alps and into Italy. They show that the homeward component of the vanishing bearings improves almost as soon as a region south of the Alps is reached. At the same time the strength of the preferred compass direction (PCD) decreases.

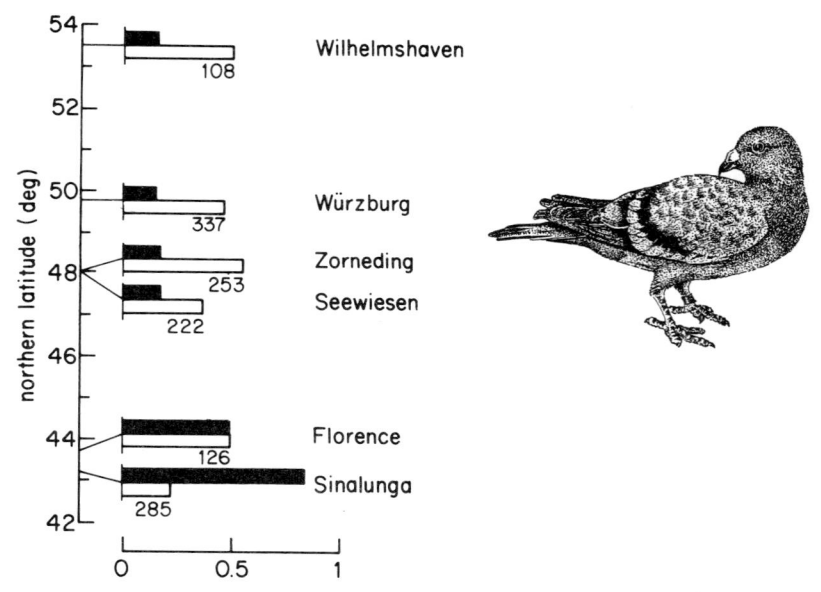

Fig. 11.9 Orientation of the vanishing bearings of inexperienced pigeons from six lofts in Germany and Italy

Dark columns, homeward component. Open columns, mean compass vectors. In each case, four release sites at distances of 20–31 km in directions approximately north, east, south and west were used.

[Simplified from Foà et al. (1982), with additional data from Wallraff and Meschini]

11.3.1. *Preferred compass direction*

Kramer (1957, 1959) seems to have been the first to suggest that the consistent biases away from the home direction shown by pigeons may give an important clue to their use of maps and the form of the navigation process. Matthews (1968) drew attention to the tendency of pigeons from certain lofts to fly in particular directions upon release and emphasized the importance of taking this into account in the design of navigation experiments (Chapter 3). He has also, of course, made great use of the PCD (I prefer this term to his 'nonsense' orientation) in his studies of orientation mechanisms. The phenomenon became a subject for study in its own right, however, when Wallraff (1974, 1978b, 1982) systematically examined the biases shown by pigeons from a variety of lofts.

Statistically, the PCD can be established by pooling data relative to north rather than relative to home. It is important, of course, that the releases from which the data derive should be in all directions from the loft. When Wallraff (1978b) surveyed in this way the results of releases from eleven lofts in Germany and the United States, he found that for ten of them birds showed a PCD that was between west-southwest and north-northwest with only a loft at Wilhelmshaven in northern Germany showing a PCD, for inexperienced birds, of east-southeast. Wallraff showed for a wide range of releases that the mean vanishing bearing is a compromise between the home direction and the PCD.

In releases from some lofts, vanishing bearings are actually more likely to be oriented in the preferred compass direction than in the home direction. The strength of orientation in these two directions can be compared by pooling the same two sets of data (for releases in different directions) first with respect to north and then with respect to the home direction from each site. When Foà et al. (1982) did this for their transect of lofts from northern Germany down into Italy, they found that in Germany the vector in the direction of the PCD was actually longer than the homeward vector. In Italy, the converse was true. Grüter et al. (1982), however, could find no PCD for pigeons from the loft at Frankfurt.

Wallraff (1978b) showed that the PCD was still present even if the birds: (1) had been restrained in aviaries with no opportunity to fly around the loft before their first release; and (2) were unable to smell when released. It was only not present if the birds were housed when young and until first release in an aviary with solid walls. However, under such conditions the ability to orient toward home also disappears. The only treatments to manipulate the PCD rather than destroy it were, on the one hand, clock-shifting and, on the other hand, keeping the birds housed in a corridor that was aligned east-west and open at both ends. The fact that clock-shifting has a predictable effect suggests that the PCD really is a compass preference and that it involves at least a sun compass. Birds housed in corridor aviaries aligned east-west tend to show a reversal in PCD, the meaning of which still lies in the realms of mystery.

More recently, Wallraff (1982) has been studying the striking PCD to the west-northwest shown by his birds at the Würzburg loft in northern Bavaria. The area is ideal for such work because releases can be made from any direction in a region with no major topographical boundaries. Thirty-four release sites between 7 and 180 km from the loft have so far been used. At distances less than 30 km from home, inexperienced birds show orientation that is virtually a pure reflection of their PCD. As with earlier results, a PCD is still shown by birds unable to smell. Indeed, pure orientation in the PCD is found in birds unable to smell at all distances from 7 to 500 km from home. This, Wallraff argues, shows that the PCD is a characteristic of the loft, not of individual release sites. Other authors, however, would not agree.

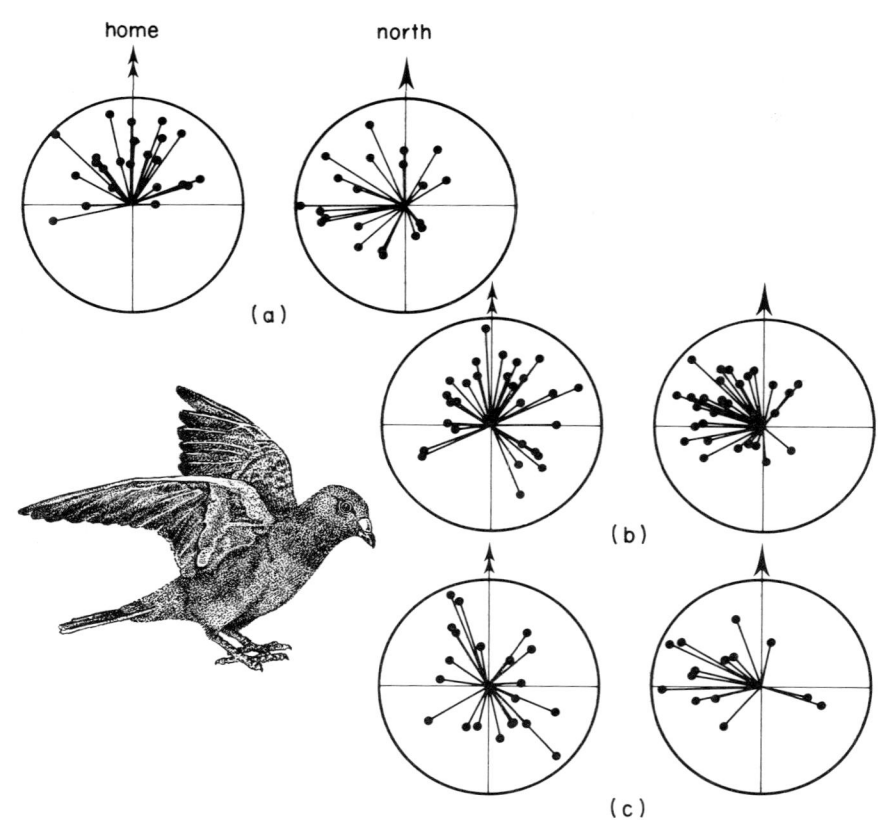

Fig. 11.10 Mean vanishing bearings of homing pigeons from the loft at Würzburg relative to home and to compass bearing

Each vector is the mean vector for a single experiment shown in the left-hand column relative to home and in the right-hand column relative to north. Release sites were evenly distributed around the home loft.

(a) Experienced, untreated birds that were more than one year old and had returned from at least 12 preceding releases. (b) Inexperienced, untreated birds that had been allowed to fly around the loft but had never been displaced. (c) Inexperienced birds unable to smell due to surgical bisection of the olfactory nerves a few weeks before release.

[Simplified from Wallraff (1982)]

11.3.2. Release-site bias

Kramer believed that consistent biases in the vanishing bearings of pigeons at particular sites were a response to the release sites themselves and might, therefore, reflect distortion of the bird's map at these sites. Keeton also subscribed to this view and referred to these biases as 'release-site biases'. On the basis of this thesis, Keeton carried out a series of detailed investigations of one such site, Castor Hill Fire Tower in New York State, 143 km north-northwest of the home loft at Ithaca (Keeton 1973).

Pigeons released at Castor Hill showed a consistent bias of 50–90 degrees clockwise of the true home direction. The same deflection was shown by both experienced and inexperienced birds, whether released from Castor Hill before or not. It was also independent of the degree of overcast and was shown not only by the birds from Cornell University but also by pigeons from other lofts around Ithaca. Moreover, birds from lofts in areas other than south-southwest from Castor Hill showed, relative to their own home direction, the same clockwise bias. Finally, even bank swallows (*Riparia riparia*) nesting in Ithaca, showed the same bias when released at Castor Hill.

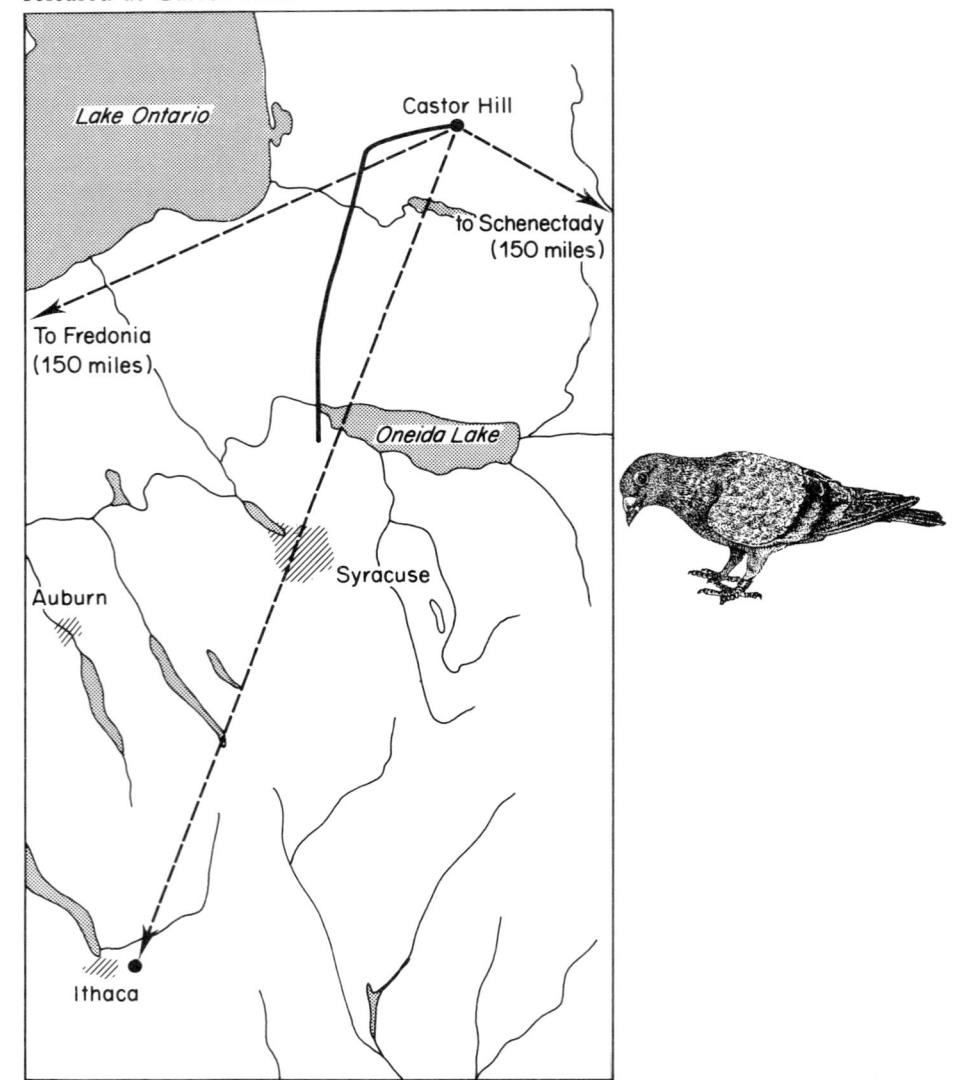

Navigation in action: short-distance homing 189

(a) sunny day

(b) overcast day

● familiar to site
○ new to site

(c) magnetic disruption

(d) clock-shifting

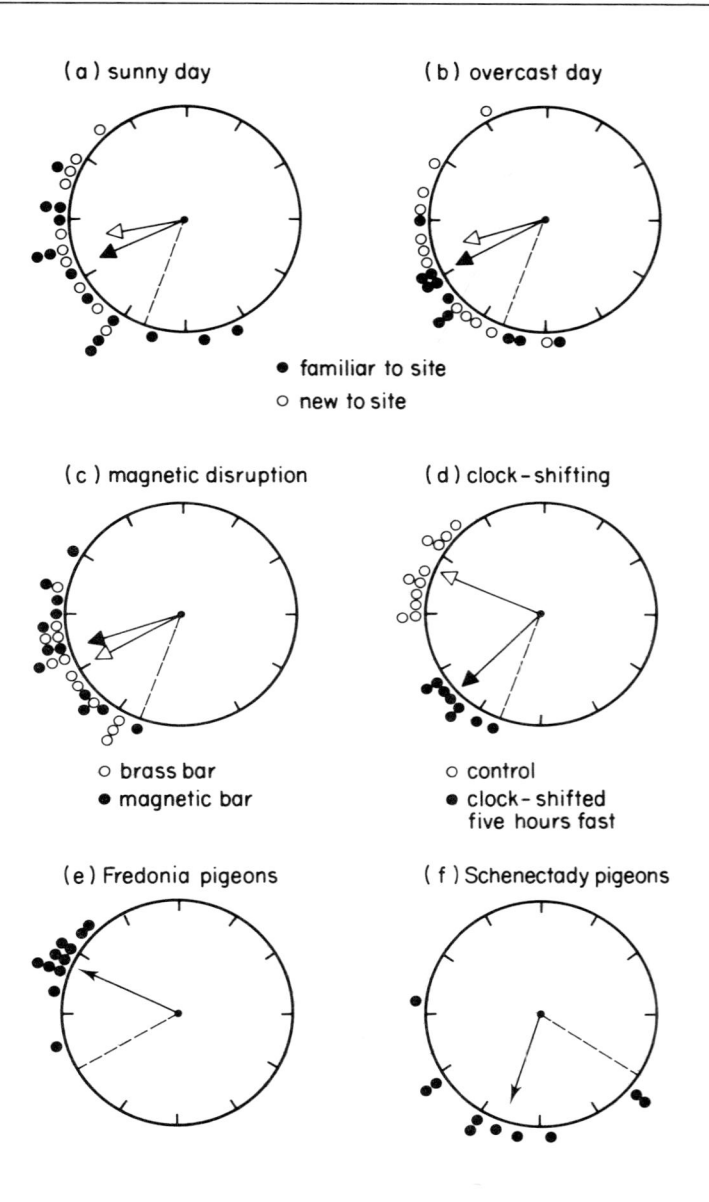

○ brass bar
● magnetic bar

○ control
● clock-shifted five hours fast

(e) Fredonia pigeons

(f) Schenectady pigeons

Fig. 11.11 Release-site bias at Castor Hill

Heavy line on map shows the actual flight path of normal Ithaca pigeons released at Castor Hill and tracked by aeroplane.

[Modified from Keeton (1974b)]

Keeton then tried clock-shifting pigeons in such a way that, when released, they actually set off in the home direction. Such birds in fact homed less well than birds that set off in the 'wrong' direction. Attempts to follow normally released pigeons by plane produced few tracks, but enough to suggest that the birds continued on their biased course for about 20–30 km before turning to the south and setting off more or less directly toward home.

11.3.3. Release-site bias, preferred compass direction and landscapes

My own feeling is that many, if not most, of the biases we have been considering could be explained if we had a better understanding of the way that birds read and respond to the landscape. Previously, the only support I could offer for this view was eclectic (Baker 1978, 1981a). Now, I see rather more direct support from an interesting and perhaps important paper by Dornfeldt (1982) on the release-site biases of pigeons, mainly from the loft at Göttingen. Using more than 2000 individual releases as data, accumulated over 153 different journeys, Dornfeldt subjected vanishing bearings to multivariate analysis in relation to a variety of different topographical, meteorological and geophysical variables.

The Göttingen loft is near the main railway station where there are a number of electrical transformers and overhead feed lines. Dornfeldt's analysis showed that, at the different release sites, the main factors associated with the deviation of vanishing bearings from the home direction were topographical cues (railway stations, buildings, transmission lines and electrical transformers) and particularly the downhill slope of the ground. Vanishing interval increased with the degree of cloud cover and the distance of the release site from home. There was no support at all for the involvement of any geophysical factor in the deviation from the true home direction.

The downslope bias of the Göttingen pigeons suggests that, from this loft at least, pigeons prefer to home over lower ground rather than higher. Pigeons released at Castor Hill from different lofts effectively took an initial route that could be interpreted as the best compromise between downslope and the true compass direction of home. Pigeons from Ithaca, in effect, flew west until they attained both low ground and a view of Lake Ontario, whereupon they turned homeward.

Keeton (1974b) suggests that visual features could not be involved in the Castor Hill bias because it was still shown by birds wearing frosted contact lenses. However, even though such lenses may have prevented a detailed view of the topography, they may not have prevented detection of high and low horizons. We know (Chapter 5) that pigeons with frosted-lenses can work out the compass direction of home (by route-based navigation?) and then fly in that direction using a compass. If they can also detect high and low horizons, they have the two elements necessary to produce the characteristic vanishing bearing at Castor Hill. The data could well fit a model in which the birds first determine the true compass direction of home and then fly on a bearing that is the best compromise between this direction and some feature of the topography, perhaps the slope of the land.

Windsor (1975) analyzed the biases shown by Ithaca pigeons at a variety of release sites in New York State. He showed that there seems to be an axis running through the loft from north-northwest to south-southeast. To the west of this axis, pigeons deviate counter-clockwise from the home direction whereas to the east they deviate in a clockwise direction. Wallraff (1978b) points out that the bias at Castor Hill is part of the regional trends found by Windsor and should not, therefore, be considered in any way exceptional. The results also indicate a PCD for Ithaca birds of north-northwest.

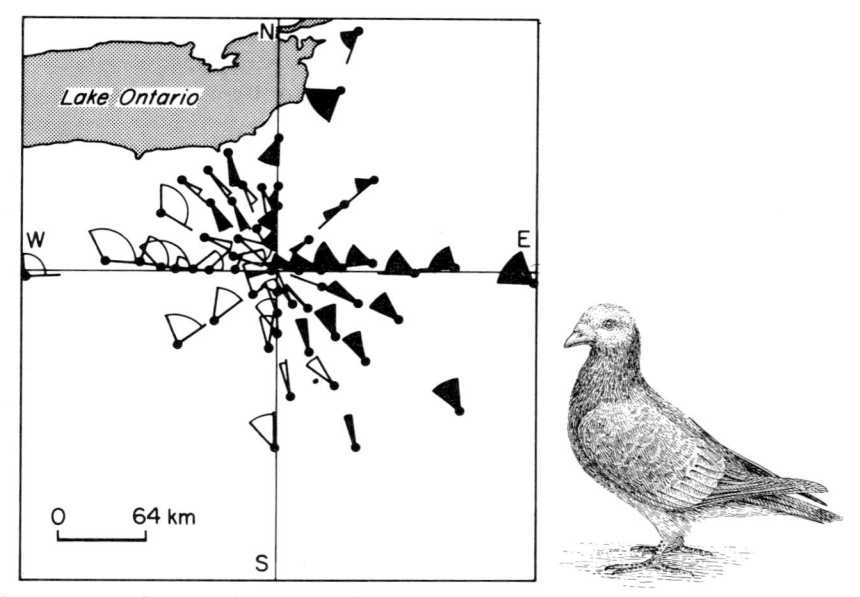

Fig. 11.12 Release-site biases around Ithaca

Clockwise biases are shown black; counter-clockwise biases, white.

[Modified from Windsor (1975)]

Looking at Windsor's diagram for the biases around Ithaca (Fig. 11.12), I could easily convince myself that they were in some way related to topography. A multivariate analysis for these data of the type carried out by Dornfeldt for the Göttingen loft could be instructive. In the case of the Ithaca birds, however, perhaps the direction of the nearest lake or river might be more appropriate than the direction of the nearest railway station or power lines.

Keeton found that, even after pigeons had homed from Castor Hill on previous occasions, they nevertheless continued to leave the site for the west instead of the more direct homeward route to the south-southwest.

This suggests that either the birds really could not read their map properly at Castor Hill or, having homed from the site, they were satisfied that the west then south route was the best. There is reason to believe that the latter is more likely.

Foà and Albonetti (1980) have shown that birds which take an uneconomical route from a site can learn on successive releases to correct for this and to take a more economical route. The direction taken by their pigeons from a particular site was changed by clock-shifting the birds so that they showed a bias comparable to that shown at Castor Hill. The birds were then released from the same site virtually every day, and gradually the birds set off more and more in the home direction. This could, of course, have been due to a recalibration of their sun compass, but when the birds were then taken to an unfamiliar site, the clock-shift was found still to be effective. Evidently, therefore, increased familiarity with the site was the most likely explanation for their correction of vanishing bearings.

I have concentrated in this section on the possibility that biases away from the home direction may be due to the way birds read and react to topographical features on their landscape. Acoustic landmarks could well generate similar effects, particularly if they are infrasonic, located hundreds of kilometres from the release site. Pigeons may need to fly toward, or away from, such sources for some distance, beyond their vanishing point, before being able to read their acoustic landscape. On the other hand, I have avoided discussion of whether the biases could be generated by the way birds read and react to familar smells because Wallraff (1978b, 1982) has shown that birds which cannot smell nevertheless still show a PCD, and I see no reason at present for assuming that a PCD and release-site biases are in any way different phenomena.

In an analysis of vanishing bearings around the Frankfurt loft, Grüter et al. (1982) see an irregular mosaic of release-site biases. When there is such a mosaic, there would be no great difficulty in imagining that deviation from the true home direction at each site was a response to the surrounding topography. When the trends are wider-spread, as around Ithaca and Würzburg, emerging in the form of a PCD, a role for topography may be less obvious. Nevertheless, topography does show regional trends, as when the ground slopes in a more or less consistent direction for tens or hundreds of kilometres or when rivers all drain in the same direction. Alternatively, a major landmark, such as a large lake to the north or a major river to the west, may provide such an easy navigational clue that it is more economical for a bird, when in some doubt over home direction, to fly, say, north or west respectively until encountering the landmark, whereupon homeward orientation is simple. Such a navigational strategy would emerge as a PCD.

Wallraff (1978b) has shown that even birds restrained in an aviary from birth until release show the PCD appropriate to their loft. If their PCD is really due to an interpretation of the landscape, therefore, it must in these cases be a reflection of those landscape features the birds can perceive from inside the aviary. We are not forced to accept that the same is true at those lofts for which aviary birds have not been tested. Even here, however, the arguments I have presented require the birds to learn particular features of the landscape around the loft. This is well illustrated by the response of the Göttingen birds to railway stations and power lines. It may even be the case that Ithaca birds learn to make use of Lake Ontario and the finger lakes to its south in developing their reaction to the landscape. In other words, as pointed out by Grüter et al. (1982), for release site biases to express themselves, they have to be related to map factors learned at or around the home loft.

Whereas Keeton (1973) found that birds from different lofts released at Castor Hill all showed similar biases, Wallraff (1967, 1978b) has also shown that birds from different lofts, when released at the same site, may show different biases. We should expect such variation if a bird's reaction to the landscape is in part a reflection of the area surrounding its home loft.

Attention turns, therefore, to the factors perceived at and around the home loft.

11.4. Information at the home

Experiments to examine the influence on pigeon navigation of restricting or changing the information available to them at the home site began with a set of palisade experiments started by Kramer (1959) and completed by Wallraff (1966, 1970a, b, 1979). Kramer's original experiments used an aviary in a crater so that birds could see only the sky. Wallraff used the more manipulable situation of an aviary with walls made of a variety of materials designed and changed so as to allow the birds to perceive some cues but not others. Palisade birds, contrary to the normal procedure, are never allowed exercise flights or practice releases.

Early results showed that if the horizon and the lowest 3 degrees of the sky are blocked off, homeward orientation upon release is reduced. This was true whether the material used was wood or glass. Consequent upon the use of various degrees and forms of blocking of cues by a mixture of materials and forms of louvres, Wallraff concluded that the birds needed access to some horizontal 'atmospheric' factor if they were to be able to orient towards home upon release. Gould (1982), however, has subjected Wallraff's data to a critical reanalysis and can see little pattern to the results except that both glass and louvres increase the amount of scatter to the vanishing bearings.

Since these pioneer experiments, and given the impetus of Papi's olfaction theory, the past decade has seen a mushrooming of various designs of loft, each intended to manipulate the cues reaching the pigeons at home in a particular way. When the loft is constructed in the form of a corridor so that wind can blow through along one axis but not another (Wallraff 1979, Ioalé 1982), pigeons are only able to orient toward home

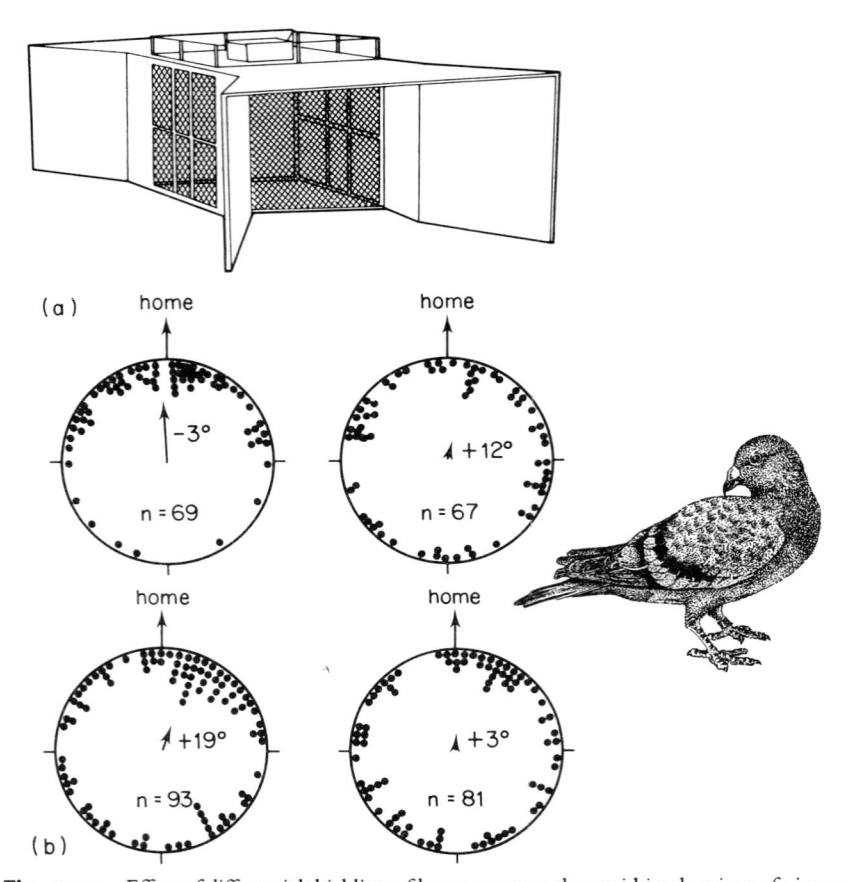

Fig. 11.13 Effect of differential shielding of home cages on the vanishing bearings of pigeons

Birds were housed in 'corridor' aviaries open at two ends. Controls (left) were birds released in directions from which they had been exposed to the wind while in the loft. Experimentals (right) were birds released at right-angles to the open axis of their home cage.
(a) Data from Italy; (b) data from Germany.

[Modified from Ioalé (1982), partly after Wallraff]

when released at sites along this axis, but not on the orthogonal axis (Fig. 11.13). Moreover, Ioalé et al. (1978) and Ioalé (1982) have shown that if the direction of natural airflow through the corridor is reversed by means of fans, birds when released fly away from home. Finally, it the loft is only open to air from one direction, the birds cannot orient even from that direction (Benvenuti and Ioalé in Papi 1982). Some form of comparison of winds from different directions may, therefore, be necessary for birds to learn that the wind carries useful information, though Papi interprets the result as showing that the pigeons may be learning gradients of smells. The question of whether deflector lofts are also relevant to the accumulation of map information at the home site has already been discussed (Chapter 5).

All of these results, of course, make sense in terms of the bird's learning of its olfactory landscape and show how much learning of this landscape may occur even before the pigeon begins to explore around its home. As Wallraff (1978b, 1982) has shown, however, an ability to smell when released seems unnecessary for the manifestation of PCDs or release-site biases. Although the studies discussed in this section are clearly relevant to the way birds build up their map of familiar smells while at the home site, they probably aid us little in understanding the biases that were the subject of the previous section.

11.5. Navigation and distance

We have nearly finished with pigeons for this book, yet still there has been no pressing need to abandon the thought that all pigeon navigation can be explained in terms of landscapes and compasses, route-based navigation and familiar area maps. The last chance for this to be forced on us is if there is evidence that pigeons, rendered incapable of route-based navigation, are still able to set off towards home even when displaced so far away that a familiar area map would be useless. Does such evidence exist?

The answer seems to be that it does not. The distance from which pigeons can orient toward home has never really been clarified (Wallraff and Graue 1973; Wallraff 1974, 1981a; Wallraff et al. 1981; Papi 1982). Up to 400 km from home, most recoveries of pigeons en route after displacement show evidence of homeward orientation (Schmidt-Koenig 1966; Wallraff 1970b, 1981a). More ambitious experiments, such as transatlantic displacements (Wallraff and Graue 1973), failed to control for the possibility of route-based navigation and in any case did not produce convincing evidence for homeward orientation (Baker 1978).

Early experiments suggested, in fact, that disorientation occurred not at longer distances but at shorter ones. Matthews (1968), with support even as late as 1979 (Schmidt-Koenig 1979), suggested that there was a 'dead zone' of disorientation between 20 and 60 km from home. Matthews interpreted this in terms of the distance at which navigation by landmarks disappeared but navigation by grid map was not yet accurate because differences in the relevant coordinates between home and the release site were not yet large enough to detect. However, Ithaca birds showed no such zone (Keeton 1970; Windsor 1975) and Grüter et al. (1982) find no such dead zone round the Frankfurt loft, nor does Wallraff (1982) around the Würzburg loft. Wallraff suggests that such a zone is not a general phenomenon but is generated in particular homing series by the particular homing histories of groups of birds. I have shown for humans that in bus experiments a similar dead zone can appear, almost certainly due to the distance around Manchester from which familiar ranges of hills can be seen (Baker 1981a). A similar argument could be applied to birds (Baker 1978).

Returning, then, to the outer limit of homeward orientation for

196 *Bird navigation: the solution of a mystery?*

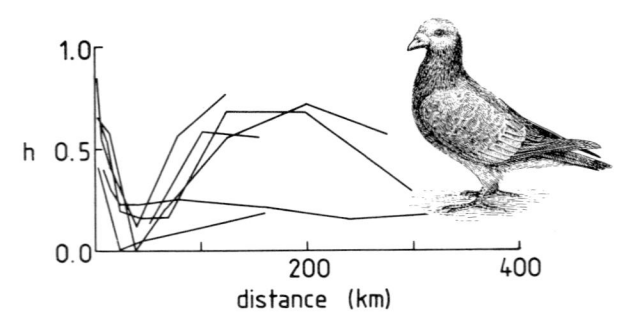

Fig. 11.14 Apparent zone of disorientation 20–60 km from the loft as observed from the vanishing bearings of homing pigeons

Homeward orientation expressed as the homeward component (h) (see Fig. 3.4).

[From Baker (1981a), modified from Schmidt-Koenig after Wallraff, Graue, Schmidt-Koenig and Matthews]

pigeons, Wallraff and his colleagues (Wallraff 1981a; Wallraff *et al.* 1981) have begun a series of experiments designed to find the spatial limit of pigeon navigation. It is worth bearing in mind that we know from pigeon races that birds are physically capable of homing from further than 1000 km. In such races, however, the birds are trained to fly in a particular direction, from whatever distance they are released. Homing pigeons by contrast have to work out the direction in which to fly.

Wallraff and his colleagues displaced pigeons from Florence, Italy, to release sites well beyond the Alps in northern Bavaria, nearly 700 km from the home loft. Inexperienced birds, trained previously at distances no greater than 30 km from home, showed no homeward orientation, but instead showed a strong tendency to fly west when released and to continue in that direction even after vanishing from view. Later groups of birds were therefore subjected to an intensive training programme, not in a single direction, as are racing pigeons, but in both directions from home along a north–south axis. These training flights took them as far as 170 km north and 150 km south.

During displacement to Bavaria, the birds were divided into control and experimental groups, the experimental birds having their nostrils plugged throughout the journey so that they could not smell. The plugs were removed before release. Both groups showed vanishing bearings to the west, as before, but this time the recoveries of the control birds showed that after vanishing from view they had eventually turned to the south in the direction of home. The experimentals, on the other hand, had continued on towards the west. One of the controls actually returned home to its loft in Florence after crossing the Alps.

As well as birds being released from northern Bavaria, some were also released 493 km north of home, just 60 km north of the Alps. Both control and experimental birds disappeared to the west-southwest, but for both

(a) controls

(b) experimentals

Fig. 11.15 Recoveries of homing pigeons from a loft near Florence (F) released in Germany at Landsberg (L) and Würzburg (W)

Experimentals (b) were prevented from smelling during the outward journey by plugging their nostrils.

[Modified from Wallraff et al. (1981)]

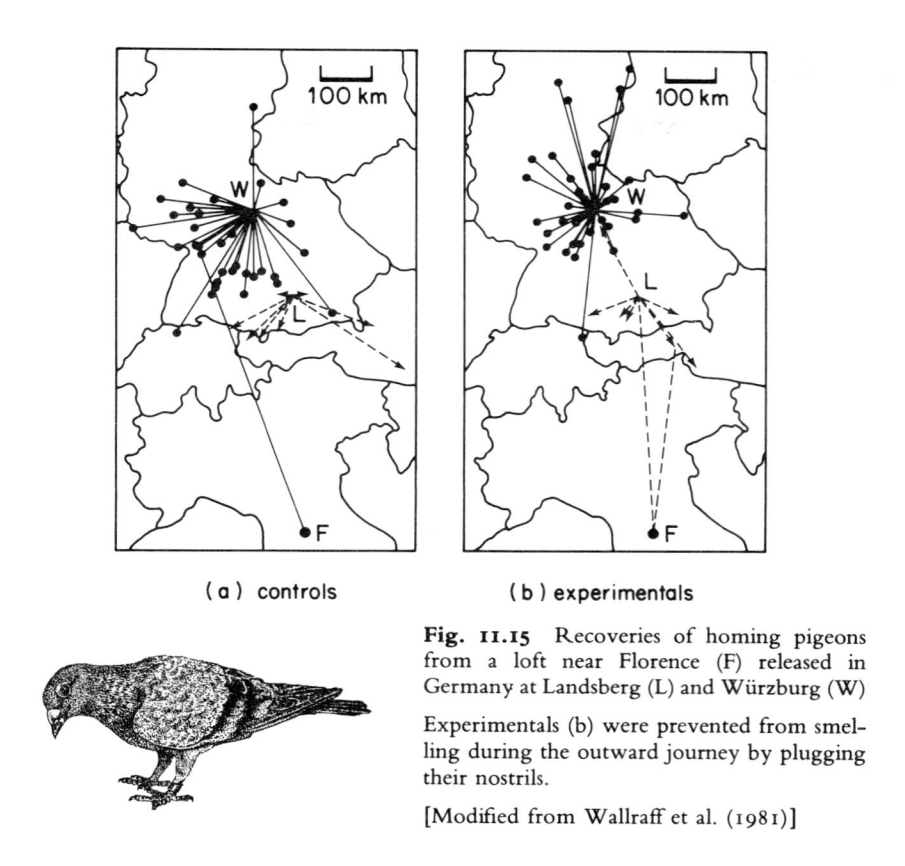

groups recoveries were oriented towards home and two of the experimental birds returned to the home loft.

11.6. Correcting mistakes: navigation while homing

These results bring us back to an important point that was made earlier. Most of the experimental treatments described in this chapter influence only the direction in which the birds vanish after release, which we assume to reflect the bird's first guess as to the direction of home. Most experimental groups in fact seem to realize their mistake fairly soon after vanishing from view and manage to return homing peformances little if at all inferior to those of controls. In the experiment just described, this was certainly the case from the release point 493 km from home but was not the case from the further release site, 680 km from home. From here, the experimental birds seemed never to recognize home direction.

Homing experiments suggest, therefore, that even when pigeons are unable to work out home direction immediately upon release, they may still be able to do so later, even from distances as great as hundreds of kilometres. In the last experiment it seems possible that from the nearer release site, a view of the Alps, familiar to the birds from their northernmost training releases, could offer sufficient information for both control and experimental birds to decide that home was to the south. From the further release site the local unfamiliar topography was of little assistance and an ability to smell during the journey was essential to work out home direction during the period after vanishing from sight. We have no way of knowing whether the birds should have been able to use their magnetic compass during displacement.

Even from long-distance displacements, therefore, there is no indication that pigeons use anything other than route-based navigation and familiar landscapes in their eventual determination of home direction, even if they are disoriented upon release. Indeed, the evidence indicated that if the birds are denied these two elements, navigation is impossible, both immediately upon release and thereafter. In most pigeon experiments, however, usually from distances less than 100 km from home, even if the birds' first estimate is wrong, as represented by their vanishing bearings, many are still able to recognize their mistake given sufficient exploration of the release site and its surrounding area or even after flight tens of kilometres in the 'wrong' direction (Fig. 4.5).

Obviously, it is more difficult to study the factors influencing a bird's reassessment of home direction once it has disappeared from sight. However, we do know that many birds that set off initially in their preferred compass direction home perfectly well. So, too, do many birds with a distorted map of familiar smells and, except that they rarely find the loft itself, so do many birds wearing frosted contact lenses. Neither the distortion of magnetic information during the outward journey nor the release of birds at a magnetic anomaly prohibit homing. Indeed, Kiepenheuer (1982a) found that upon their second release at an anomaly, his birds were as well oriented as birds released outside of the anomaly (Fig. 10.5).

In all of these cases, there are a variety of other landscape features and compasses to which the birds are able to refer in their eventual decision over the direction of home. The fact that birds, heading in the wrong direction, may cross familiar tracts of land before correcting their bearing is not evidence against their using the landscape to reassess their location. I have done the same myself on many occasions. When approaching a familiar place from an unfamiliar direction, I can pass many familiar landmarks before realizing that they are familiar. This does not mean I do not normally make use of such landmarks.

The most potent methods of preventing pigeons from reassessing their location after disappearing from view, seem to be plugging the nostrils, surgery on the olfactory nerve, and clock-shifting. When using the first two of these methods, as pointed out by Gould (1982), we cannot separate the possible influence of the treatment on navigational ability from the certain influence on the birds' energy, endurance and motivation. In clock-shifting experiments, however, few birds succeed in homing on the day of release, before they have had some opportunity to rephase their internal clock. It seems, therefore, that even after disappearing from view, if clock-shifted birds do reassess their location from the landscape beneath them, they continue to translate this location into the compass direction of home and misorient themselves accordingly. Perhaps those few birds that do home on the same day are the only ones to change their technique to a point-to-point steeplechase from one familiar landmark to the next. Of 69 clock-shifted pigeons wearing frosted contact lenses all failed to home on the day of release (Schmidt-Koenig and Keeton 1977).

The ability of many homing pigeons to correct their first mistake at some time after disappearing from view should not, however, necessarily be taken to mean that map information eventually always gives correct information that was lacking from a route-based estimate. It is possible that, upon release, a bird overrides its initial route-based estimate of home direction in favour of a map estimate. On perceiving its mistake, however, it may well revert to its original route-based estimate.

My conclusion, from the evidence as it exists so far, would be that the tremendous effort and ingenuity that has gone into research into pigeon navigation over the past few decades has taught us that birds have two major weapons in their armoury of navigational techniques: (1) a familiar area map based on a rich and varied landscape; and (2) a variety of efficient compasses. The map is stored as a mosaic of familiar sites with learned compass relationships, along with vague topographical and perhaps olfactory gradients filling the void between the important landmarks. Upon exploration or experimental displacement, the bird decides on the compass direction of home first by the use of route-based navigation during the outward journey and then, upon release, by reference to the landscape beneath. It then homes by a form of vector navigation, flying on the decided compass bearing until it meets a target area of more detailed familiarity around its home.

Release-site biases are a response to local topography, either the result of mistaken familiarity with topographical features or a response to topographical features that lead to more economical routes home. The preferred compass direction reflects release-site biases common to large areas around the loft because of regional trends in topography.

The outer reaches of the familiar area map are built up by distant perception from the furthest limits of the familiar area. In the first instance, this map begins to be established, even before the bird begins to explore away from home, using visual, olfactory and acoustic information perceived from the home loft.

200 Bird navigation: the solution of a mystery?

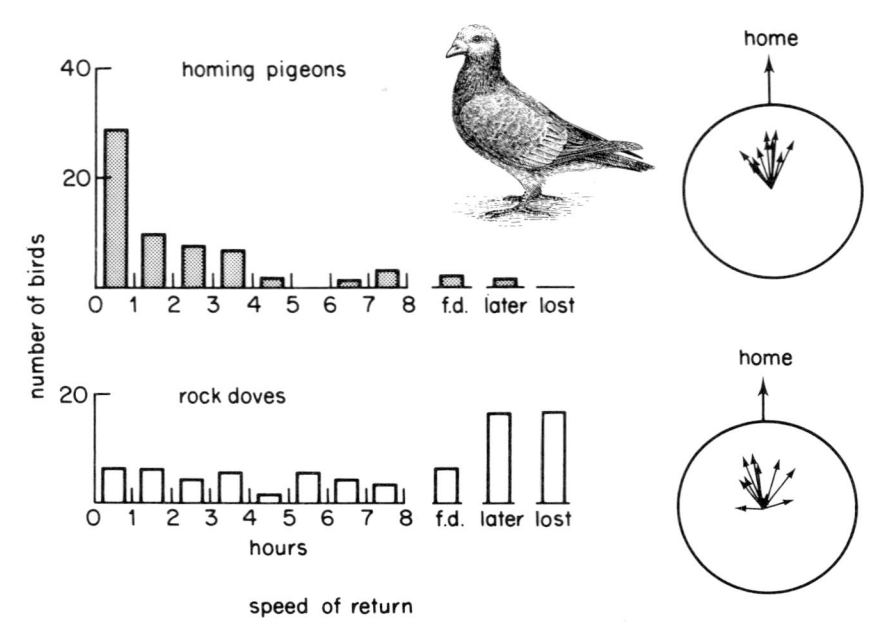

Fig. 11.16 Relative performance of homing pigeons and rock doves (**Columbia livia**) in homing experiments

Each arrow inside the circles is the mean vector of vanishing bearings in a single experiment. Histograms show the number of birds that returned home at different intervals after release. f.d. = following day. The vanishing bearings of homing pigeons are not significantly better than those of rock doves, but homing speeds are dramatically faster. The main impact of artificial selection during domestication seems to have been on the motivation to return home following displacement and release rather than on navigational ability.

[Drawn from data in Visalberghi et al. (1978)]

With this picture as a paradigm for bird navigation, we can move on to the final challenge. Will this system of navigation and orientation suffice to take a young bird of a seasonally migrant species from its birthplace to some unknown wintering grounds and back again, perhaps over distances of thousands of kilometres, or are there still fresh factors for which we must search? This is the subject of the final chapter.

12 Navigation in action: long-distance migrants

The aim of this chapter is to see how far the paradigm of pigeon navigation that has emerged during this book can explain the orientation and navigation mechanisms of long-distance migrants. This is done by focusing on a single, new-born individual of some hypothetical 'typical' passerine and following it through its first two years of life. At each stage, our picture of pigeon navigation is applied and examined in the light of what little direct evidence is available for migrant birds.

At the end, we have to ask ourselves whether the exercise was successful. If it was, we may begin to feel that an understanding of the enduring mystery of bird navigation is almost within our grasp. If the exercise fails, we can only assume that there are still major discoveries to be made.

The place to begin is when the young and helpless bird first struggles out of its egg, lies prostrate in its nest, and perceives the world for the first time. The first time, that is, unless it has already learned while in the egg that different directions have different magnetic 'feels' to them. What inborn package of behaviour patterns and predispositions does this bird possess, as a result of its evolutionary past, to help it as it travels through time and space on its way from birth to death?

12.1. Inborn programs

The only way to answer this question is to hand-raise young nestlings, maintain them in conditions as constant as possible and to see how they behave. This is the approach used for the past decade by Gwinner and Berthold and, as a result of their efforts, we now seem to know a great deal about the inborn programs of migrant passerines.

All of the early work on inborn programs was carried out in Germany using Old World warblers of the genera *Phylloscopus* and *Sylvia* (e.g. Gwinner 1972, 1981; Berthold 1973; Berthold and Querner 1981; see reviews by Berthold 1981; Gwinner 1981), but experiments have begun recently on muscicapids, the pied flycatcher (*Ficedula hypoleuca*) and collared flycatcher (*F. albicollis*) (Gwinner and Schwabl-Benzinger 1982).

Taken as a whole, the work has shown convincingly that birds are born

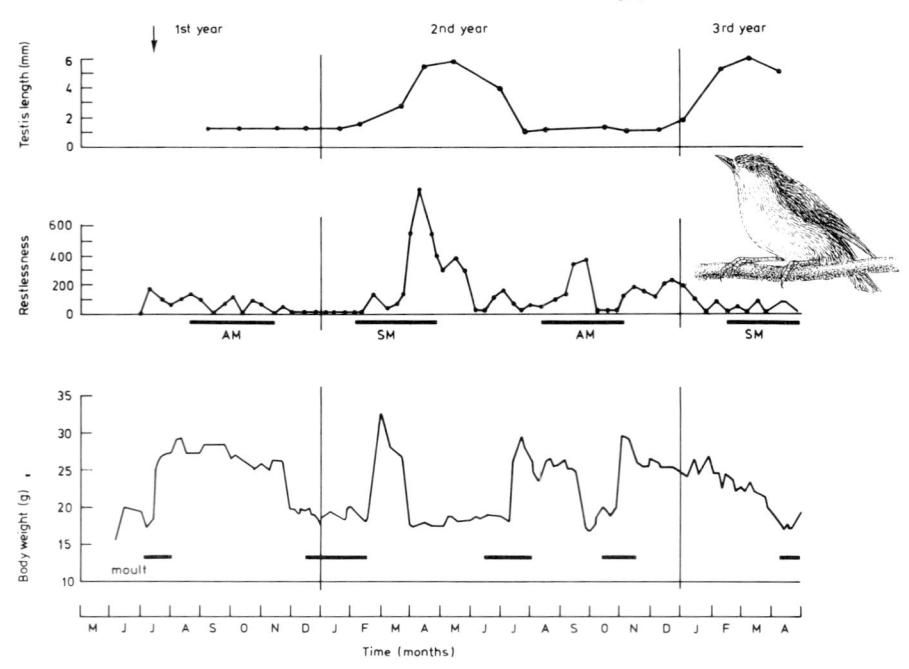

Fig. 12.1 Programmed sequences of moulting, changes in body weight, migratory restlessness and testis length in a garden warbler (**Sylvia borin**)

The bird was hand-raised before being transferred (arrow) to constant photoperiodic conditions of 10 hours light and 14 hours dark. Solid bars beneath the graph of restlessness show the timing of migration (AM, autumn migration; SM, Spring migration) for free-living garden warblers. [From Baker (1982), after Berthold] with an internal program that, even in the absence of changing day-length and temperature, will organize the sequence and timing of moult, deposition of fat, development and regression of the gonads, and migratory restlessness. In constant conditions, the program does not have a perfect, year-long cycle (i.e. it is circannual rather than annual), but it does persist for at least the first two to three years of a bird's life. Normally, of course, the program would be kept strictly in phase with the seasons by such phase-setters as day-length and temperature. Of major interest to this chapter, however, is the fact that migratory restlessness is part of this program, for it raises the question of just how much of the migratory behaviour is inborn.

All of the species studied so far are nocturnal migrants. Primarily, this is because migratory restlessness in such birds is so easy to separate from normal activity. In non-migration seasons, the birds are inactive at night. During the seasons for migration, they are very active (Fig. 3.7). In day-migrating birds, however, migratory restlessness is less easy to separate from other activity.

Glück (1982) has recently described experiments with the day-migrating goldfinch (*Carduelis carduelis*). He obtained a measure for migratory restlessness by first calculating a regression line for activity on day-length. The base-line given by this regression could then be subtracted from measured activity levels. Seasonal excesses of activity coincided with the normal migration seasons for the species, and presumably, therefore, reflect migratory restlessness. Perhaps, before long, we shall have information on the inborn programs of diurnal migrants. For the moment, however, all data relate to nocturnal migrants.

In this chapter, our main interest is in the extent to which the direction and distance of migration are coded into the bird at birth. We have already seen in Chapter 9 that the direction of migration is part of the inborn, circannual program. Not only do the young of migrant birds have an instinctive preference to travel south in autumn and north in spring but some species show more detailed preferences even within a single season. Garden warblers (*Sylvia borin*), for example, prefer to fly southwest early on in autumn migration but later they prefer to fly south (Fig. 9.4).

Insofar as warblers that migrate further are more restless than warblers that migrate less far, there is some evidence that the distance of migration is also reflected in the inborn program. However, the evidence is less clear for the pair of muscicapids that have been studied. Even less convincing is the evidence that details of the migration behaviour over different stages of the route may also be programmed into the bird at birth. Nevertheless, it has been suggested that changes in the intensity of restlessness through the late summer and autumn period mimic changes in the migratory speed of free-living birds. Thus, at a time when some warblers are expected to be crossing the Mediterranean Sea and Sahara Desert, caged birds may show more restlessness than when they are expected to be migrating more slowly through Europe or Africa.

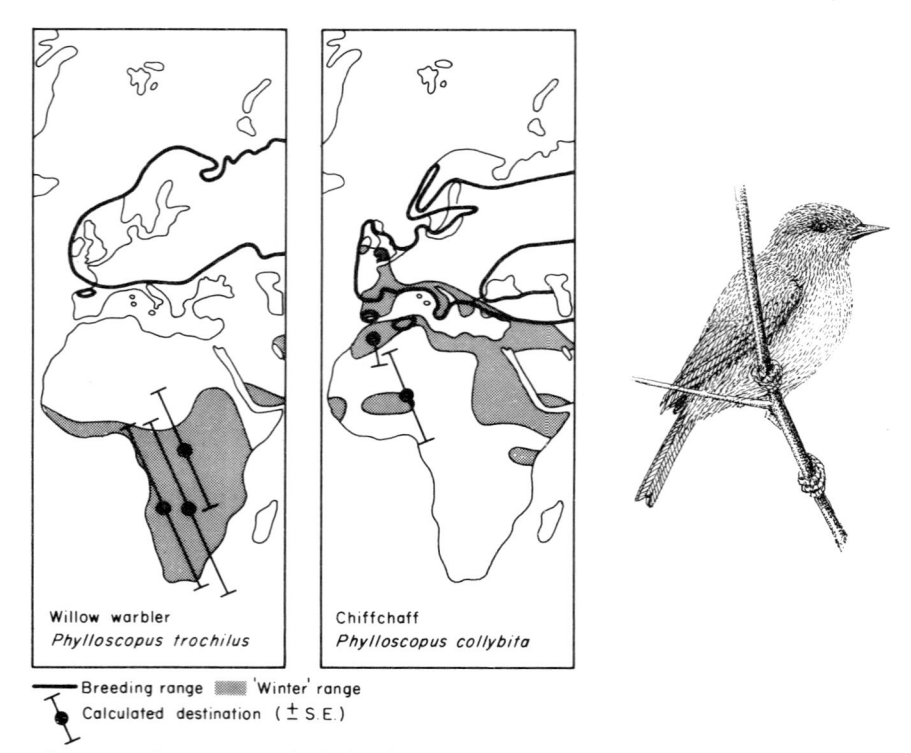

Fig. 12.2 Comparison of calculated and observed end-points of autumn migration for two species of **Phylloscopus**

A 'conversion factor' was calculated from observed migration rates over one section of the migration route and used to convert measured amounts of migratory restlessness into migration distance. On the basis of this factor, the end-points of migration for 20 willow warblers and 24 chiffchaffs were calculated from their total amount of restlessness during autumn. These calculated destinations are shown in relation to the winter ranges of the two species. [From Baker (1978), after Gwinner]

It seems, therefore, that birds are predisposed at birth to show a given amount of migratory activity and to orient in a particular sequence of preferred compass directions. Moreover, the internal program from which these predispositions derive is geared roughly to the seasons, even in the absence of a changing day-length and temperature. Just how we imagine this program to be manifest in the real world depends on which theory of bird migration we accept.

12.2. Models of migration

There are three major theories of bird migration: (1) clock-and-compass; (2) goal-area navigation; and (3) exploration (reviewed in Baker 1978, 1982; Rabøl 1978).

Navigation in action: long-distance migrants 205

The clock-and-compass model grew out of the experiments by Perdeck (1958), described in Chapter 9 and Fig. 2.2, on young starlings (*Sturnus vulgaris*) which were captured on migration through The Netherlands. When displaced to Switzerland, these starlings continued to migrate to the west-southwest, parallel to their original course, and ended up in Spain instead of in Britain and northern France.

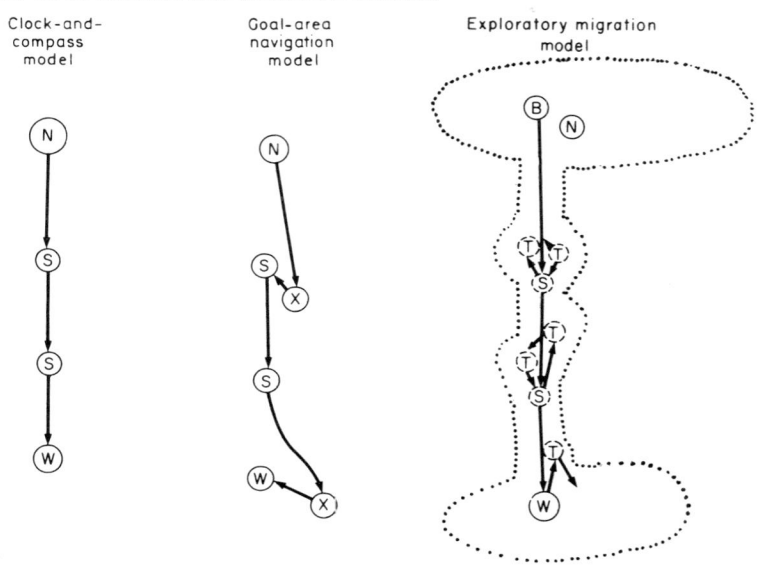

Fig. 12.3 Three models of the first autumn migration of a young bird

N, natal site; S, place at which a migrant lands after one leg of the migration; T, potential staging sites at which the bird can feed and rest; W, winter home range; B, potential breeding site for following year. Dashed lines, limits of familiar area; dotted lines, limits to the familiar area map. [Modified from Baker (1978)]

The clock-and-compass model suggests that the first autumn migration of a bird progresses in a series of preprogrammed stages. At each stage, the bird migrates for a programmed length of time (using an internal clock) in a programmed direction (using a compass). Between each stage of migration the bird rests and feeds in a particular staging area. Either the number of such stages, or the total time spent in migration, are determined by the inborn program. When this program has finally run its course, the bird is in effect deposited in its ancestral wintering area.

Before a bird sets off on its autumn migration and again upon arrival at the wintering grounds, it 'imprints' on their coordinates. Thereafter, perhaps even during its first spring migration, but certainly in subsequent years, the bird navigates to and fro between these two sets of coordinates. Alternatively, the first spring migration may be a repeat of the autumn sequence, but in the opposite direction. When its spring clock-and-compass program runs out, the bird should be deposited near to its breeding range. Spring migration then ends with a final exercise in navigation as the bird regains more precisely the coordinates of the breeding site upon which it imprinted the previous autumn.

The goal-area navigation model was proposed and elaborated by Rabøl (1970, 1978). He suggested that a bird is born with the coordinates of its ancestral migration track already encoded within its genetic material. As the seasons progress, the coordinates preferred by the bird gradually change to those appropriate to successive positions along the migration track. Young and adults alike, therefore, live their lives in a continual process of navigation, forever moving to regain the values of those geographical coordinates that have been generated by the program within them.

Different variants on what is essentially an exploration model for bird migration were proposed independently by Baker (1978) and W. and R. Wiltschko (1978). Briefly, both versions suggest that the young fledgling first explores in various directions around its birth site. During these explorations, it does two things: (1) it builds up a large familiar area map; and (2) it searches for a suitable breeding site for use the following year. At a given time in autumn, the young bird's internal program switches on a preference for the compass direction appropriate to seasonal migration.

From this point on, the Wiltschkos' model does not differ in any major way from the clock-and-compass model except insofar as they suggest that the bird learns a map at each staging post. They still see the distance and direction of migration being determined in a fairly inflexible way by the internal program for migratory restlessness.

My model is more extreme and has its emphasis on flexibility. I interpret each stage of the autumn migration as an exploratory foray and suggest that the bird will often, if circumstances are favourable, explore either back along major sections of its track or sideways, at right angles to the track, before moving on to the next stage of the migration journey. Such large-scale explorations during the autumn migration are suggested to be aimed primarily at finding the best staging posts for use in future years, and eventually at finding the best wintering grounds. If reverse and sideways explorations from the main migration axis are as flexible and large-scale as I propose, the bird will not simply be deposited in its traditional wintering grounds when a program for time and distance has run its course, as on the clock-and-compass model. Some other mechanism by which the young bird recognizes the end point of its journey is necessary. The mechanism suggested is discussed later, along with other details of this version of the exploration model.

Evidence against the clock-and-compass model is threefold: (1) birds displaced sideways from their migration track and then caged often seem to orient either back in the direction from which they were displaced or at some compromise angle between that and the standard direction for autumn migration (Evans 1966, 1972; Rabøl 1970, 1972); (2) birds blown off-course during the night may compensate by reorientation the following morning (Gauthreaux 1978) or on a later night (Evans 1972); and (3) birds respond to the direction of the wind during their migration in complex ways (discussed in Section 12.4) that are inconsistent with a simple time-and-distance program.

Evidence against the goal-area navigation model is twofold: (1) when young birds are displaced sideways from their migration track and then released, they continue on a course parallel to their original track and spend the winter at coordinates different from those traditional for the population (Perdeck 1958); (2) when young birds are displaced from their wintering grounds and released elsewhere, they often adopt the release site as their winter home range (see Section 12.5). In none of these cases is there an indication that the birds are preprogrammed to be at particular coordinates.

Further evidence against clock-and-compass and goal-area navigation models and, at the same time support for the more extreme version of the exploration model, comes from an analysis of ringing returns for lesser black-backed gulls (*Larus fuscus*) breeding in Britain (Baker 1980a). Over the past few decades, the gulls have shifted their wintering grounds from Northwest Africa and Iberia to Britain and northern France. Both clock-and-compass and goal-area navigation models would predict that such a change in migration pattern would be brought about by a change in the internal program with which the gulls are born. The exploration model, on the other hand, would predict that the internal program for exploration by the young bird during its first year could remain unchanged. The new migration pattern could instead result from the birds judging different habitats now from previously to be the most suitable for wintering. In the past, sites in the south of the range explored by the young birds were judged to be most suitable; more recently, sites in the north were chosen.

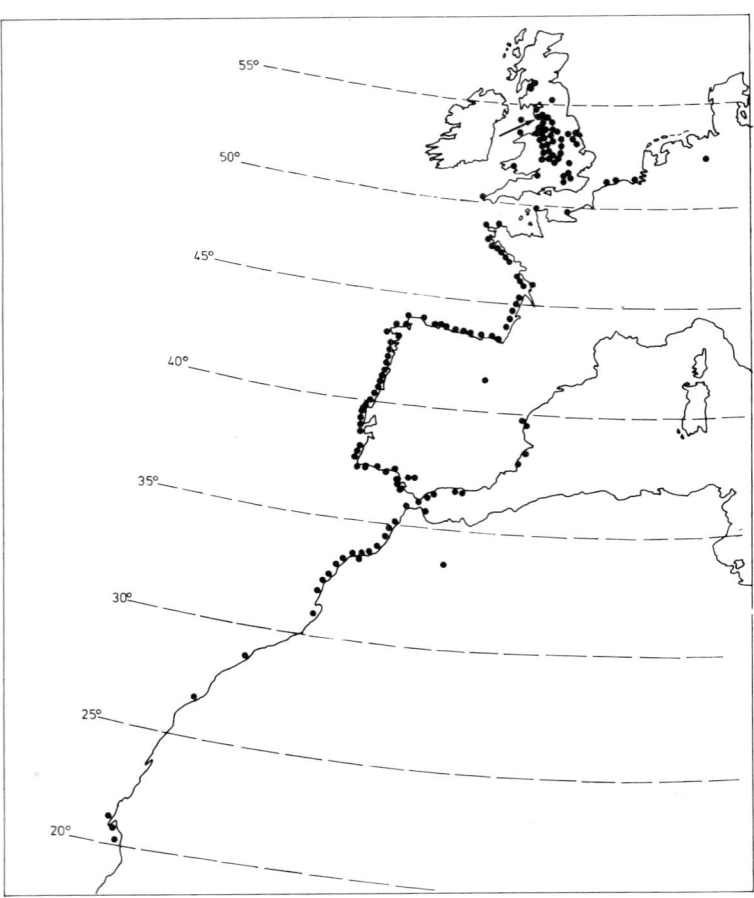

Fig. 12.4 Migration route of lesser black-backed gulls (**Larus fuscus**) born on Walney Island, England, (arrowed) as indicated by ringing recoveries.

Each dot shows the recovery of a dead bird ringed as a pullus on Walney Island between 1962 and 1975. The migration route is overland across England but primarily coastal thereafter. [From Baker (1982). Photo (p. 207) by Eric Hosking]

On the exploration model, therefore, first-year birds could show the same migration pattern now as previously, with the change being shown primarily by older birds. An analysis of ringing returns for lesser black-backed gulls ringed as pulli in the nest and then later recovered over the period 1962–1975 (Fig. 12.5) supports the predictions of the exploration model.

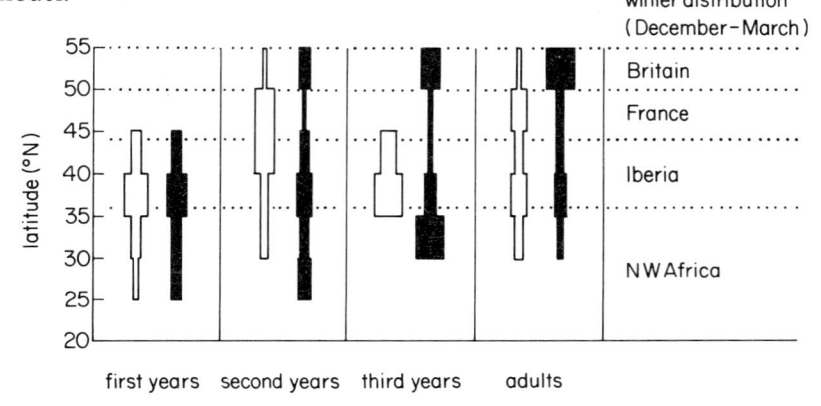

Fig. 12.5 Change in winter distribution of lesser black-backed gulls (**Larus fuscus**) born on Walney Island, England, as indicated by ringing recoveries

Histograms show the distribution of ringing recoveries for each 5° of latitude for birds during each of their first three years of life and for adults. Open histograms, recaptures 1962–68; solid black, recaptures 1969–75.

[Modified from Baker (1982). Photo (p. 210) by Gösta Håkansson, courtesy of Frank W. Lane]

In the remainder of this chapter, the navigation of birds during seasonal migration is discussed in the context of my version of the exploration model. First, however, we have to consider how to interpret the migratory restlessness of caged birds, for at first sight the behaviour seems much more compatible with a clock-and-compass than with exploration.

The answer may be that such restlessness is simply a reflection of a bird's urge to explore rather than an inflexible program for time and direction (Baker 1978, 1982). It seems to be comparable to the wheel-running behaviour of caged rodents (Baker 1978). There is now evidence that the amount of wheel-running reflects the level of motivation of a rodent to acquire resources not available in the cage (Mather 1981). Moreover, the direction of wheel-running reflects the direction in which particular resources are most likely to be found (Mather in prep.) Mather (1981) concludes that, in most laboratory conditions, wheel-running indeed reflects the level of motivation for the rodent to explore for new sites.

If we apply the same interpretation to caged birds, it follows that the amount of migratory restlessness over the whole migration season will reflect the amount of exploration needed to explore for a succession of staging posts en route from the traditional breeding grounds to the traditional wintering grounds. On this interpretation, the amount of restlessness should still correlate with the traditional migration distance, but the relationship is much more flexible than the relationship required by the clock-and-compass model.

With this picture of the internal program for migration with which the bird is born, we can now return to our newly-born nestling and begin to follow it during its first year of life.

12.3. Exploration and navigation after fledging

We might expect that, even while in its nest, our young bird could refine its appreciation of magnetic compass bearings. Perhaps, also, depending on the location of the nest, the nestling may even begin to develop star and sun compasses and to appreciate their relationship to wind direction and the direction of movement of clouds. It may even begin to build up its landscape of familiar smells.

All of these elements in its package of navigational techniques, along with the placement of topographical and acoustic features on its landscape, will be developed still further around the time of fledging when the bird leaves the nest and begins to survey its environment from higher vantage points. If infrasonic landmarks are to form part of the landscape, they seem unlikely to be detected until the bird begins to fly.

Even before our bird leaves the immediate vicinity of its nest, therefore, it should have begun to develop an integrated compass system and already to have a well-developed centre to its ever-growing map. Some of the more conspicuous olfactory, visual and acoustic landmarks should by this stage be stored in the bird's spatial memory in terms of their compass direction from the nest site.

It follows that, when the young bird first begins to explore the area surrounding the nest, it is doing so within a familiar map that already extends tens of kilometres, perhaps more, in different directions. First explorations are likely to cover distances that are relatively short and should serve primarily to fill in details around areas of immediate importance, such as feeding, drinking, roosting and escape sites. Once the bird has found, or is shown by its parents, enough sites to provide all of its immediate requirements, and once it can find its way from one to another through orientation on the compass bearings that link them, the stage is set for it to begin to explore over longer distances.

The distance the bird can travel on its first explorations beyond its initial map is determined partly by the size of the map, partly by how easily familiar landscapes can be recognized when approached from unfamiliar directions, and partly by the accuracy with which the bird can perform route-based navigation. Suppose, for example, that initially it can perform route-based navigation only crudely, perhaps to within, say, 45 degrees. If the bird's familiar area map extends only 20 km in all directions from its nest site, it can travel up to about 20 km beyond the previous limits of its map. Within this distance it could return simply by flying in the compass direction determined by route-based navigation during the outward flight. If it were more accurate, say to within 10 degrees, it could have flown over 100 km beyond its previous map.

Suppose that each time a bird explores, it extends its mosaic map of major landmarks and that, at the same time, it gradually refines the accuracy with which it is able to use its various compasses. Within a very short time the bird could indeed have a wide-ranging map of its landscape. I calculate that, even with relatively modest levels of visibility, time spent exploring and accuracy of route-based navigation, a bird could still obtain over the course of about a week a mosaic map of an area roughly the size of the southern half of England and Wales. If infrasonic landmarks were also part of the landscape, the feat would be simple. Such an area would make a very suitable target area when the bird returns from migration the following spring.

Route-based navigation may well be of primary importance whenever the bird explores beyond the previous limits of its familiar landscape. Once its mosaic map takes on such large dimensions, however, and much of the bird's explorations take place within the outer limits of this familiar mosaic, the bird is likely to make increasing use of map-based mechanisms (Wiltschko and Wiltschko 1982).

There is abundant evidence that young birds are very mobile during the phase of post-fledging exploration (e.g. Mead and Harrison (1979) for sand martins (*Riparia riparia*); see discussion by Baker (1982)). There is also evidence that, apart from developing a large target area at which to aim when returning from migration the following spring, one of the main reasons for a bird to explore so widely during this post-fledging period is to find a potential nesting site suitable for the young bird's own use the following year.

Experiments in Russia (Shcherbakov and Palivanov in Mauersberger 1957) and southwest Germany (Löhrl 1959) involved the release of hand-raised pied flycatchers (*Ficedula hypoleuca*) at sites up to 400 km from their birthplace. Each of these sites seemed to be suitable for the species but had not supported breeding populations for some years. When displaced before or just after fledging, over 5 per cent of the birds returned in later years and successfully established a breeding colony. Berndt and Winkel (1981),

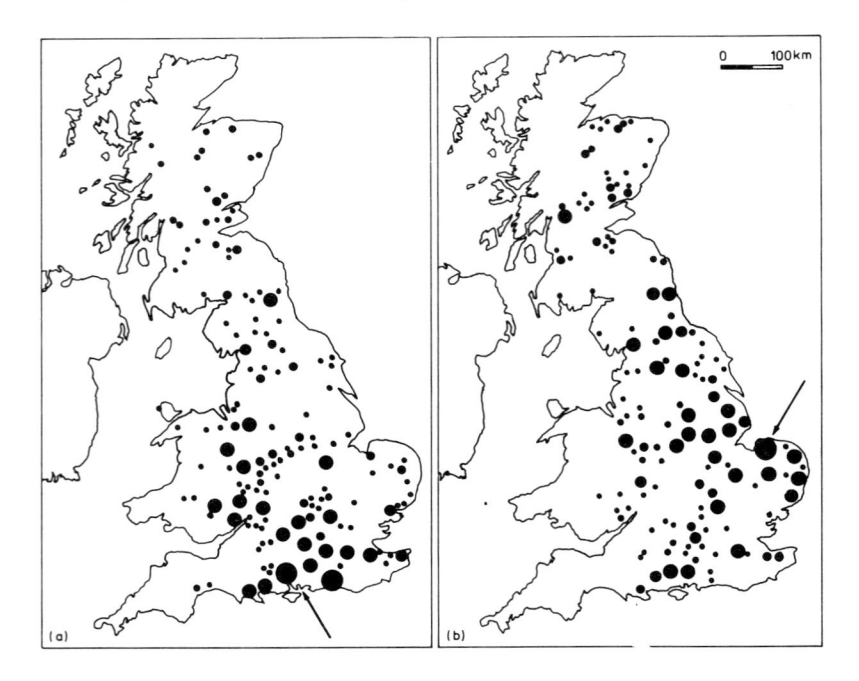

Fig. 12.6 Post-fledging migrations of sand martins / bank swallows (**Riparia riparia**)

Arrows show the position of two roosts: Chichester (a), and Fenland (b). Dots show colonies visited during the period of post-fledging migration by birds that use each roost. Size of dot reflects number of individuals.

[From Baker (1982), (after Mead and Harrison)]

using the same species, exchanged eggs, nestlings and juveniles between sites 250 km apart in northern Germany. Without exception, of those that returned to one or other of the two sites, all returned to their foster site, not to their birth site. This does not mean, of course, that no other birds settled elsewhere, but it does indicate that the coordinates of the breeding site are not encoded within the bird at birth as required by Rabøl's goal-area navigation model.

As discussed in Chapter 9, it is possible that each individual bird has an inborn preference for exploring along a particular compass axis during post-fledging exploration. In sand martins, for example, most of the population seem to prefer to explore along a north–south axis during this phase (Fig. 12.6).

For many 'non-migrant' birds, such as rock doves/homing pigeons, this period soon after fledging may reflect almost the entirety of their inborn program for exploration. In seasonal migrants, however, the compass direction preferred for post-fledging exploration changes at some stage to that preferred for the beginning of autumn migration. At the same time, most individuals adopt more or less the same direction, and in many species the time of day or night at which the bird carries out most of its exploration also changes.

12.4. Autumn migration

On the exploration model, each stage of the autumn migration serves two major functions: (1) to displace the young bird in the direction that evolution has dictated is optimum for that particular lineage (Baker 1978); (2) to enable the young bird to locate the best staging post along that particular stretch of the migration route. In addition, of course, we should expect height, time, speed and length of flight also to be optimum (see review by Alerstam 1981).

The best way to compromise between displacement in a particular direction (say to the south) and exploration for the most suitable staging post, is first to fly a transect to the south and, during fly-over, to observe potential staging areas, albeit at a very general level (e.g. forest, lake, moorland). Among several factors to influence the optimum length of this transect, two in particular are relevant to orientation and navigation: (1) distance travelled over the ground; and (2) the state of wind, cloud and visibility. The first of these factors, distance, has already been discussed and is a function of the size of the target area last vacated and the accuracy with which the bird can employ route-based navigation (Fig. 4.3). The bird has to be able, upon landing, to estimate with enough accuracy the compass direction of the breeding or last staging area that it has just vacated. The second factor, the state of wind, cloud and visibility, leads us into an area that has been the subject of much research and discussion over the past decade.

In Chapter 4 we considered some of the ways in which route-based navigation might function in the real world, as opposed to during experimental displacement. The possibility was mooted that birds may well make use during their outward journey of features of the landscape, as well as of a compass. A bird with a preferred compass direction to the south, for example, encountering during a particular stage of its migration a major river or mountain range aligned north–south would gain many advantages from using that feature as a leading line, flying along it and memorizing its location for use in future years. There is also a disadvantage, of course, in that if many birds follow the same leading line there will be much greater competition for any suitable staging areas to which it might lead. Against this, however, is the advantage that it provides a line that is easy both to perceive and follow. This makes it much easier for a bird to compensate for any tendency of a cross-wind to displace it sideways from its track. Often, also, such features provide favourable air-currents that reduce the energy cost of migration.

Evidence that migrating birds, by day and by night, are aware of the landscape as it passes beneath them was presented in Chapter 5. It is likely that a bird makes use of this awareness of the landscape in three ways: (1) it enables the bird to extend its mosaic map as it migrates, though presumably only with respect to the more major features of the landscape; (2) it gives the bird an opportunity to scan for potential staging areas; (3) it allows the bird to monitor and perhaps correct and compensate for any tendency to be drifted by cross-winds.

The way in which birds may wait to migrate with a tail-wind were described in Chapter 9, but in the absence of such winds some migration may occur with cross-winds, when the risk of drift is real, or even with head winds. An example of the influence of wind direction on the number

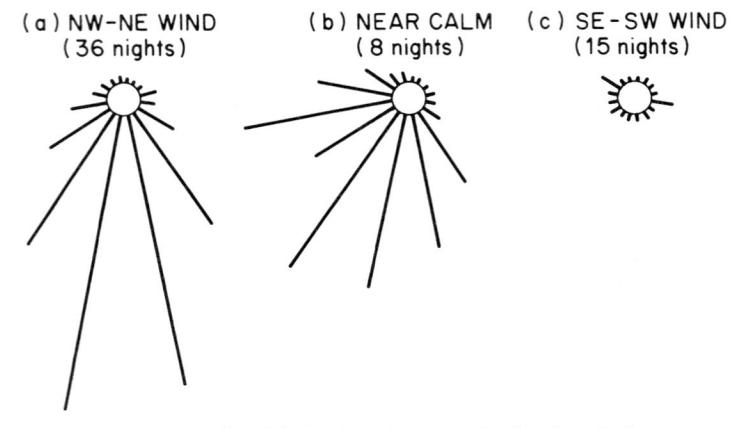

Fig. 12.7 Mean number of landbirds migrating in each direction during autumn nights with northerly (a), near calm (b) and southerly (c) winds

[Simplified from Richardson (1982a)]

of birds migrating is shown in Fig. 12.7 (also reviewed by Alerstam 1981). Excessive drift off the preferred track can, of course, be costly to a bird in terms of the total time and energy it ultimately expends on migration. As far as orientation by a first-year bird is concerned, however, it may be less important to correct for drift than to be aware of it. The important factor is that the bird is able to use route-based navigation so that it can work out the compass direction of the area it has just left.

Most authors agree that the best way for a bird to judge drift is to watch landmarks (Griffin 1969; Emlen 1975) and the evidence that birds do so has been presented in Chapter 5. Theoretically, it would also be possible for birds to judge wind direction even in the absence of landmarks by detecting patterns of turbulence in the lower atmosphere (Nisbet 1955; Bellrose 1967; see also Larkin 1982). Finally, Williams and Teal (1973) found that some herring gulls (*Larus argentatus*) could maintain their position, even when blindfolded, against a 25 km/h wind. They suggested that the birds managed to do this by attention to acoustic cues from the ground beneath.

One of the features that distinguishes my model of bird migration from others concerns the behaviour of a young bird on completion of each southward transect forming one stage of its autumn migration. On landing in a potential staging area, the bird first links this with the area most recently vacated back along the migration track. This mental link on the mosaic map should involve memorizing the compass axis by which the two areas are joined along with any leading lines that aided migration between them. According to the exploration model, the bird then has to set about finding the best feeding and resting sites for this stage of its journey. My own feeling is that the exploration involved may be extensive, though perhaps less so than that in the breeding area. In particular, such exploration may routinely involve flights back along the migration track to potential sites first noticed as the bird flew over during the previous southward transect.

There are abundant data from radar and ringing studies that such reverse migrations do occur and are performed primarily by first-year birds (reviewed by Baker 1978, 1982). The important point here, however, is that whether reverse migration over any distance occurs or not depends almost certainly on the weather, and in particular on wind direction.

Gulls are particularly useful birds for the study of migration behaviour because they can be aged while flying overhead. My own, as yet unpublished, observations on gulls passing through the Manchester region suggest a major difference in orientation of birds of different ages. First-year birds, particularly in autumn, migrate primarily downwind, irrespective of its direction. Adults on the other hand, wait for winds in directions favourable for migration to the south before travelling. It is as if young birds take advantage of winds of any direction to explore in that direction. Adults, on the other hand, await winds favourable for migration to a particular destination.

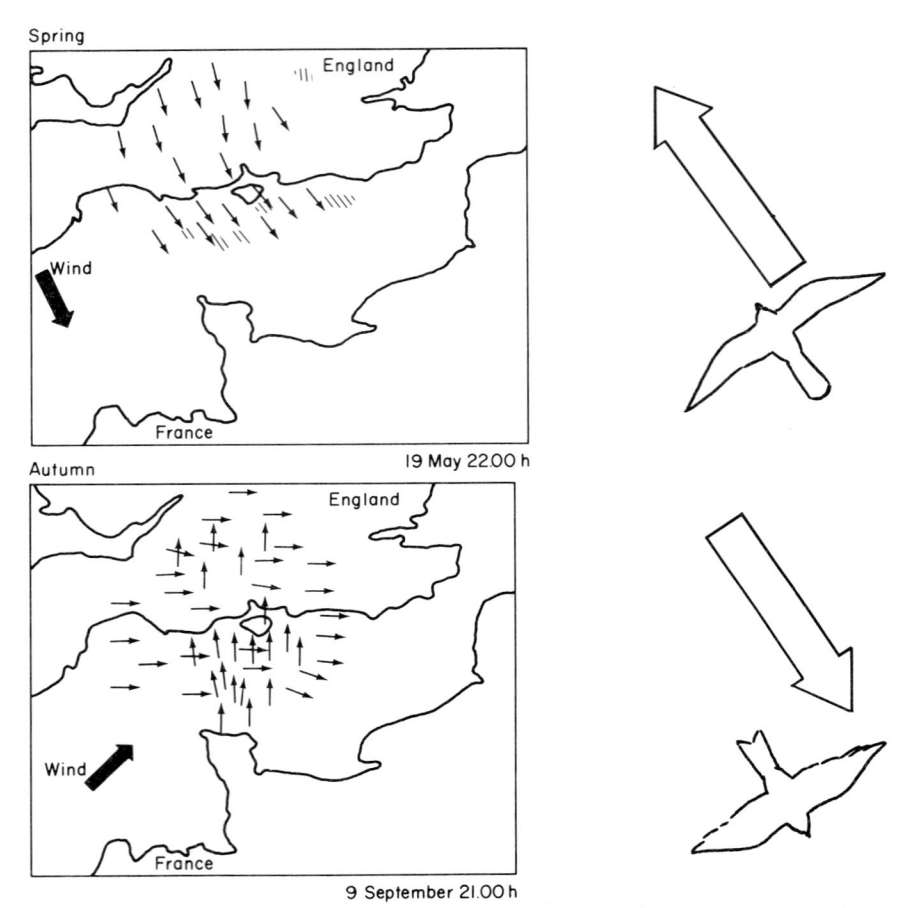

Fig. 12.8 Two examples of 'reverse' migration by nocturnal migrants over southern England as detected by radar

Small arrows show the direction of migrating birds. Large black arrows show wind direction. Large white arrows show the standard migration direction for the time of year indicated. Hatched areas in the spring diagram show rain showers.

[From Baker (1978), after Parslow]

Gauthreaux (1982), in studies of caged passerines, has also found that adults are much more loyal than first-year birds to the expected seasonal direction. He warns that age differences must be taken into account in trying to rationalize observed patterns of migration. This warning seems particularly apt when we survey data on the response of nocturnal migrants to wind and overcast.

Most radar studies of the direction of migration encounter night to night variation that seems to relate to changes in wind direction (Evans 1966; Nisbet and Drury 1968; Alerstam 1976). Alerstam suggested that much of this was 'pseudodrift' due to populations with different preferred compass directions settling out from, and taking off into, the wind whenever its direction changed. Few authors, however, have encountered the dependence on wind direction discovered by Gauthreaux and Able (1970) and Able (1974) in the southeastern United States. There they found nocturnal songbirds flying downwind irrespective of wind direction. Moreover, migration direction kept phase with downwind direction even when the wind changed during the course of a single night. Flocked waterfowl and shorebirds, however, maintained seasonally appropriate directions throughout.

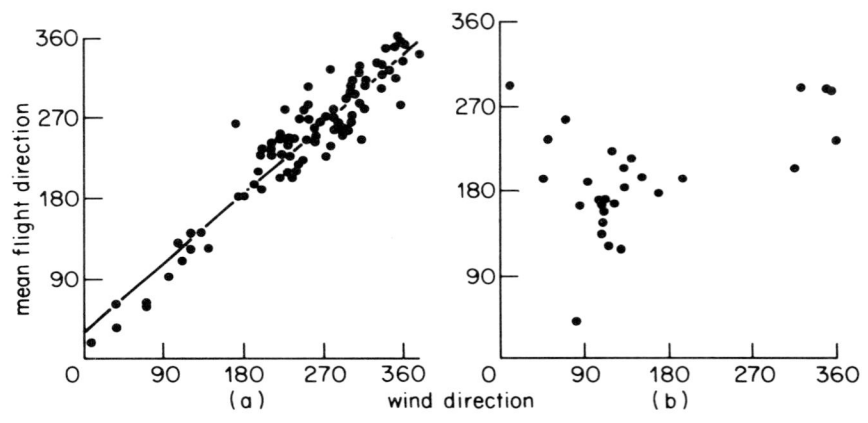

Fig. 12.9 Flight directions of passerine nocturnal migrants as a function of wind direction in northeast and southeast USA

Each point gives a one hour mean for the flight direction of the birds. Wind directions are those toward which the winds were blowing.
(a) Autumn migration at Lake Charles, Louisiana and Athens, Georgia, southeast USA.
(b) Autumn migration near Albany, New York, northeast USA.

[Redrawn from Able (1980)]

Able (1980) expressed doubts that there are sufficient populations with different preferred directions to account for this phenomenon in terms of pseudodrift. In the conventional sense, that pseudodrift relates to populations from different source areas, we are forced to agree with him. If the exploration model is accepted, however, we can suggest that downwind migrants are primarily first-year birds taking advantage of wind direction to explore economically in directions that would otherwise be barred to them. Evans (1972) has already shown that migration in seasonally inappropriate directions tends to be associated with an influx of young birds.

In the northeastern United States, downwind flight in seasonally inappropriate directions occurs less often (Drury and Nisbet 1964; Richardson 1972; Able 1978). The same is true in nearby southern Ontario (Richardson 1982a), though headwinds during autumn migration cause considerable reduction in migration traffic and a more or less equal division of birds migrating north and south. Yet, by and large, the birds studied in northeastern and southeastern United States will be from the same populations but at different stages in their migration flight. Perhaps, in the early stages of autumn migration, southward progress is more important to young birds than finding perfect staging areas. Later on, however, when in the southeast, many are near their wintering grounds and others are preparing for a long flight to South America, often over the Atlantic and Caribbean (Williams and Williams 1978). Here, perhaps, it is of more immediate importance to find good staging areas then to make further southward progress. On the Florida peninsula in autumn (Williams et al. 1977) and in the Carribbean, downwind flight in inappropriate directions is also rare, for obvious reasons.

In a radar study of reverse migration over Nova Scotia, Canada, in autumn, Richardson (1982b) found small numbers of landbirds flying northeast at all hours of the day and night. As usual, most but not all reverse migrations occurred with tail-winds, in this case from the southwest. The birds maintained straight courses and were oriented, not simply being blown downwind. It seemed that the birds were engaged in oriented movements over distances of at least tens and sometimes hundreds of kilometres. Such reverse migrations remained common during October and even into November. Richardson accepted that such late flights might be consistent with exploration during autumn migration, but doubted that the November flights are adaptive because weather conditions deteriorate rapidly at this time. This does not, however, seem to be a major objection for in the lesser black-backed gull, ringing returns indicate an apparently adaptive but nevertheless long-distance reverse migration by second-year birds in the middle of winter (Baker 1980a).

The reaction of birds to landscape and wind during migration is influenced by the degree of overcast. In a careful comparison of the flight paths of individual birds tracked by radar on clear and overcast nights, Able (1982a) came to a number of interesting conclusions. He found that birds flying below cloud on overcast nights seemed totally unaffected by the absence of stars and Moon. Tracks were as straight, level and fast as on clear nights, and the spread of headings was similar. However, when flying within or between continuous cloud layers, headings were random and tracks were slightly less straight. This contrasted with an earlier study by Griffin (1973), who found birds to be well oriented in such situations.

Earlier, Hebrard (1971) and Emlen and Demong (personal communication to Able 1982a), found that in periods of uninterrupted overcast lasting several days, the headings of nocturnal migrants were random within 24 hours of the onset of overcast conditions. Able (1982a), however, found that birds continued to be well-oriented as long as they were flying below the cloud. If overcast skies set in during late afternoon so that birds had no view of the sun late in the day, or of the stars at night, the birds headed downwind. When the winds were opposed to the standard direction for the season, the birds flew in the inappropriate direction. This influence of conditions at dusk was discussed in Chapter 9.

In an attempt to study the influence of wind experimentally, Able et al. (1982) adopted a technique pioneered by Emlen and Demong (1978). White-throated sparrows were taken aloft in a cage by balloon, then released and followed by tracking radar. In Able's study, the birds were fitted with frosted contact lenses and released only on clear nights with winds opposed to the normal migration direction. Birds fitted with frosted lenses flew downwind, whereas control birds with a clear view of their surroundings flew in the appropriate seasonal direction. The birds fitted with frosted lenses did not just drift, passively, but were actually oriented downwind. How they managed to orient with respect to the wind, in the absence of visual cues and within a few seconds of release, is of considerable interest in its own right.

Alerstam (1979a, b) has suggested that migrants may save time by allowing themselves partially to be drifted by strong cross-winds at high altitude at night and then, at low altitude in weaker winds, perhaps during the following day, to correct for their resulting displacement. In the southeastern United States, Gauthreaux (1978) indeed seemed to find that early morning flights by migrants showed a tendency to correct for possible displacement from the standard migration direction the previous night. In the northeastern United States, however, Bingman (1980) found no indication of such behaviour.

The exploration model proposes that a young, first-year bird does not embark on the next stage of its autumn migration until it has found a suitable staging post for the current stage, plus perhaps several alternatives for contingencies. It should also memorize, as a mosaic map, a large area of landscape to serve as a target at which it can aim on future visits. The process of alternating a long flight in the standard direction with shorter, but often substantial, exploratory flights in other directions, is then repeated as many times as is necessary for the bird to arrive at its winter range. The next question is: how does it recognize the winter range, and what does it do on arrival?

12.5. Arrival at the winter range

The clock-and-compass and goal-area navigation models had no problem in depositing first-year birds in their traditional winter range. The clock-and-compass program simply runs out after a given time, whereupon the bird should be in more or less the correct wintering area. Goal-area navigation assumes that the coordinates of the winter range are encoded into the bird's genetic material at birth and awaits only the correct time of year to have the bird navigate to these coordinates.

Unfortunately, a clock-and-compass program runs into trouble if the bird encounters a succession of head- or, even worse, cross-winds during migration. Similarly, evidence presented below shows it to be most unlikely that precise winter coordinates are coded into the bird. We are left, then, with the exploration model which, at first sight, would seem to be the theory with the greatest difficulty in placing a bird in the correct area for winter.

My suggestion (Baker 1978, 1982) is that the internal program for migratory restlessness does not, in fact, simply run out. We concluded earlier (Chapter 9) that one of the predispositions with which a migrant bird is born was to migrate in autumn in the direction that felt right relative to the magnetic field. In a similar way, we can imagine that a bird has programmed within it a mental image of how a suitable winter area should feel as perceived by its various senses (Baker 1982). Day-length, temperature, food availability, even type of vegetation, may all be built into this mental image. Indeed, we have experimental evidence that some birds are born with a preference for particular types of vegetation (Partridge 1978). We can propose, therefore, that a bird will continue to explore, ever further in its preferred direction, until it finds an area that provides a suitable match to this inborn mental image. Put another way, the bird continues on autumn migration until its threshold for further migration is no longer exceeded.

Two problems with this thesis might come to mind. Firstly, why should the end of migration be interpreted as a threshold response instead of simply a termination of the autumn phase of the migration program? Secondly, why does migratory restlessness cease in caged birds?

Evidence on the first point comes from the phenomenon of hard-weather movements in winter. Further migration in the standard direction can be triggered in most birds at any time during the winter by the sudden onset of adverse conditions (Fig. 12.10). It follows, therefore, that the end point of migration is not fixed and further migration does occur if the appropriate threshold is exceeded.

A possible answer to the second point is that the migratory restlessness of caged birds ends because the cage environment, with its warmth, light and abundant food, is interpreted by the bird as a very favourable winter environment. Evidence that it is not inevitable for migratory restlessness to end is provided by the behaviour of caged birds in spring (data from Gwinner and Czeschlik 1978; interpretation from Baker 1982). At this time of year, restlessness can continue well beyond the migration season of free-living birds and, indeed, may even extend until the time of moulting in late summer. If the above interpretation is correct, it follows that the birds do not consider the cage environment to be as suitable for summer as they do for winter, and the motivation to explore remains strong. So what resources are missing? Anecdotal evidence suggests that it is the absence of a mate that is important. If a restless bird in spring is placed within sight and sound of a potential mate, migratory restlessness stops.

On the exploration model, therefore, a bird ceases to migrate in the

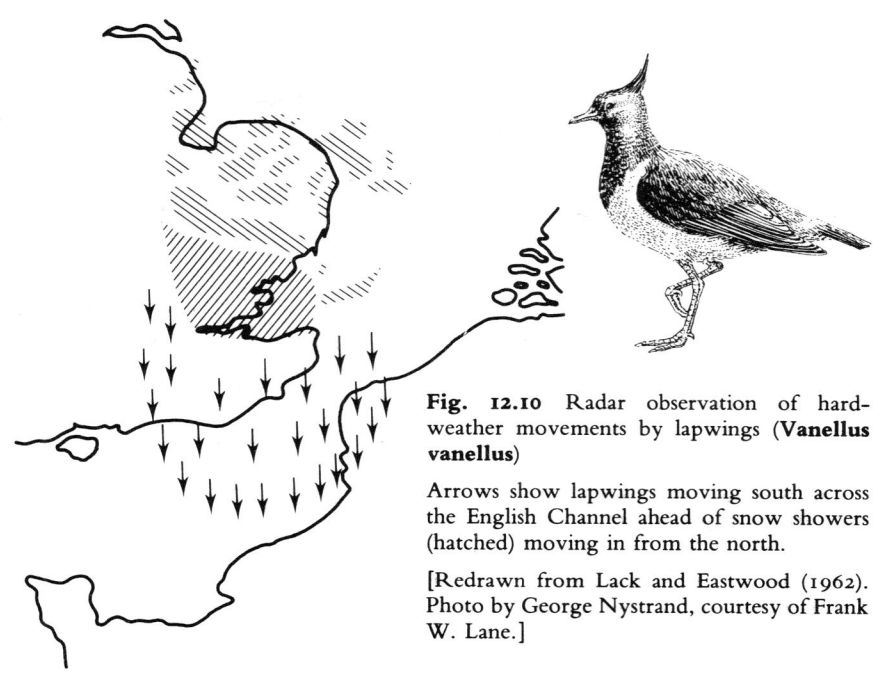

Navigation in action: long-distance migrants 221

Fig. 12.10 Radar observation of hard-weather movements by lapwings (**Vanellus vanellus**)

Arrows show lapwings moving south across the English Channel ahead of snow showers (hatched) moving in from the north.

[Redrawn from Lack and Eastwood (1962). Photo by George Nystrand, courtesy of Frank W. Lane.]

standard direction for autumn as soon as it arrives in an area that matches its mental image of how a suitable winter range should feel. When this happens, directed movement in one main direction gives way to exploration of the type described first in relation to post-fledging exploration. We can assume that at this far end of its range the bird builds up yet another large mosaic map. On this map eventually are sites that provide the range of resources necessary for winter survival. At the same time, the map itself provides a large target area for navigation in future years.

A number of homing experiments have been carried out on birds displaced from their winter range. Peterson (1953) with gulls (*Larus* spp), Schwärtz (1963) with waterthrushes (*Seiurus noveboracensis*), Ralph and Mewaldt (1975) with sparrows (*Zonotrichia* spp) and Benvenuti and Ioalé (1980) with robins (*Erithacus rubecula*) and other passerines, all found that displaced birds would return to their place of capture. In addition, the studies by Ralph and Mewaldt and by Benvenuti and Ioalé both showed that adults were much more likely to return than first-year birds.

Ralph and Mewaldt captured 900 white crowned (*Z. leucophrys*) and golden-crowned (*Z. atricapilla*) sparrows and displaced them distances of 4–160 km from winter sites in California. Whereas the adults returned to their original site the same winter, most first-years stayed at the release site, apparently treating the experimental displacement as free exploration. No adults, but many of the ex-immatures, returned to the release site the following winter. Of those immatures that did return, however, all had been displaced before mid-January of the previous winter. None of those displaced after mid-January returned to the release site the following winter. This suggests that, by mid-January, first-year birds have found suitable feeding and other sites and, presumably, have built up their mosaic map, whereas before that time they are quite likely to decide that the release site is the most suitable area they have found.

A similar trend can be seen in even longer displacements (Mewaldt 1963, 1964; Mewaldt *et al.* 1973). The same species of sparrows were displaced from winter sites around San José in California to Baton Rouge, Louisiana, (2900 km) and to Laurel, Maryland, (3860 km). None returned to California the same winter, but significant numbers returned there to winter the following year, presumably after having spent the summer at their breeding grounds in Alaska. Sparrows displaced from San José to Seoul, Korea, (9060 km Great Circle), failed to produce any returns. In the displacements to sites within the United States, there was the same tendency for immature birds to remain around their release site and to return there the following winter.

Fig. 12.11 Breeding and wintering grounds of sparrows displaced by Mewaldt from wintering grounds in San José, California, to Baton Rouge and to Laurel

[From Baker (1978)]

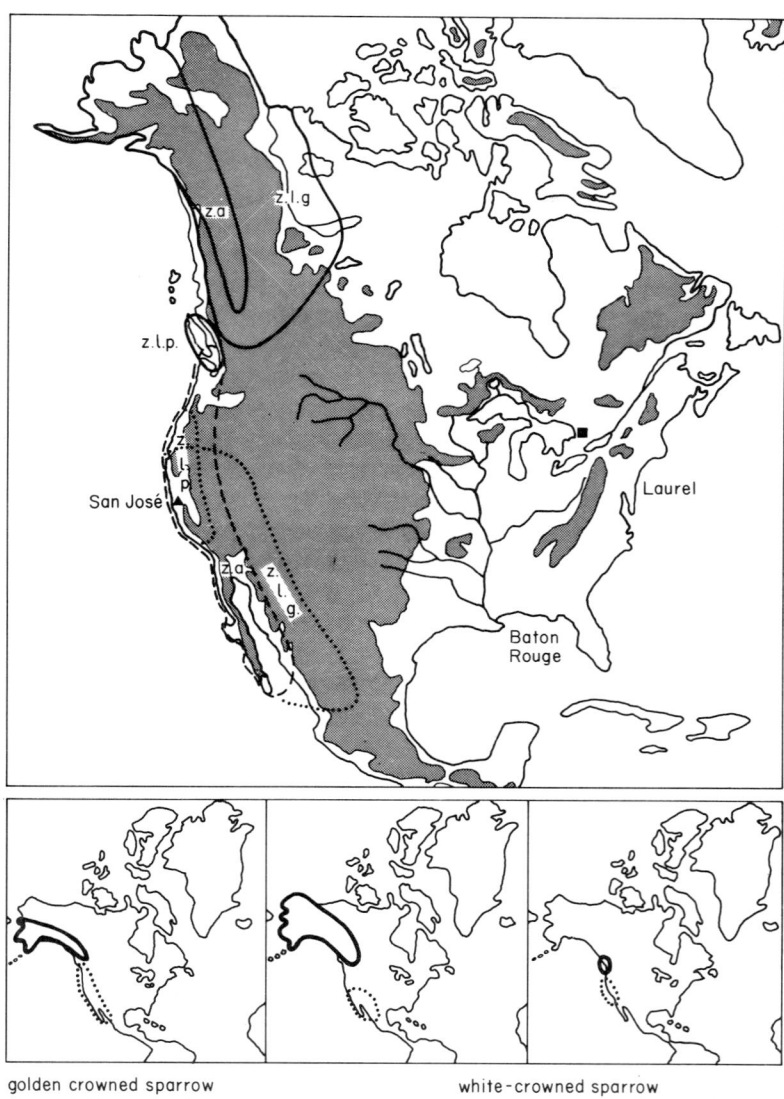

Navigation in action: long-distance migrants 223

golden crowned sparrow
Zonotrichia atricapilla

white-crowned sparrow
Zonotrichia leucophrys

Z.l. gambelii   *Z.l. pugetensis*

12.6. Spring return

Whether a bird, while in its winter range, is aware of the compass direction of its breeding range or whether its map is simply a consecutive sequence of staging areas and their relative compass directions has not been examined. Nor is it known if most birds during spring migration simply navigate back along their autumn migration route, retracing their steps from one staging area to another. Some, however, certainly do not, for they travel back in a direction different from the axis of the autumn migration. We must assume that such birds have a program for exploration comparable to that which organized their migration in autumn. Whether such birds actually link up the two mosaic maps they must produce, as with a zip fastener, is unknown, but my own prediction is that probably they do.

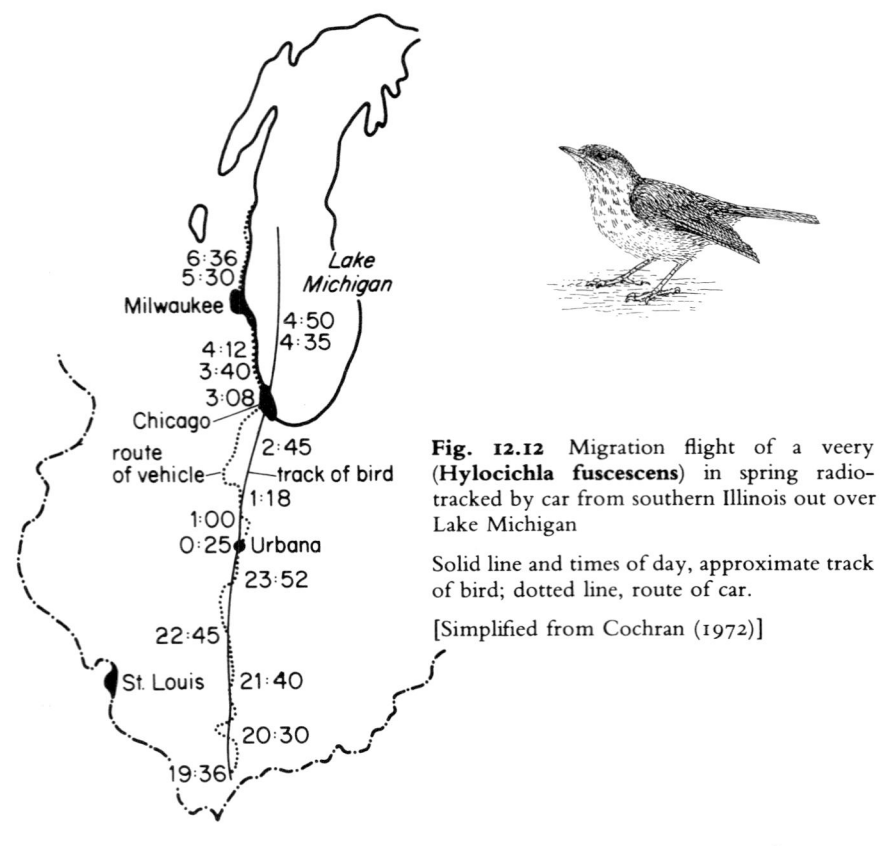

Fig. 12.12 Migration flight of a veery (**Hylocichla fuscescens**) in spring radio-tracked by car from southern Illinois out over Lake Michigan

Solid line and times of day, approximate track of bird; dotted line, route of car.

[Simplified from Cochran (1972)]

There will often be some advantage in carrying out separate explorations in autumn and spring, for areas that provide suitable staging posts in autumn do not necessarily do the same the following spring. Moreover, wind fields that favour one route in autumn may well favour a different route in spring. Given that the preferred compass directions governing the spring migration flights (Chapter 9) have evolved to be part of the internal program with which the bird is born (Gwinner and Wiltschko 1978), and given that each stage of spring migration probably follows the same principles as described for autumn migration, there is nothing new to say about spring migration, until, that is, we come to consider the final return of the bird to its breeding area.

Experiments in which young birds were displaced laterally from their migration track in either autumn (e.g. Perdeck 1958, 1967 with the starling; Fig. 2.2) or spring (e.g. Rüppell 1944 with hooded crows; Fig 9.3) suggest that birds may regain their breeding range but only if the distance of lateral displacement is not too great and does not take them the wrong side of a geographic barrier. Many of the starlings displaced during autumn migration from Holland to Switzerland seemed to regain their original breeding area (Fig. 12.13). When displaced from Holland to Spain,

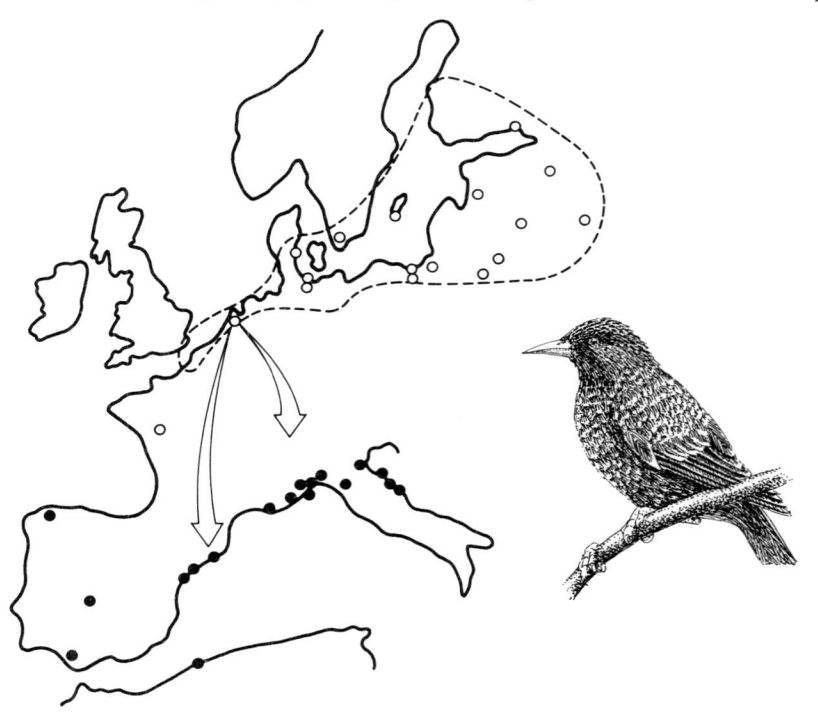

Fig. 12.13 Subsequent grounds of young starlings (**Sturnus vulgaris**) displaced during their first autumn migration

Starlings were captured during autumn migration through The Hague and displaced by aeroplane to Switzerland or Spain. Dashed line show the limits of the breeding range of birds that migrate through The Hague. Dots show recaptures of displaced birds in subsequent summers. Open symbols, birds displaced to Switzerland; solid symbols, birds displaced to Spain. [Compiled from Perdeck (1958, 1967)] however, most seemed to end up breeding south of the area from which they were likely to have originated. Hooded crows, displaced so that their traditional direction would take them north of the Baltic Sea, when their original breeding areas lie to the south, also failed to regain the area from which they were likely to have originated.

These data are yet further evidence against the goal-area navigation model. They suggest, also, that these two species may rely on their spring migration program to deposit them near enough to their breeding area to encounter the mosaic map that they built up during post-fledging exploration the previous autumn. If, because of experimental displacement, this does not happen, the birds are either unable or unwilling to find their original area.

Normally, however, we expect young birds to regain their original target area successfully and to return to the place they selected the previous autumn as the best site in which to breed.

12.7. Migration as an adult

Earlier we asked, but could not answer, the question of whether a bird in its winter range is aware of the compass direction of its breeding range, or whether it simply memorizes only the compass directions of successive staging sites. There is a tenuous indication, however, that having completed one round trip, a bird may be aware of the direction of its winter grounds from its breeding grounds.

The American golden plover (*Pluvialis dominica*) breeds on much of the Canadian tundra and winters in South America (Fig. 12.14). Young birds, in their first autumn, migrate south by an inland route that takes them more or less down the Mississippi to Louisiana and Texas, across the gulf to Central America, and then down through South America to the pampas grounds of Argentina. Spring migration is virtually the reverse of this route. On their next autumn migration, however, they fly to the coasts of Labrador and then south, out over the Atlantic, to the Lesser Antilles or even non-stop to the northern coast of South America. Return migration is by the same inland route that they followed during their first year.

This change in behaviour between the first and second years of life could, of course, simply be part of each bird's inborn program. Alternatively, it could indicate that, having once completed the round trip, and taking into account the prevailing wind fields, the plovers are able to work out a more economical route to their winter range.

The golden plover is exceptional. Most species probably take similar routes when adult to those they took during their first year. Evidence is slowly accumulating that the same individuals may indeed use the same staging areas at roughly the same time during successive years' migrations (see Baker 1978; Wiltschko and Wiltschko 1978). There is no reason, of course, why even adults should not carry out some further explorations in

Fig. 12.14 Breeding range, winter range and migration routes of the American golden plover (**Pluvialis dominica dominica**)

First-year birds take an overland route both in autumn and spring. Adults migrate south over the Atlantic Ocean but return overland in spring.

[Redrawn from Schmidt-Koenig (1979), after Salomonsen and Schüz]

subsequent years. Indeed, many species, such as redwings (*Turdus iliacus*) (Alerstam 1981) and various finches (Newton 1972) seem to pioneer different routes each year as part of their adaptation to a source of winter food that is rarely abundant in the same area two years in succession. Similarly, adults of species liable to lose breeding areas while away in their winter range seem routinely each year, after completion of breeding, to explore for further potential breeding sites. This is the explanation for the post-breeding mobility of sand martins (*Riparia riparia*) offered by Mead and Harrison (1979). In this species, landslides can destroy whole nesting banks over winter.

For most species, however, the seasonal migrations of adults can be seen as a succession of homing flights to consecutive staging areas strewn out along an elongated mosaic map that stretches from their breeding grounds at one end to their wintering grounds at the other. Adults, like young, should also migrate primarily by compass orientation. The reasons, however, are different. Whereas young migrate along a compass bearing because it is part of an inborn program, adults do so because, as for all movements within a mosaic map, they have learned the compass bearings that link successive familiar sites. Adults also respond differently to the wind. Whereas the young are opportunistic, adults await winds favourable for migraton to their next destination. Finally, a radar study by Nisbet and Drury (1968), suggests that perhaps adults select for migration those weather patterns offering good conditions for navigation by map at their destination, rather than en route, during which compass orientation alone could be sufficient to hit the target area around the destination.

12.8. Are landscapes and compasses enough?

We began this chapter by asking whether the use of compasses and landscape maps, welded together in sequences of route-based and location-based navigation, would suffice to allow a long-distance migrant to carry out its seasonal migrations. I can see nothing in the admittedly meagre evidence for such migrants to make me feel that there is still some major reference system or navigational technique that remains hidden from us. There are still many details to be uncovered, but these are details, not major principles.

I now feel, almost for the first time, that bird navigation, whether by pigeons or long-distance migrants, is no longer enigmatic. The mystery seems to have been solved, though it will probably be some time before we can bring ourselves to believe it.

If the mystery really is over, the answer was neither singular nor dramatic. Instead, it has come gradually, through a careful accumulation of evidence by many patient biologists over the past few decades. Now, in the early 1980s, we seem to have reached a stage where there are no major loose ends, only a number of frustrating but fascinating inconsistencies.

13 Epilogue: the next ten years

If the major part of the mystery of bird navigation really has been solved, we may wonder what are the most important issues to be resolved over the next ten years?

Perhaps the most important question mark still hangs over the apparent phenomenon of the extreme sensitivity of birds to tiny fluctuations in the geomagnetic field. It is ironic that while we struggle experimentally to find the best way to produce a predictable rotation in magnetic orientation using field intensities of tens of thousands of nanoTesslas, tests that require a sensitivity of the animal to changes in field intensity of only a few tens of nanoTesslas nevertheless appear to succeed in obtaining results. I predict that these small intensity effects will all be found to be in some way related to the functioning of the magnetic compass. If I am wrong and proof is found that such extreme sensitivity does indeed reflect the fact that birds normally read their location on a magnetic grid map, then the conclusion of the last chapter must be rescinded. Landscapes and compasses would have proved not to be the sum total of the navigational armoury of birds.

As it is, however, I envisage that the primary research over the next ten years will be an attempt to understand better the nature of the avian landscape and the development and interaction of the various compass systems.

As far as landscapes are concerned, it is to be hoped that the next ten years will clarify the presence and prominence of infrasonic landmarks. The controversy over whether smells can form familiar landmarks seems as good as over. The main drive should now be toward an understanding of how such smells are perceived at the home site so as to be incorporated into a functional map of the landscape. I doubt that deflector lofts will contribute greatly to this understanding. Perhaps the major debate over the next ten years will try to resolve whether consistent biases in the vanishing bearings of homing pigeons are specific to the release-site or to the way that map information is perceived while at the loft. If such biases are due to the release-site, the question must be asked whether the factors causing them are often topographical in origin. I foresee more analyses of the type carried out by Dornfeldt for the Göttingen loft.

Progress in our understanding of the way that birds are born with, and develop, their different compass systems may well depend on the discovery of experimental situations in which the magnetic compass can express itself with full force. Many details of the internal programs with which seasonal migrants are born remain to be identified as do details of the ways that adults and first-year birds respond differently to their environment, particularly with regard to wind direction. All of these details will influence the final decision about whether the exploration model is indeed the most realistic description of bird migration.

If this book were to be written again in the early 1990s, it would, I am sure, present a much more complete picture of bird navigation. In particular, there should be fewer major inconsistencies and most controversy should concern details of landscapes and compasses rather than conflict between major theories. I should be surprised, however, if such a book could see any further element of mystery to navigational behaviour, and suspect that, in retrospect, it would conclude that the mystery of bird navigation died a very quiet death, scarcely noticed, in the late 70s and early 80s.

References and author index

The most recent review on each of the major aspects of bird navigation is shown by an asterisk (*). Figures in brackets after each reference give the page-numbers on which the publication is quoted.

ABLE, K. P. (1974) Environmental influences on the orientation of freeflying nocturnal bird migrants. *Anim. Behav.* **22**, 224–238. (217)

ABLE, K. P. (1978) Field studies of the orientation cue hierarchy of nocturnal songbird migrants. In: Schmidt-Koenig, K. and Keeton, W. T. (eds) *Animal Migration, Navigation and Homing.* Springer, Heidelberg. pp. 228–238. (146, 217)

*ABLE, K. P. (1980) Mechanisms of orientation, navigation, and homing. In: Gauthreaux, S. A. Jr. (ed.) *Animal Migration, Orientation, and Navigation.* Academic Press, New York. pp. 284–374. (59, 60, 92, 149, 217)

ABLE, K. P. (1982a) The effects of overcast skies on the orientation of free-flying nocturnal migrants. In: Papi, F. and Wallraff, H. G. (eds) *Avian Navigation.* Springer, Heidelberg. pp. 38–49. (60, 146, 150, 160, 218)

ABLE, K. P. (1982b) Field studies of avian nocturnal migration I. Interaction of sun, wind and stars as directional cues. *Anim. Behav.* **30**, 761–767. (146, 150)

ABLE, K. P. (Personal Communication). 85

ABLE, K. P., BINGMAN, V. P., KERLINGER, P. and GERGITS, W. (1982) Field studies of avian nocturnal migratory orientation II. Experimental manipulation of orientation in white-throated sparrows (*Zonotrichia albicollis*) released aloft. *Anim. Behav.* **30**, 768–773. (219)

ADLER, H. E. (ed.) (1971) *Orientation: Sensory Basis.* Vol. 188. N.Y. Academy of Science, New York. (6)

ADLER, K. and TAYLOR, D. H. (1973) Extraocular perception of polarized light by orienting salamanders. *J. Comp. Physiol.* **87**, 203–212. (117)

ALDRICH, J. W., GRABER, R. R., MUNRO, D. A., WALLACE, G. J., WEST, G. C. and GAHALANE, V. H. (1966) Mortality at ceilometers and towers. *Auk* **83**, 465–467. (94)

ALERSTAM, T. (1976) *Bird migration in relation to wind and topography.* Ph.D. Thesis, University of Lund. (216)

ALERSTAM, T. (1979a) Wind as selective agent in bird migration. *Ornis Scand.* **10**, 76–93. (219)

ALERSTAM, T. (1979b) Optimal use of wind by migrating birds: combined drift and overcompensation. *J. Theor. Biol.* **79**, 341–353. (219)

*ALERSTAM, T. (1981) The course and timing of bird migration. In: Aidley, D. J. (ed.) *Animal Migration.* University Press, Cambridge. pp. 9–54. (213, 215, 227)

ALERSTAM, T. and PETTERSSON, S.-G. (1976) Do birds use waves for orientation when migrating across the sea? *Nature, Lond.* **259**, 205–207. (60)

ALEXANDER, J. R. and KEETON, W. T. (1974) Clock-shifting effect on initial orientation of pigeons. *Auk* **91**, 370–374. (141)

ANESHANSLEY, D. J. and LARKIN, T. S. (1981) V-test is not a statistical test of 'homeward' direction. *Nature, Lond.* **293**, 239. (25)

BAKER, R. R. (1978) *The Evolutionary Ecology of Animal Migration.* Hodder and Stoughton, London. (6, 14, 15, 42, 52, 56, 126, 131, 133, 190, 195, 204, 206, 209, 213, 215, 220, 226)

BAKER, R. R. (1980a) The significance of the Lesser Black-Backed Gull, *Larus fuscus*, to models of bird migration. *Bird Study* **27**, 41–50. (207, 218)

BAKER, R. R. (ed.) (1980b) *The Mystery of Migration.* Macdonald, London. (3, 4)

BAKER, R. R. (1980c) Goal orientation by blindfolded humans after long-distance displacement: possible involvement of a magnetic sense. *Science* **210**, 555–557. (172)

BAKER, R. R. (1981a) *Human Navigation and the Sixth Sense.* Hodder and Stoughton, London. (4, 6, 95, 121, 145, 172, 179, 190, 195)

BAKER, R. R. (1981b) Man and other vertebrates: a common perspective to migration and navigation. *In:* Aidley, D. J. (ed.) *Animal Migration.* University Press, Cambridge. pp. 241–260. (6, 14)

*BAKER, R. R. (1982) *Migration: paths through time and space.* Hodder and Stoughton, London. (6, 14, 52, 131, 133, 204, 209, 212, 215, 220)

BAKER, R. R. (1984a) Sinal magnetite and direction finding. *Physics in Technology.* (in press). (106, 116)

BAKER, R. R. (1984b) Magnetoreception by humans and other primates. *In:* Kirschvink, J. L., Jones, D. S. and MacFadden, B. J. (eds) *Magnetite Biomineralization and Magnetoreception in Organisms: a new magnetism.* Plenum, New York. (in press). (110, 116, 172)

BAKER, R. R. and MATHER, J. G. (1982a) Magnetic compass sense in the large yellow underwing moth, *Noctua pronuba* L. *Anim. Behav.* **30**, 543–548. (145)

BAKER, R. R. and MATHER, J. G. (1982b) A comparative approach to bird navigation: implications of parallel studies on mammals. *In:* Papi, F. and Wallraff, H. G. (eds) *Avian Navigation.* Springer, Heidelberg. pp. 308–312. (68, 115, 178)

BAKER, R. R. and SADOVY, Y. J. (1978) The distance and nature of the light-trap response of moths. *Nature, Lond.* **276**, 818–821. (94)

BAKER, R. R., MATHER, J. G. and KENNAUGH, J. H. (1982) The human compass? *EOS,* **63**, 156. (68, 122)

BAKER, R. R., MATHER, J. G. and KENNAUGH, J. H. (1983) Magnetic bones in human sinuses. *Nature,* **301**, 78–80. (68, 122, 123)

BALDACCINI, N. E., BENVENUTI, S., FIASCHI, V. and PAPI, F. (1975) Pigeon navigation: effects of wind deflection at home cage on homing behaviour. *J. Comp. Physiol.* **99**, 177–186. (64)

*BALDACCINI, N. E., BENVENUTI, S., FIASCHI, V., IOALÉ, P. and PAPI, F. (1982) Pigeon orientation: experiments on the role of the olfactory stimuli perceived during the outward journey. *In:* Papi, F. and Wallraff, H. G. (eds) *Avian Navigation.* Springer, Heidelberg. pp. 160–169. (180, 181)

BANG, B. J. (1971) Functional anatomy of the olfactory system in 23 orders of birds. *Acta Anat. Suppl.* **58**, **79**, 1–76. (74)

*BARLOW, J. S. (1964) Inertial navigation as a basis for animal navigation. *J. Theoret. Biol.* **6**, 76–117. (39, 172)

*BATSCHELET, E. (1981) *Circular Statistics in Biology.* Academic Press, London. (24, 97)

BECK, W. and WILTSCHKO, W. (1982) The magnetic field as a reference system for the genetically encoded migratory direction in Pied Flycatchers (*Ficedula hypoleuca Pallas*). *Z. Tierpsychol.* (in press). (131)

BELLROSE, F. C. (1958) Celestial orientation in wild Mallards. *Bird Band.* **29**, 75–90. (88)

BELLROSE, F. C. (1963) Orientation behaviour of four species of waterfowl. *Auk*, **80**, 257–289. (88)

BELLROSE, F. C. (1967) Radar in orientation research. *Proc. XIV Int. Orn. Congr., Oxford.* pp. 281–309. (215)

BELLROSE, F. C. (1972) Possible steps in the evolutionary development of bird navigation. In: Galler, S. R., Schmidt–Koenig, K., Jacobs, G. J. and Belleville, R. E. (eds) *Animal Orientation and Navigation*. NASA SP-262 US Govt. Printing Office, Washington D.C. pp. 223–257. (4, 75)

BENVENUTI, S. (1976) Homing pigeons with prism goggles: an experiment for testing the sun navigation hypothesis. *Monitore zool. ital. (N.S.)* **10**, 219–227. (155)

BENVENUTI, S. and IOALÉ, P. (1980) Homing experiments with birds displaced from their wintering ground. *J. Orn.* **121**, 281–286. (222)

BENVENUTI, S. and IOALÉ, P. in Papi (1982). (194)

BENVENUTI, S., FIASCHI, V., FIORE, L. and PAPI, L. (1973) Homing performances of inexperienced and directionally trained pigeons subjected to olfactory nerve section. *J. Comp. Physiol.* **83**, 81–92. (67)

BENVENUTI, S., FIASCHI, V. and FOÀ, A. (1977) Homing behaviour of pigeons disturbed by application of an olfactory stimulus. *J. Comp. Physiol.* **120**, 173–179. (69)

*BENVENUTI, S., BALDACCINI, N. E. and IOALÉ, P. (1982) Pigeon homing: effect of altered magnetic field during displacement on initial orientation. In: Papi, F. and Wallraff, H. G. (eds) *Avian Navigation*. Springer, Heidelberg. pp. 140–148. (177, 179, 184)

BERGMAN, G. (1964) Zur Frage der Abtriftskompensation des Vogelzuges. *Orn. Fenn.* **41**, 106–110. (60)

BERNDT, R. and WINKEL, W. (1981) Field experiments on problems of imprinting to the birthplace in the Pied Flycatcher *Ficedula hypoleuca. Proc. XVII Int. Orn. Congr., Berlin.* pp. 156–171. (212)

BERTHOLD, P. (1973) Relationships between migratory restlessness and migration distance in six *Sylvia* species. *Ibis* **115**, 594–599. (202)

*BERTHOLD, P. (1981) Die endogene Steuerung der Jahresperiodik: eine kurze Übersicht. *Proc. XVII Int. Orn. Congr., Berlin.* pp. 473–478. (202)

BERTHOLD, P. and QUERNER, U. (1981) Genetic basis of migratory behavior in European warblers. *Science* **212**, 77–79. (202)

BINGMAN, V. P. (1980) Inland morning flight behavior of nocturnal passerine migrants in eastern New York. *Auk* **97**, 465–472. (219)

BINGMAN, V. P. (1981) Savannah Sparrows have a magnetic compass. *Anim. Behav.* **29**, 962–963. (97, 131, 149)

BINGMAN, V. P. (personal communication). (149)

BINGMAN, V. P. and ABLE, K. P. (1979) The sun as a cue in the orientation of the white-throated sparrow, a nocturnal migrant bird. *Anim. Behav.* **27**, 621–622. (145)

*BINGMAN, V. P., ABLE, K. P. and KERLINGER, P. (1982) Wind drift, compensation, and the use of landmarks by nocturnal bird migrants. *Anim. Behav.* **30**, 49–53. (59, 60)

BLAKEMORE, R. P. (1975) Magnetotactic bacteria. *Science* **190**, 377–379. (120)

*BOOKMAN, M. A. (1978) Sensitivity of the homing pigeon to an earth-strength magnetic field. In: Schmidt-Koenig, K. and Keeton, W. T. (eds) *Animal Migration, Navigation and Homing.* Springer, Heidelberg, pp. 127–134. (105)

BRANDENSTEIN, C. G. von (1972) The symbolism of the north-western Australian zigzag design. *Oceania* **42**, 227. (35)

BRINES, M. (1980) Dynamic patterns of skylight polarization as clock and compass. *J. Theor. Biol.* **86**, 507–512. (83)

*BRUDERER, B. (1982) Do migrating birds fly along straight lines? In: Papi, F. and Wallraff, H. G. (eds) *Avian Navigation.* Springer, Heidelberg. pp. 3–14. (59, 60)

BUDDENBROCK, W. von (1937) *Gundriss der vergleichenden Physiologie.* Berlin. (92)

CAVÈ, A. J., BOL, C. and SPEEK, G. (1974) Experiments on discrimination by the starling between geographical locations. *Inst. Roy. Netherlands Acad. Arts and Sci.* **63**, 82. (58)

CHERFAS, J. in Baker (1980b). (3, 4)

CLARKE, W. E. (1912) *Studies in Bird Migration.* Gurny and Jackson, London. (94)

*COCHRAN, W. W. (1972) Long-distance tracking of birds. In: Galler, S. R., Schmidt-Koenig, K., Jacobs, G. J. and Belleville, R. E. (eds) *Animal Orientation and Navigation.* NASA SP-262 US Govt. Printing Office, Washington D.C. pp. 39–59. (224)

DANILOV, V., DEMIROCHAGLYAN, G., AVETYSAN, Z., AELAKHNYERDYAN, M., GRIGORYAN, S. and SARIBEKHYAN, G. (1970) Possible mechanisms of magnetic sensitivity in birds (in Russian). *Biol. Zh. Arm.* **23**, 26–34. (117)

DECOURSEY, P. J. (1962) Effect of light on the circadian activity rhythm of the flying squirrel (*Glaucomys volans*). *Z. Vergl. Physiol.* **44**, 331–354. (82)

DELIUS, J. D. and EMMERTON, J. (1978) Sensory mechanisms related to homing in pigeons. In: Schmidt-Koenig, K. and Keeton, W. T. (eds) *Animal Migration, Navigation and Homing.* Springer, Heidelberg. pp. 35–41. (39)

DELIUS, J. D., PERCHARD, R. J. and EMMERTON, J. (1976) Polarized light discrimination by pigeons and an electroretinographic correlate. *J. Comp. Physiol. Psychol.* **90**, 560–571. (84, 117)

DEMONG, N. J. and EMLEN, S. T. (1978) Radar tracking of experimentally released migrant birds. *Bird Band.* **49**, 342–59. (150)

*DORNFELDT, K. (1982) Dependence of the homing pigeons' initial orientation on topographical and meteorological variables: a multivariate study. In: Papi, F. and Wallraff, H. G. (eds) *Avian Navigation.* Springer, Heidelberg. pp. 253–264. (58, 190)

DRURY, W. H. (1959) Orientation of gannets. *Bird Band.* **30**, 118–119. (60)

DRURY, W. H. and NISBET, I. C. T. (1964) Radar studies of orientation of songbird migrants in south-eastern New England. *Bird Band.* **35**, 69–119. (217)

EDRICH, W. and KEETON, W. T. (1977) A comparison of homing behavior in feral and homing pigeons. *Z. Tierpsychol.* **44**, 389–401. (141)

EMLEN, S. T. (1967a) Migratory orientation in the Indigo Bunting, *Passerina cyanea*. Part I: The evidence for use of celestial cues. *Auk* **84**, 309–342. (87, 166)
EMLEN, S. T. (1967b) Migratory orientation in the Indigo Bunting, *Passerina cyanea*. Part II. Mechanisms of celestial orientation. *Auk* **84**, 463–489. (87, 166)
EMLEN, S. T. (1969) The development of migration orientation in young Indigo Buntings. *Living Bird* **1969**, 113–126. (90)
EMLEN, S. T. (1970) Celestial rotation: its importance in the development of migratory orientation. *Science* **170**, 1198–1201. (90)
EMLEN, S. T. (1972) The ontogenetic development of orientation capabilities. In: Galler, S. R., Schmidt-Koenig, K., Jacobs, G. J. and Belleville, R. E. (eds) *Animal Orientation and Navigation*. NASA SP-262 US Govt. Printing Office, Washington D.C. pp. 191–210. (90)
EMLEN, S. T. (1975) Migration: orientation and navigation. In: Farner, D. S. and King, J. R. (eds) *Avian Biology*. Vol. V. Academic Press, London. pp. 129–219. (59, 60, 88, 215)
EMLEN, S. T. and DEMONG, N. J. (1978) Orientation strategies used by free-flying bird migrants: a radar tracking study. In: Schmidt-Koenig, K. and Keeton, W. T. (eds) *Animal Migration, Navigation and Homing*. Springer, Heidelberg. pp. 283–293. (150, 219)
EMLEN, S. T. and DEMONG, N. J. in Able (1982a). (218)
EMLEN, S. T. and EMLEN, J. T. (1966) A technique for recording migratory orientation of captive birds. *Auk* **83**, 361–367. (88)
EMLEN, S. T., WILTSCHKO, W., DEMONG, N. J., WILTSCHKO, R. and BERGMAN, S. (1976) Magnetic direction finding: evidence for its use in migratory Indigo Buntings. *Science* **193**, 505–508. (97, 138)
ENRIGHT, J. T. (1972) When the beachhopper looks at the moon: the moon-compass hypothesis. In: Galler, S. R., Schmidt-Koenig, K., Jacobs, G. J. and Belleville, R. E. (eds) *Animal Orientation and Navigation*. NASA SP-262 US Govt. Printing Office, Washington D.C. pp. 523–555. (92)
EVANS, P. R. (1966) Migration and orientation of passerine night migrants in northeast England. *J. Zool., Lond.* **150**, 319–369. (207, 216)
EVANS, P. R. (1972) Information on bird navigation obtained by British long-range radars. In: Galler, S. R., Schmidt-Koenig, K., Jacobs, G. J. and Belleville, R. E. (eds) *Animal Orientation and Navigation*. NASA SP-262 US Govt. Printing Office, Washington D.C. pp. 139–149. (207, 217)
FERGENBAUER, I. in Wiltschko (1982). (139)
FERGUSON, D. E. (1971) The sensory basis of orientation in amphibians. *Ann. N.Y. Acad. Sci.* **188**, 30–36. (69)
FIASCHI, V. and WAGNER, G. (1976) Pigeon homing: some experiments for testing the olfactory hypothesis. *Experientia* **32**, 991–993. (176)
FOÀ, A. and ALBONETTI, E. (1980) Does familiarity with the release site influence the initial orientation of homing pigeons? Experiments with clock-shifted birds. *Z. Tierpsychol.* **54**, 327–338. (192)
FOÀ, A., WALLRAFF, H. G., IOALÉ, P. and BENVENUTI, S. (1982) Comparative investigations of pigeon homing in Germany and Italy. In: Papi, F. and Wallraff, H. G. (eds) *Avian Navigation*. Springer, Heidelberg. pp. 232–238. (184, 186)
FRANKEL, R. B., BLAKEMORE, R. P. and WOLFE, R. S. (1979) Magnetite in freshwater magnetic bacteria. *Science* **203**, 1355–1357. (120)

FREI, U. (1982) Homing pigeons' behaviour in the irregular magnetic field of western Switzerland. In: Papi, F. and Wallraff, H. G. (eds) *Avian Navigation*. Springer, Heidelberg. pp. 129–139. (161)

FREI, U. and WAGNER, G. (1976) Die Anfangsorientierung von Brieftauben im erdmagnetisch gestörten Gebiet des Mont Jorat. *Rev. Suisse Zool.* **83**, 891–897. (161)

FRISCH, K. von (1967) *The Dance Language and Orientation of Bees*. Oxford University Press, London. (82)

GATTY, H. (1958) *Nature is your guide*. Collins, London. (49, 54)

GAUTHREAUX, S. A. Jr. (1978) Importance of the daytime flights of nocturnal migrants: redetermined migration following displacement. In: Schmidt-Koenig, K. and Keeton, W. T. (eds) *Animal Migration, Navigation and Homing*. Springer, Heidelberg. pp. 219–227. (207, 219)

GAUTHREAUX, S. A. Jr. in Able (1980). (59)

GAUTHREAUX, S. A. Jr. (1982) Age-dependent orientation in migratory birds. In: Papi, F. and Wallraff, H. G. (eds) *Avian Navigation*. Springer, Heidelberg. pp. 68–74. (216)

GAUTHREAUX, S. A. Jr. and ABLE, K. P. (1970) Wind and the direction of nocturnal songbird migration. *Nature, Lond.* **228**, 476–477. (217)

GERRARD, E. C. (1981) *The Instinctive Navigation of Birds*. The Scottish Research Group, Pabay. (7, 95, 130)

GLÜCK, E. (1982) Locomotory activity of day-migrating finches. In: Papi, F. and Wallraff, H. G. (eds) *Avian Navigation*. Springer, Heidelberg. pp. 90–95. (203)

GOODWIN, D. (1967) *Pigeons and Doves of the World*. Trustees of the British Museum (Natural History), London. (6)

GOULD, J. L. (1980) The case for magnetic sensitivity in birds and bees (such as it is). *Am. Sci.* **68**, 256–267. (73, 158, 160, 161, 164)

*GOULD, J. L. (1982) The map sense of pigeons. *Nature, Lond.* **296**, 205–211. (48, 50, 67, 69, 158, 160, 161, 162, 164, 165, 166, 193, 198)

GOULD, J. L., KIRSCHVINK, J. L. and DEFFEYES, K. S. (1978) Bees have magnetic remanence. *Science* **102**, 1026–1028. (120)

GRAUE, L. C. (1963) The effects of phase shifts in the day-night cycle on pigeon homing at distances of less than one mile. *Ohio J. Sci.* **63**, 214–217. (36, 57, 174)

GRAUE, L. C. (1965) Initial orientation in pigeon homing related to magnetic contours (abstract). *Am. Zool.* **5**, 704. (161)

GRIFFIN, D. R. (1969) The physiology and geophysics of bird navigation. *Q. Rev. Biol.* **44**, 255–276. (60, 70, 215)

GRIFFIN, D. R. (1973) Oriented bird migration in or between opaque cloud layers. *Proc. Amer. Phil. Soc.* **117**, 117–141. (61, 218)

GRIFFIN, D. R. (1981) *The Question of Animal Awareness*. William Kaufmann, New York. (95)

GRIFFIN, D. R. and BUCHLER, E. R. (1978) Echolocation of extended surfaces. In: Schmidt-Koenig, K. and Keeton, W. T. (eds) *Animal Migration, Navigation and Homing*. Springer, Heidelberg. pp. 201–208. (75)

GRIFFIN, D. R. and HOPKINS, C. D. (1974) Amphibian choruses as a cue for night-migrating birds in spring. *Anim. Behav.* **22**, 672–678. (75)

GRONAU, J. and SCHMIDT-KOENIG, K. (1970) Annual fluctuation in pigeon homing. *Nature, Lond.* **226**, 87–88. (183)

*GRÜTER, M., WILTSCHKO, R. and WILTSCHKO, W. (1982) Distribution of release-site biases around Frankfurt â. M., Germany. *In*: Papi, F. and Wallraff, H. G. (eds) *Avian Navigation*. Springer, Heidelberg. pp. 222–231. (186, 192, 193, 195)

GWINNER, E. (1972) Adaptive functions of circannual rhythms in warblers. *Proc. XV Int. Orn. Congr., Berlin.* pp. 218–236. (202)

*GWINNER, E. (1981) Circannual systems. *In*: Aschoff, J. (ed.) *Handbook of Behavioral Neurobiology*, Vol. V. Plenum, New York. pp. 391–410. (202)

GWINNER, E. and CZESCHLIK, D. (1978) On the significance of spring migratory restlessness in caged birds. *Oikos* **30**, 364–372. (220)

GWINNER, E. and SCHWABL-BENZINGER, I. (1982) Adaptive temporal programming of molt and migratory disposition in two closely related long-distance migrants, the pied flycatcher (*Ficedula hypoleuca*) and the collared flycatcher (*Ficedula albicollis*). *In*: Papi, F. and Wallraff, H. G. (eds) *Avian Navigation*. Springer, Heidelberg. pp. 75–89. (202)

GWINNER, E. and WILTSCHKO, W. (1978) Endogenously controlled changes in the migratory direction of the Garden Warbler, *Sylvia borin. J. Comp. Physiol.* **125**, 267–273. (225)

GWINNER, E. and WILTSCHKO, W. (1980) Circannual changes in the migratory orientation of the Garden Warbler, *Sylvia borin. Behav. Ecol. Sociobiol.* **7**, 73–78. (131)

HARTWICK, R. F., KIEPENHEUER, J. and SCHMIDT-KOENIG, K. (1978) Further experiments on the olfactory hypothesis of pigeon navigation. *In*: Schmidt-Koenig, K. and Keeton, W. T. (eds) *Animal Migration, Navigation and Homing*. Springer, Heidelberg. pp. 107–118. (69, 176)

HASLER, A. D. and SCHWASSMANN, H. O. (1960) Sun orientation of fish at different latitudes. *Cold Spring Harb. Symp.* **25**, 429–441. (80)

HEBRARD, J. J. (1971) Fall nocturnal migration during two successive overcast days. *Condor* **74**, 106–107. (218)

HOFFMANN, K. (1954) Versuche zu der im Richtungsfinden der Vögel enthaltenen Zeitschätzung. *Z. Tierpsychol.* **11**, 453–475. (77, 80)

HOFFMANN, K. (1958) Repetition of an experiment on bird orientation. *Nature. Lond.* **181**, 1435–1437. (155)

HOFFMANN, K. (1959) Die Richtungsorientierung von Staren unter der Mitternachtssonne. *Z. Vergl. Physiol.* **41**, 471–480. (78)

HUTCHISON, L. V. and WENZEL, B. M. (1982) Activity of central olfactory neurons in the pigeon. *In*: Papi, F. and Wallraff, H. G. (eds) *Avian Navigation*. Springer, Heidelberg. pp. 362–372. (74)

HUTH, H. and BURKHARDT, D. (1972) Der spektrale Sehbereich eine Violettohr-Kolibris. *Naturwiss.* **59**, 650. (85)

IOALÉ P. (1980) Further investigations on the homing behaviour of pigeons subjected to reverse wind direction at the loft. *Monitore zool. ital. (N. S.)* **14**, 77–87. (194)

IOALÉ, P. (1982) Pigeon homing: effects of differential shielding of home cages. *In*: Papi, F. and Wallraff, H. G. (eds) *Avian Navigation*. Springer, Heidelberg. pp. 170–178. (193, 194)

IOALÉ, P., PAPI, F., FIASCHI, V. and BALDACCINI, N. E. (1978) Pigeon navigation: effects upon homing behaviour by reversing wind direction at the loft. *J. Comp. Physiol.* **128**, 285–295. (63, 194)

JONES, D. S. and MACFADDEN, B. J. (1981) Induced magnetization in the monarch butterfly, *Danaus plexippus* L. (Insecta, Lepidoptera). *J. exp. Biol.* **96**, 1–9. (120)

KALMIJN, A. (1978) Experimental evidence of geomagnetic orientation in elasmobranch fishes. *In*: Schmidt-Koenig, K. and Keeton, W. T. (eds) *Animal Migration, Navigation and Homing.* Springer, Heidelberg. pp. 345–353. (118)

KALMIJN, A. and BLAKEMORE, R. P. (1978) Magnetic behaviour of mud bacteria. *In*: Schmidt-Koenig, K. and Keeton, W. T. (eds) *Animal Migration, Navigation and Homing.* Springer, Heidelberg. pp. 354–355. (120)

KALMUS, H. (1956) Sun navigation of *Apis mellifica* L. in the southern hemisphere. *J. Exp. Biol.* **33**, 554–565. (80)

KEETON, W. T. (1969) Orientation by pigeons: is the sun necessary? *Science* **165**, 922–928. (155)

KEETON, W. T. (1970) Do pigeons determine latitudinal displacement from the sun's altitude? *Nature, Lond.* **227**, 626–627. (155, 195)

KEETON, W. T. (1971) Magnets interfere with pigeon homing. *Proc. Nat. Acad. Sci.* **68**, 102–106. (126, 134, 136, 140, 163)

KEETON, W. T. (1972) Effects of magnets on pigeon homing. *In*: Galler, S. R., Schmidt-Koenig, K., Jacobs, G. J. and Belleville, R. E. (eds) *Animal Orientation and Navigation.* NASA SP-262 US Govt. Printing Office, Washington D.C. pp. 579–594. (125, 126)

KEETON, W. T. (1973) Release site bias as a possible guide to the "map" component in pigeon homing. *J. Comp. Physiol.* **86**, 1–16. (187, 193)

KEETON, W. T. (1974a) The orientational and navigational basis of homing in birds. *In*: Lehrman, D., Rosenblatt, J. S., Hinde, R. A. and Shaw, E. (eds) *Advances in the Study of Behaviour, Vol. 5.* Academic Press, London. pp. 47–132. (95, 155, 168)

KEETON, W. T. (1974b) The mystery of pigeon homing. *Sci. Am.* **231**, 96–107. (155, 190)

KEETON, W. T. (1974c) Pigeon homing: no influence of outward-journey detours on initial orientation. *Monitore zool. ital. (N. S.)* **8**, 227–234. (176)

KEETON, W. T. (1981) The orientation and navigation of birds. *In*: Aidley, D. J. (ed.) *Animal Migration.* University Press, Cambridge. pp. 81–104. (73, 165)

KEETON, W. T. and GOBERT, A. (1970) Orientation by untrained pigeons requires the sun. *Proc. Nat. Acad. Sci, USA* **65**, 853–856. (134, 138)

KEETON, W. T., LARKIN, T. S. and WINDSOR, D. M. (1974) Normal fluctuations in the earth's magnetic field influence pigeon orientation. *J. Comp. Physiol.* **95**, 95–103. (160, 164)

KEETON, W. T., KREITHEN, M. L. and HERMAYER, K. L. (1977) Orientation of pigeons deprived of olfaction by nasal tubes. *J. Comp. Physiol.* **114**, 289–299. (68, 69)

KIEPENHEUER, J. (1978a) Pigeon homing: a repetition of the deflector loft experiment. *Behav. Ecol. Sociobiol.* **3**, 393–395. (64, 176)

KIEPENHEUER, J. (1978b) Pigeon navigation and the magnetic field: information collected during the outward journey is used in the homing process. *Naturwiss.* **65**, 113. (176)

KIEPENHEUER, J. (1978c) Inversion of the magnetic field during transport: its influence on the homing behaviour of pigeons. *In*: Schmidt-Koenig, K. and Keeton, W. T. (eds) *Animal Migration, Navigation and Homing.* Springer, Heidelberg. pp. 135–142. (177)

KIEPENHEUER, J. (1979) Pigeon homing: deprivation of olfactory information does not affect the deflector loft effect. *Behav. Ecol. Sociobiol.* **6**, 11–22. (64, 69)

KIEPENHEUER, J. (1982a) The effect of magnetic anomalies on the homing behaviour of pigeons: an attempt to analyse the possible factors involved. *In*: Papi, F. and Wallraff, H. G. (eds) *Avian Navigation*. Springer, Heidelberg. pp. 120–128. (161, 162, 198)

KIEPENHEUER, J. (1982b) Pigeon orientation: a preliminary evaluation of factors involved or not involved in the deflector loft effect. *In*: Papi, F. and Wallraff, H. G. (eds) *Avian Navigation*. Springer, Heidelberg. pp. 203–210. (64, 65)

KIEPENHEUER, J., BALDACCINI, N. E. and ALLEVA, E. (1979) A comparison of orientational and homing performances of homing pigeons of German and Italian stock raised together in Germany and Italy. *Monit. zool. ital.* (N. S.) **13**, 159–171. (184)

KING, J. M. B. (1959) Orientation of migrants over sea in fog. *Brit. Birds* **52**, 125–126. (60)

KIRSCHVINK, J. L. (1981) Ferromagnetic crystals (magnetite?) in human tissue. *J. exp. Biol.* **92**, 333–335. (123)

KIRSCHVINK, J. L. and GOULD, J. L. (1981) Biogenic magnetite as a basis for magnetic field detection in animals. *BioSystems* **13**, 181–201. (120, 123)

KNOX, E. G., ARMSTRONG, E., LANCASHIRE, R., WALL, M. and HYNES, R. (1979) Magnetic storms and admissions to coronary intensive care units. *Nature, Lond.* **281**, 564–567. (115)

KÖHLER, K. L. (1978) Do pigeons use their eyes for navigation? A new technique! *In*: Schmidt-Koenig, K. and Keeton, W. T. (eds) *Animal Migration, Navigation and Homing*. Springer, Heidelberg. pp. 57–64. (57)

KRAMER, G. (1949) Über Richtungstendenzen bei der nächtlichen Zugunruhe gekäfigter Vögel. *In*: Mayr, E. and Schuz, E. (eds) *Ornithologie als Biologische Wissenschaft*. Springer, Heidelberg. (32, 145)

KRAMER, G. (1950) Orientierte Zugaktivität gekäfigter Singvögel. *Naturwiss.* **37**, 188. (77)

KRAMER, G. (1953) Die Sonnenorientierung der Vögel. *Verh. Deut. Zool. Ges. Freiburg*, **1952**, 72–84. (32)

KRAMER, G. (1957) Experiments in bird orientation and their interpretation. *Ibis* **99**, 196–227. (154, 168, 185)

KRAMER, G. (1959) Recent experiments on bird orientation. *Ibis*, **101**, 399–416. (185, 193)

KRAMER, G. and RIESE, E. (1952) Die Dressur von Brieftauben auf Kompassrichtung im Wahlkäfig. *Z. Tierpsychol.* **9**, 245–251. (78)

KRAMER, G. and SAINT PAUL, U. von (1950) Ein wesentlicher Bestandteil der Orientierung der Reisetaube: die Richtungsdressur. *Z. Tierpsychol.* **7**, 620–631. (77)

KREITHEN, M. L. (1978) Sensory mechanisms for animal orientation—can any new ones be discovered? *In*: Schmidt-Koenig, K. and Keeton, W. T. (eds) *Animal Migration, Navigation and Homing*. Springer, Heidelberg. pp. 25–34. (70, 72, 84, 149)

*KREITHEN, M. L. (1979) The sensory world of the homing pigeon. *In*: Granda, A. M. and Maxwell, J. H. (eds) *Neural Mechanisms of Behavior in the Pigeon*. Plenum, Washington D.C. pp. 21–33. (85, 117)

KREITHEN, M. L. and EISNER, T. (1978) Ultraviolet light detection by the homing pigeon. *Nature, Lond.* **272**, 347–348. (84)

KREITHEN, M. L. and KEETON, W. T. (1974a) Detection of changes in atmospheric pressure by the homing pigeon, *Columba livia. J. Comp. Physiol.* **89**, 73–82. (71, 149)

KREITHEN, M. L. and KEETON, W. T. (1974b) Detection of polarized light by the homing pigeon, *Columba livia. J. Comp. Physiol.* **89**, 83–92 (83, 84, 117)

KREITHEN, M. L. and KEETON, W. T. (1974c) Attempts to condition homing pigeons to magnetic stimuli. *J. Comp. Physiol.* **91**, 355–362. (105)

KREITHEN, M. L. and QUINE, D. B., (1979) Infrasound detection by the homing pigeon: a behavioural audiogram. *J. Comp. Physiol.* **129**, 1–4. (70, 71)

LACK, D. and EASTWOOD, E. (1962) Radar films of migration over eastern England. *Brit. Birds* **55**, 388–414. (221)

LANDSBOROUGH THOMSON, A. (1974) *A New Dictionary of Birds.* Nelson, London. (5, 6, 14)

LARKIN, R. P. (1982) Spatial distribution of migrating birds and small-scale atmospheric motion. In: Papi, F. and Wallraff, H. G. (eds) *Avian Navigation.* Springer, Heidelberg. pp. 28–37. (215)

LARKIN, T. S. and KEETON, W. T. (1976) Bar magnets mask the effect of normal magnetic disturbances on pigeon orientation. *J. Comp. Physiol.* **110**, 227–231. (160)

LARKIN, T. S. and KEETON, W. T. (1978) An apparent lunar rhythm in the day-to-day variations in the initial bearings of homing pigeons. In: Schmidt-Koenig, K. and Keeton, W. T. (eds) *Animal Migration, Navigation and Homing.* Springer, Heidelberg. pp. 92–106. (73, 165)

LEASK, M. J. M. (1977) A physico-chemical mechanism for magnetic field detection by migratory birds and homing pigeons. *Nature, Lond.* **267**, 144–146. (118)

*LEDNOR, A. J. (1982) Magnetic navigation in pigeons: possibilities and problems. In: Papi, F. and Wallraff, H. G. (eds) *Avian Navigation.* Springer, Heidelberg. pp. 109–119. (48, 162, 166, 179)

LEWIS, D. (1972) *We, the Navigators.* Australian National University Press, Canberra. (35, 46, 49, 75)

LIPP, H. P. and FREI, U. (1982) Variations of nocturnal performance in pigeons. In: Papi, F. and Wallraff, H. G. (eds) *Avian Navigation.* Springer, Heidelberg. pp. 271–280. (56)

LÖHRL, H. (1959) Zur Frage des Zeitpunkts einer Prägung auf die Heimatregion beim Halsbandschnäpper *(Ficedula albicollis). J. Orn.* **100**, 132–140. (212)

LOWENSTAM, H. A. (1962) Magnetite in denticle capping in recent chitons. *Geol. Soc. Am. Bull.* **73**, 435–438. (120)

LOWERY, G. H. and NEWMAN, R. J. (1966) A continent-wide view of bird migration on four nights in October. *Auk* **83**, 547–586. (59)

MATHER, J. G. (1981) Wheel-running activity: a new interpretation. *Mammal Review.* **11**, 41–51. (210)

MATHER, J. G. and BAKER, R. R. (1980) A demonstration of navigation by rodents using an orientation cage. *Nature, Lond.* **284**, 259–262. (23)

MATHER, J. G. and BAKER, R. R. (1981) Magnetic sense of direction in woodmice for route-based navigation. *Nature, Lond.* **291**, 152–155. (120, 122)

MATHER, J. G., BAKER, R. R. and KENNAUGH, J. H. (1982) Magnetic field detection by small mammals. EOS **63**, 156. (68, 120, 122)

MATTHEWS, G. V. T. (1951) The sensory basis of bird navigation. *J. Inst. Nav.* **4**, 260–275. (49, 153)

MATTHEWS, G. V. T. (1953) Sun navigation in homing pigeons. *J. exp. Biol.* **30**, 243–267. (49, 153)

MATTHEWS, G. V. T. (1955a) An investigation of the 'chronometer' factor in bird navigation. *J. exp. Biol.* **32**, 39–58. (167)

MATTHEWS, G. V. T. (1955b) *Bird Navigation*, 1st Ed. University Press, Cambridge. (49, 52, 163)

MATTHEWS, G. V. T. (1963) The astronomical bases of 'nonsense' orientation. *Proc. XIII Int. Orn. Cong.*, Ithaca pp. 415–429. (88)

MATTHEWS, G. V. T. (1968) *Bird Navigation*, 2nd Ed. University Press, Cambridge. (26, 78, 82, 95, 96, 126, 153, 165, 175, 176, 185, 195)

MATTHEWS, G. V. T. *in* Landsborough Thomson (1974) (5, 6, 14)

MATTHEWS, G. V. T. (1973) Biological clocks and bird migration. *In*: Mills, J. N. (ed.) *Biological Aspects of Circadian Rhythms*. Plenum, London. pp. 281–311. (92)

MATTHEWS, G. V. T. and COOK, W. A. (1982) Further complexities in the fixed "nonsense" orientations of mallard. *In*: Papi, F. and Wallraff, H. G. (eds) *Avian Navigation*. Springer, Heidelberg. pp. 283–289. (26)

MAUERSBERGER, G. (1957) Umsiedlungsversuche am Trauerschnäpper (*Muscicapa hypoleuca*), durchgeführt in der Sowjetunion. Ein Sammelreferat. *J. Orn.* **98**, 445–447. (212)

MCDONALD, D. L. (1972) Some aspects of the use of visual cues in directional training of homing pigeons. *In*: Galler, S. R., Schmidt-Koenig, K., Jacobs, G. J. and Belleville, R. E. (eds) *Animal Orientation and Navigation*. NASA SP-262 US Govt. Printing Office, Washington D.C. pp. 293–304. (82)

MEAD, C. J. and HARRISON, J. D. (1979) Sand Martin movements within Britain and Ireland. *Bird Study* **26**, 73–86. (211, 227)

MERKEL, F. W. (1978) Angle sense in painted quails—a parameter of geodetic orientation? *In*: Schmidt-Koenig, K. and Keeton, W. T. (eds) *Animal Migration, Navigation and Homing*. Springer, Heidelberg. pp. 269–274. (42)

MERKEL, F. W. and FROMME, H. G. (1958) Untersuchungen über das Orientierungsvermögen nächtlich ziehender Rotkehlchen, *Erithacus rubecula*. *Naturwiss.* **45**, 499–500. (96)

MERKEL, F. W. and WILTSCHKO, W. (1965) Magnetismus und Richtungsfinden zugunruhiger Rotkehlchen (*Erithacus rubecula*). *Vogelwarte* **23**, 71–77. (96, 97)

MEWALDT, L. R. (1963) California crowned sparrows return from Louisiana. *West. Bird band.* **38**, 1–4. (222)

MEWALDT, L. R. (1964) California sparrows return from displacement to Maryland. *Science* **146**, 941–942. (222)

MEWALDT, L. R., COWLEY, L. T. and PYONG-OH WON (1973) California sparrows fail to return from displacement to Korea. *Auk* **90**, 857–861. (222)

MICHENER, M. and WALCOTT, C. (1967) Homing of single pigeons—an analysis of tracks. *J. Exp. Biol.* **47**, 99–131. (26, 56)

MIDDENDORF, A. von (1855) Die Isepipetsen Russlands; Grundlagen zur Erforschung der Zugzeiten und Zugrichtungen der Vögel Russlands. *Mem Acad. Sci. St. Petersbourg* **8**, 1–143. (4, 95)

MITTLESTAEDT, H. and MITTLESTAEDT, M.-L. (1982) Homing by path integration. In: Papi, F. and Wallraff, H. G. (eds) *Avian Navigation*. Springer, Heidelberg. pp. 290–297. (172)

MONTGOMERY, K. C. and HEINEMANN, E. G. (1952) Concerning the ability of homing pigeons to discriminate patterns of polarized light. *Science* **116**, 454–456. (83)

MOORE, B. R. (1980) Is the homing pigeon's map geomagnetic? *Nature, Lond.* **285**, 69–70. (158)

MOORE, F. R. (1977) Geomagnetic disturbance and the orientation of nocturnally migrating birds. *Science* **196**, 682–684. (160)

MOORE, F. R. (1978) Sunset and the orientation of a nocturnal migrant bird. *Nature, Lond.* **274**, 154–156. (145)

*MOORE, F. R. (1980) Solar cues in the migratory orientation of the Savannah Sparrow (*Passerculus sandwichensis*). *Anim. Behav.* **28**, 684–704. (145)

NAZARCHUK, G. K., KISTYAKOVSKII, A. B., SMOGORZHEVSKII, L. A. and SHULMAN, L. M. (1969) Solar navigation of birds. *Vestnik Zoologii* **6**, 3–15. (153)

NEWTON, I. (1972) *Finches*. Collins, London. (227)

NISBET, I. C. T. (1955) Atmospheric turbulence in bird flight. *Brit. Birds.* **48**, 557–559. (215)

NISBET, I. C. T. and DRURY, W. H. (1968) Short-term effects of weather on bird migration: a field study using multivariate statistics. *Anim. Behav.* **16**, 496–530. (216, 228)

O'KEEFE, J. and NADEL, L. (1979) *The Hippocampus as a Cognitive Map*. Oxford University Press, London. (34)

PAPI, F. (1976) The olfactory navigation system of homing pigeons. *Verh. Deut. Zool. Ges.* **1976**, 184–205. (42)

*PAPI, F. (1982) Olfaction and homing in pigeons: ten years of experiments. In: Papi, F. and Wallraff, H. G. (eds) *Avian Navigation*. Springer, Heidelberg. pp. 149–159. (v, 42, 67, 68, 194, 195)

PAPI, F. and PARDI, L. (1959) Nuovi reperti sull'orientamento lunare di *Talitrus saltator* Montagu (Crustacea Amphipoda). *Z. vergl. Physiol.* **41**, 583–596. (92)

PAPI, F. and PARDI, L. (1963) On the lunar orientation of sandhoppers. *Biol. Bull.* **124**, 97–105. (92)

PAPI, F. and SYRJÄMÄKI, J. (1963) The sun-orientation rhythm of Wolf Spiders at different latitudes. *Arch. ital. Biol.* **101**, 59–77. (79)

*PAPI, F. and WALLRAFF, H. G (eds) (1982) *Avian Navigation*. Springer, Heidelberg. (v)

PAPI, F., FIORE, L., FIASCHI, V and BENVENUTI, S. (1972) Olfaction and homing in pigeons. *Monit. zool. ital. (N. S.)* **6**, 85–95. (63, 67)

PAPI, F., FIASCHI, V., BENVENUTI, S. and BALDACCINI, N. E. (1973) Pigeon homing: outward journey detours influence the initial orientation. *Monitore zool. ital. (N. S.)* **7**, 129–133. (67, 175)

PAPI, F., IOALÉ, P., FIASCHI, V., BENVENUTI, S. and BALDACCINI, N. E. (1974) Olfactory navigation of pigeons: the effect of treatment with odorous air currents. *J. Comp. Physiol.* **94**, 187–193. (63)

PAPI, F., IOALÉ, P., FIASCHI, V., BENVENUTI, S. and BALDACCINI, N. E. (1978a) Pigeon homing: cues detected during the outward journey influence initial orientation. In: Schmidt-Koenig, K. and Keeton, W. T. (eds) *Animal Migration, Navigation and Homing*. Springer, Heidelberg. pp. 65-77. (176, 177, 180)

PAPI, F., KEETON, W. T., BROWN, A. I. and BENVENUTI, S. (1978b) Do American and Italian pigeons rely on different homing mechanisms? *J. Comp. Physiol.* **128**, 303-317. (67, 68, 69, 176)

PAPI, F., MARIOTTI, G., FOÀ, A. and FIASCHI, V. (1980) Orientation of anosmatic pigeons. *J. Comp. Physiol.* **135**, 227-232. (67)

PARTRIDGE, L. 1978) Habitat selection. In: Krebs, J. R. and Davies, N. B. (eds) *Behavioural Ecology: an evolutionary approach*. Blackwell, London. pp. 351-376. (220)

PENNYCUICK, C. J. (1960a) The physical basis of astronavigation in birds: theoretical considerations. *J. exp. Biol.* **37**, 573-593. (153)

PENNYCUICK, C. J. (1960b) Sun navigation in birds. *Nature, Lond.* **188**, 1128. (153)

PENNYCUICK, C. J. (1961) Sun navigation in birds? *Nature, Lond.* **190**, 1026. (153)

PERDECK, A. C. (1958) Two types of orientation in migrating starlings, *Sturnus vulgaris* L., and chaffinches, *Fringilla coelebs* L., as revealed by displacement experiments. *Ardea* **46**, 1-37. (130, 205, 207, 225)

PERDECK, A. C. (1967) Orientation of starlings after displacement to Spain. *Ardea* **55**, 194-202. (225)

PERRY, A., BAUER, G. B. and DIZON, A. E. (1981) Magnetite in the green turtle. *EOS*, **62**, 849. (120)

PETERSEN, E. (1953) Orienteringsforsog med Haettemage (*Larus r. ridibundus* L.) og Stormmage (*Larus c. canus* L.) i vinterkvarteret. *Dansk. Orn. Foren. Tids.* **47**, 133-178. (222)

PHILLIPS, J. B. and ADLER, K. (1978) Directional and discriminatory responses of salamanders to weak magnetic fields. In: Schmidt-Koenig, K. and Keeton, W. T. (eds) *Animal Migration, Navigation and Homing*. Springer, Heidelberg. pp. 325-333. (32)

PHILLIPS, J. B. and WALDVOGEL, J. A. (1982) Reflected light cues generate the short-term deflector-loft effect. In: Papi, F. and Wallraff, H. G. (eds) *Avian Navigation*. Springer, Heidelberg. pp. 190-202. (65, 84)

PRESTI, D. and PETTIGREW, J. D. (1980) Ferromagnetic coupling to muscle receptors as a basis for geomagnetic field sensitivity in animals. *Nature, Lond.* **285**, 99-101. (120)

QUINE, D. B. (1979) *Infrasound detection and frequency discrimination by the homing pigeon*. Ph.D. Thesis, Cornell University. (72)

*QUINE, D. B. (1982) Infrasounds: a potential navigational cue for homing pigeons. In: Papi, F. and Wallraff, H. G. (eds) *Avian Navigation*. Springer, Heidelberg. pp. 373-376. (72)

QUINE, D. B. and KREITHEN, M. L. (1981) Frequency shift discrimination: can homing pigeons locate infrasounds by Doppler Shifts? *J. Comp. Physiol.* **141**, 153-155. (72)

RABØL, J. (1970) Displacement and phaseshift experiments with night-migrating Passerines. *Ornis Scand.* **I**, 27-43. (52, 206, 207)

RABØL, J. (1972) Displacement experiments with night-migrating Passerines (1972). *Z. Tierpsychol.* **30**, 14–25. (207)
*RABØL, J. (1978) One-direction orientation versus goal area navigation in migratory birds. *Oikos* **30**, 216–223. (52, 204, 206)
RALPH, C. J. and MEWALDT, L. R. (1975) Timing of site fixation upon wintering grounds in sparrows. *Auk* **92**, 698–705. (222)
RANVAUD, R. (1982) Study of homing pigeons in Brazil: some preliminary results. In: Papi, F. and Wallraff, H. G. (eds) *Avian Navigation*. Springer, Heidelberg. pp. 265–270. (80)
RAWSON, K. S. and RAWSON, A. M. (1955) The orientation of homing pigeons in relation to change in sun declination. *J. Orn.* **96**, 168–172. (154)
RICHARDSON, W. J. (1972) Autumn migration and weather in eastern Canada: a radar study. *Am. Birds* **26**, 10–17. (217)
RICHARDSON, W. J. (1974) Autumn migration over Puerto Rico and the Western Atlantic: a radar study. *Ibis* **118**, 309–32. (160)
RICHARDSON, W. J. (1978) Reorientation of nocturnal landbird migrants over the Atlantic Ocean near Nova Scotia in autumn. *Auk* **95**, 717–732. (59)
RICHARDSON, W. J. (1982a) Nocturnal landbird migration over Southern Ontario, Canada: Orientation vs. wind in autumn. In: Papi, F. and Wallraff, H. G. (eds) *Avian Navigation*. Springer, Heidelberg. pp. 15–27. (217)
*RICHARDSON, W. J. (1982b) Northeastward reverse migration of birds over Nova Scotia, Canada, in autumn. A radar study. *Behav. Ecol. Sociobiol.* **10**, 193–206. (218)
ROSENBLUM, B. and JUNGERMAN, R. (1981) Induction-based magnetoreception (theoretical analysis). *EOS* **62**, 849. (107, 118)
ROWAN, W. (1946) Experiments in bird migration. *Trans. Roy. Soc. Canada* **40**, 123–135. (128)
RÜPPELL, W. (1944) Versuche über Heimfinden ziehender Nebelkrähen nach Verfrachtung. *J. Orn.* **92**, 106–133. (225)
SAILA, S. B. and SHAPPY, R. A. (1963) Random movement and orientation in salmon migration. *J. Cons. perm. int. Explor. Mar.* **128**, 153–166. (20)
SAINT PAUL, U. von (1953) Nachweis der Sonnenorientierung bei nächtlich ziehenden Vögeln. *Behaviour* **6**, 1–7. (145)
SAINT PAUL, U. von (1956) Compass directional training of western meadowlarks (*Sturnella neglecta*). *Auk* **73**, 203–210. (78)
SAINT PAUL, U. von (1982) Do geese use path integration for walking home? In: Papi, F. and Wallraff, H. G. (eds) *Avian Navigation*. Springer, Heidelberg. pp. 298–307. (174)
SALMONSEN, F. (1969) *Vogelzug*. BLV Verlagsgesellschaft, Munchen. (227)
SARGENT, T. D. (1962) A study of homing in the bank swallow (*Riparia riparia*). *Auk* **79**, 234–246. (22, 57)
SAUER, E. G. F. (1957) Die Sternorientierung nächtlich ziehender Grasmücken (*Sylvia atricapilla, borin* und *curruca*). *Z. Tierpsychol.* **14**, 29–70. (87, 157)
SAUER, E. G. F. (1961) Further studies on the stellar orientation of nocturnally migrating birds. *Psychol. Forschung.* **26**, 224–244. (87, 166)
SAUER, E. G. F. and SAUER, E. M. (1955) Zur Frage der nächtlichen Zugorientierung von Grasmücken. *Rev. Suisse Zool.* **62**, 250–259. (87)

SAUER, E. G. F. and SAUER, E. M. (1960) Star navigation of nocturnal migrating birds. The 1958 planetarium experiments. *Cold Spring Harb. Symp. Quant. Biol.* **25**, 463–473. (87, 157, 166)

SAUER, E. G. F. and SAUER, E. M. (1962) Richtungsfinden und Raumbeherrschung von Zugvogeln nach Gestirnen. *Nach. d. Obers-gessellschaft Bremen* **50**, 12–16. (166)

SCHMIDT-KOENIG, K. (1958) Experimentelle Einflussnahme auf die 24-Stunden-Periodik bei Brieftauben und deren Auswirkungen unter besonderer Berücksichtigung des Heimfindevermögens. *Z. Tierpsychol.* **15**, 301–331. (78)

SCHMIDT-KOENIG, K. (1960) Internal Clocks and homing. *Cold Spring Harb. Symp. Quant. Biol.* **25**, 389–393. (78)

SCHMIDT-KOENIG, K. (1961) Die Sonne als Kompass im Heim-Orientierungssystem der Brieftauben. *Z. Tierpsychol.* **68**, 221–244. (78, 166)

SCHMIDT-KOENIG, K. (1963a) Sun compass orientation of pigeons upon displacement north of the arctic circle. *Biol. Bull.* **127**, 154–158. (78)

SCHMIDT-KOENIG, K. (1963b) Sun compass orientation of pigeons upon equatorial and trans-eqeatorial displacement. *Biol. Bull.* **124**, 311–321. (79)

SCHMIDT-KOENIG, K. (1966) Über die Entfernung als Parameter bei der Anfangsorientierung der Brieftaube. *Z. vergl. Physiol.* **52**, 33–55. (195)

SCHMIDT-KOENIG, K. (1972) New experiments on the effects of clock shifts on homing pigeons. In: Galler, S. R., Schmidt-Koenig, K., Jacobs, G. J. and Belleville, R. E. (eds) *Animal Orientation and Navigation*. NASA SP-262 US Govt. Printing Office, Washington D.C. pp. 275–282. (166)

SCHMIDT-KOENIG, K. (1979) *Avian Orientation and Navigation*. Academic Press, London. (79, 80, 81, 119, 156, 179, 195)

SCHMIDT-KOENIG, K. and KEETON, W. T. (1977) Sun compass utilization by pigeons wearing frosted contact lenses. *Auk* **94**, 143–145. (199)

SCHMIDT-KOENIG, K. and KEETON, W. T. (eds) (1978) *Animal Migration, Navigation and Homing*. Springer, Heidelberg. (6)

SCHMIDT-KOENIG, K. and PHILLIPS, J. B. (1978) Local anaesthesia of the olfactory membrane and homing in pigeons. In: Schmidt-Koenig, K. and Keeton, W. T. (eds) *Animal Migration, Navigation and Homing*. Springer, Heidelberg. pp. 119–124. (66, 69)

SCHMIDT-KOENIG, K. and SCHLICHTE, H. J. (1972) Homing in pigeons with impaired vision. *Proc. Nat. Acad. Sci. USA* **69**, 2446–2447. (61)

SCHMIDT-KOENIG, K. and WALCOTT, C. (1978) Tracks of pigeons homing with frosted lenses. *Anim. Behav.* **26**, 480–486. (158)

SCHNEIDER, F. (1961) Die Einflussung der Aktivat des Maikafers durch Veranderung der gegenseitigen Lage magnetischer und elektrischer Felder. *Mitteil. Schweiz. Ent. Ges.* **33**, 232–237. (107)

SCHREIBER, B. and ROSSI, O. (1976) Correlation between race arrivals of homing pigeons and solar activity. *Boll. Zool.* **43**, 317–320. (160)

SCHREIBER, B. and ROSSI, O. (1978) Correlation between magnetic storms due to solar spots and pigeon homing performances. *IEEE Trans. Magn.* **14**, 961–963. (160)

SCHÜZ, E. (1949) Die Spat-Auflassung ostpreussischer Jungstorche in West-Deutschland durch die Vogelwarte Rossitten 1933. *Vogelwarte* **15**, 63–78. (128)

SCHÜZ, E. (1950) Fruh-Auflassung ostpreussischer Jungstorche in West-Deutschland durch die Vogelwarte Rossitten 1933–36. *Bonner zool. Beitr.* **1**, 239–253. (128)

SCHWÄRTZ, P. (1963) Orientation experiments with northern Waterthrushes wintering in Venezuela. *Proc. XIII Int. Orn. Congr., Ithaca* pp. 481–487. (222)

SEMM, P., SCHNEIDER, T., VOLLRATH, L. and WILTSCHKO, W. (1982) Magnetic sensitive pineal cells in pigeons. *In*: Papi, F. and Wallraff, H. G. (eds) *Avian Navigation*. Springer, Heidelberg. pp. 329–337. (124)

SHCHERBAKOV and PALIVANOV *in* Mauersberger (1957). (212)

SHUMAKOV, M. E. (1965) Preliminary results of the investigation of migrational orientation of passerine birds by the round-cage method. (In Russian). *Bionica* 371–378. (145)

SLEPIAN, J. (1948) Physical basis of bird navigation. *J. Appl. Phys.* **19**, 306. (117)

SMITH, J. N. M. and SWEATMAN, H. P. A. (1974) Food searching behaviour of titmice in patchy environments. *Ecology* **55**, 1216–1232. (81)

SOTTHIBANDHU, S. and BAKER, R. R. (1979) Celestial orientation by the large yellow underwing moth, *Noctua pronuba* L. *Anim. Behav.* **27**, 786–800. (92)

SOUTHERN, W. E. (1969) Orientation behaviour of gull chicks. *Condor* **71**, 418–425. (114)

SOUTHERN, W. E. (1970) En route behavior of homing herring gulls as determined by radio-tracking. *Wilson Bull.* **82**, 189–200. (58)

SOUTHERN, W. E. (1971) Gull orientation by magnetic cues: a hypothesis revisited. *Ann. N.Y. Acad. Sci.* **188**, 295–311. (114, 115, 164)

SOUTHERN, W. E. (1972) Magnets disrupt the orientation of juvenile Ring-Billed Gull chicks. *BioScience* **22**, 476–479. (115)

SOUTHERN, W. E. (1978) Orientation responses of Ring-Billed Gull chicks: a re-evaluation. *In*: Schmidt-Koenig, K. and Keeton, W. T. (eds) *Animal Migration, Navigation and Homing*. Springer, Heidelberg. pp. 311–317. (114, 137)

SOUTHERN, W. E., HANZELY, L., BAILEY, R. L. and MOLSEN, D. V. (1982) Is the avian eye a magnetic sensor? *In*: Papi, F. and Wallraff, H. G. (eds) *Avian Navigation*. Springer, Heidelberg. pp. 344–351. (117)

SRIVASTAVA, B. J. and SAXENA, S. (1980) Geomagnetic-Biological correlations: some new results. *Indian J. Radio Space Phys.* **9**, 121–126. (115)

STEWART, O. J. A. (1957) A bird's inborn navigational device. *Trans. Ky. Acad. Sci.* **18**, 78–84. (117)

TALKINGTON, L. (1967) Bird navigation and geomagnetism (abstract). *Am. Zool.* **7**, 199. (117, 161)

VERHEIJEN, F. J. (1980) The moon: a neglected factor in studies on collision of nocturnal migrant birds with tall lighted structures and with aircraft. *Die Vogelwarte* **30**, 305–320. (94)

VERHEIJEN, F. J. (1981) Bird kills at lighted man-made structures: not on nights close to a full moon. *Am. Birds.* **35**, 251–254. (94)

VERHEYEN, R. (1950) La Cigogne blanche dans son quartier d'hiver. *Gerfaut* **40**, 1–17. (130)

VERMA, S. D., SINGHAL, K. P. and SINHA, V. (1977) Geomagnetic–Biological correlations. *Indian J. Radio Space Phys.* **6**, 12–17. (116)

VERMA, S. D., SINHA, V. and NIGAM, M. (1978) Geomagnetic activity and admissions for myocardial infarction. *Indian J. Radio Space Phys.* **7**, 119—125. (115)

VIEHMANN, W. (1979) The magnetic compass of Blackcaps (*Sylvia atricapilla*). *Behaviour* **68**, 24—30. (97, 101)

VIEHMANN, W. (1982) Interrelation of magnetic compass, star orientation, and the sun in the orientation of blackcaps and robins. In: Papi, F. and Wallraff, H. G. (eds) *Avian Navigation*. Springer, Heidelberg. pp. 59—67. (147)

VIGUIER, C. (1882) Le sens d'orientation et ses organes chez les animaux et chez l'homme. *Rev. Phil.* **14**, 1—36. (117, 151)

VISALBERGHI, E. and ALLEVA, E. (1979) Magnetic influences on pigeon homing. *Biol. Bull.* **156**, 246—256. (141, 163)

*VISALBERGHI, E., FOÀ, A., BALDACCINI, N. E. and ALLEVA, E. (1978) New experiments on the homing ability of the rock pigeon. *Monitore zool. ital. (N.S.)* **12**, 199—209. (200)

VLEUGEL, D. A. (1953) Über die wahrscheinliche Sonnen-Orientierung einiger Vogelarten auf dem Zuge. *Orn. Fen.* **30**, 41—51. (145)

VLEUGEL, D. A. (1959) Über die wahrscheinlichste Method der Wind-Orientierung ziehender Buchfinken. *Orn. Fenn.* **36**, 78—88. (145)

VLEUGEL, D. A. (1962) Über nachtlichen Zug von Drosseln und ihre Orentierung. *Vogelwarte* **21**, 307—313. (145)

WAGNER, G. (1972) Topography and pigeon orientation. In: Galler, S. R., Schmidt—Koenig, K., Jacobs, G. J. and Belleville, R. E. (eds) *Animal Orientation and Navigation*. NASA SP-262 US Govt. Printing Office, Washington D.C. pp. 249—273. (57)

WAGNER, G. (1976) Das Orientierungsverhalten von Brieftauben im erdmagnetisch gestorten Gebiete des Chasseral. *Rev. Suisse Zool.* **83**, 883—890. (161)

WAGNER, G. (1978) Homing pigeons flight over and under low stratus. In: Schmidt—Koenig, K. and Keeton, W. T. (eds) *Animal Migration, Navigation and Homing*. Springer, Heidelberg. pp. 455—470. (57)

*WALCOTT, B. and WALCOTT, C. (1982) A search for magnetic field receptors in animals. In: Papi, F. and Wallraff, H. G. (eds) *Avian Navigation*. Springer, Heidelberg. pp. 338—343. (68, 121, 123)

WALCOTT, C. (1977) Magnetic fields and the orientation of homing pigeons under sun. *J. exp. Biol.* **70**, 105—123. (141, 163)

WALCOTT, C. (1978) Anomalies in the earth's magnetic field increase the scatter of pigeons' vanishing bearings. In: Schmidt-Koenig, K. and Keeton, W. T. (eds) *Animal Migration, Navigation and Homing*. Springer, Heidelberg. pp. 143—151. (161, 164)

WALCOTT, C. in Schmidt-Koenig (1979). (176)

WALCOTT, C. (1980a) Homing pigeon vanishing bearings at magnetic anomalies are not altered by bar magnets. *J. exp. Biol.* **86**, 349—352. (161)

WALCOTT, C. (1980b) Magnetic orientation in homing pigeons. *IEEE Trans. Magn.* **16**, 1008—1013. (158, 164)

*WALCOTT, C. (1982) Is there evidence for a magnetic map in homing pigeons? In: Papi, F. and Wallraff, H. G. (eds) *Avian Navigation*. Springer, Heidelberg. pp. 99—108. (112, 161)

WALCOTT, C. and GOULD, J. L. in Walcott (1982). (163, 164)

WALCOTT, C. and GREEN, R. P. (1974) Orientation of homing pigeons is altered by a change in the direction of an applied magnetic field. *Science* **184**, 180–182. (141, 163)

WALCOTT, C. and MICHENER, M. C. (1971) Sun navigation in homing pigeons—attempts to shift sun-coordinates. *J. exp. Biol.* **54**, 291–316. (155, 166, 168)

WALCOTT, C., GOULD, J. L. and KIRSCHVINK, J. L. (1979) Pigeons have magnets. *Science* **205**, 1027–1029. (120, 121)

WALDVOGEL, J. A. and PHILLIPS, J. B. (1982) Pigeon homing: new experiments involving permanent-resident deflector-loft birds. In: Papi, F. and Wallraff, H. G. (eds) *Avian Navigation*. Springer, Heidelberg. pp. 179–189. (65)

WALDVOGEL, J. A., BENVENUTI, S., KEETON, W. T. and PAPI, F. (1978) Homing pigeon orientation influenced by deflected winds at home loft. *J. Comp. Physiol.* **128**, 297–301. (64)

WALKER, M. M. and DIZON, A. E. (1981) Identification of magnetite in tuna. *EOS* **62**, 850. (68, 120, 123)

WALLRAFF, H. G. (1960) Können Grasmücken mit Hilfe des Sternenhimmels navigieren? *Z. Tierpsychol.* **17**, 165–177. (157, 166)

WALLRAFF, H. G. (1966) Über die Heimfineleistungen von Brieftauben nach Haltung in verschiedenartig abgeschirmten Volieren. *Z. vergl. Physiol.* **52**, 215–259. (193)

WALLRAFF, H. G. (1967) The present status of our knowledge about pigeon homing. In: Snow, D. W. (ed.) *Proc. XIV Int. Orn. Congr., Oxford.* pp. 331–358. (193)

WALLRAFF, H. G. (1969) Über das Orientierungsvermögen von Vögeln unter natürlichen und Künstlichen Sternmustern. Dressurversuche mit Stockenten. *Ver. Deut. Zool. Ges. Innsbruck*, **1968**, 348–357. (89)

WALLRAFF, H. G. (1970a) Weitere Volierenversuche mit Brieftauben: Wahrscheinlicher Einfluss dynamischer Faktoren der Atmosphäre auf die Orientierung. *Z. vergl. Physiol.* **68**, 182–201. (193)

WALLRAFF, H. G. (1970b) Über die Flugrichtungen verfrachteter Brieftauben in Abhängigkeit vom Heimatort und vom Ort der Freilassung. *Z. Tierpsychol.* **27**, 303–351. (193, 195)

WALLRAFF, H. G. (1972) An approach toward an analysis of the pattern recognition involved in the stellar orientation of birds. In: Galler, S. R., Schmidt-Koenig, K., Jacobs, G. J. and Belleville, R. E. (eds) *Animal Orientation and Navigation*. NASA SP-262 US Govt. Printing Office, Washington D.C. pp. 211–212. (89)

WALLRAFF, H. G. (1974) *Das Navigationssystem der Vogel*. Oldenbourg, Munchen. (34, 42, 52, 185, 195)

WALLRAFF, H. G. (1977) Selected aspects of migratory orientation in birds. *Vogelwarte Sonderh* **29**, 64–76. (59)

WLLRAFF, H. G. (1978a) Proposed principles of magnetic field perception in birds. *Oikos* **30**, 188–194. (103, 107, 117, 118)

WALLRAFF, H. G. (1978b) Preferred compass directions in initial orientation of homing pigeons. In: Schmidt-Koenig, K. and Keeton, W. T. (eds) *Animal Migration, Navigation and Homing*. Springer, Heidelberg. pp. 171–183. (26, 185, 186, 191, 192, 193, 195)

WALLRAFF, H. G. (1979) Goal-oriented and compass-oriented movements of displaced homing pigeons after confinement in differentially shielded aviaries. *Behav. Ecol. Sociobiol.* **5**, 201–225. (193)

WALLRAFF, H. G. (1980a) Olfaction and homing in pigeons: nerve-section experiments, critique, hypotheses. *J. Comp. Physiol.* **139**, 209–224. (67, 68)

WALLRAFF, H. G. (1980b) Does pigeon homing depend on stimuli perceived during displacement? I. Experiments in Germany. *J. Comp. Physiol.* **139**, 193–201. (184)

WALLRAFF, H. G. (1981a) The olfactory component of pigeon navigation: steps of analysis. *J. Comp. Physiol.* **143**, 411–422. (157, 195, 196)

WALLRAFF, H. G. (1981b) Clock-controlled orientation in space. In: Aschoff, J. (ed.) *Handbook of Behavioral Neurobiology*, Vol. 4. Plenum, New York. pp. 299–309. (92)

*WALLRAFF, H. G. (1982) Homing to Würzburg: an interim report on long-term analyses of pigeon navigation. In: Papi, F. and Wallraff, H. G. (eds) *Avian Navigation*. Springer, Heidelberg. pp. 211–221. (185, 186, 192, 195)

WALLRAFF, H. G. and FOÀ, A. (1981) Pigeon navigation: charcoal filter removes relevant information from environmental air. *Behav. Ecol. Sociobiol.* **9**, 67–77. (181)

WALLRAFF, H. G. and GELDERLOOS, O. G. (1978) Experiments on migratory orientation of birds with simulated stellar sky and geomagnetic field: method and preliminary results. *Oikos* **30**, 207–215. (103, 109)

WALLRAFF, H. G. and GRAUE, L. C. (1973) Orientation of pigeons after transatlantic displacement. *Behaviour* **44**, 1–35. (195)

WALLRAFF, H. G. and HUND, K. (1982) Homing experiments with starlings (*Sturnus vulgaris*) subjected to olfactory nerve section. In: Papi, F. and Wallraff, H. G. (eds) *Avian Navigation*. Springer, Heidelberg. pp. 313–318. (68)

WALLRAFF, H. G., FOÀ, A. and IOALÉ, P. (1980) Does pigeon homing depend on stimuli perceived during displacement? II. Experiments in Italy. *J. Comp. Physiol.* **139**, 203–208. (179, 184)

WALLRAFF, H. G., PAPI, F., IOALÉ, P. and FOÀ, A. (1981) On the spatial range of pigeon navigation. *Monit. zool. ital. (N.S.)* **15**, 155–161. (157, 180, 195, 196)

WENZEL, B. M. (1982) Functional status and credibility of avian olfaction. In: Papi, F. and Wallraff, H. G. (eds) *Avian Navigation*. Springer, Heidelberg. pp. 352–361. (74)

*WHITEN, A. (1978) Operant studies on pigeon orientation and navigation. *Anim. Behav.* **26**, 571–610. (156, 168)

WHITNEY, L. F. (1963) Landmarks and homing. *Amer. Racing Pigeon News* **79**, 8–10. (56)

WILKINSON, D. H. (1949) Some physical principles of bird orientation. *Proc. Linn. Soc. Lond.* **160**, 94–99. (46, 117)

WILKINSON, D. H. (1952) The random element in bird 'navigation'. *J. exp. Biol.* **29**, 532–560. (20)

WILLIAMS, T. C. and TEAL, J. T. (1973) The flight of blindfolded birds. *Bird Band.* **44**, 102–109. (215)

WILLIAMS, T. C. and WILLIAMS, J. M. (1970) Radio tracking of homing and feeding flights of a neotropical bat, *Phyllostomus hastatus*. *Anim. Behav.* **18**, 302–309. (75)

WILLIAMS, T. C. and WILLIAMS, J. M. (1978) Orientation of transatlantic migrants. In: Schmidt-Koenig, K. and Keeton, W. T. (eds) Animal Migration, Navigation and Homing. Springer, Heidelberg. pp. 239–251. (218)
WILLIAMS, T. C., WILLIAMS, J. M., TEAL, J. M. and KANWISHER, J. W. (1974) Homing flights of herring gulls under low visibility conditions. Bird Band. 45, 106–114. (58)
WILLIAMS, T. C., BERKELEY, P. and VICTOR, H. (1977) Bird Band. 48, 1–10. (218)
WILTSCHKO, R. and WILTSCHKO, W. (1978a) Relative importance of stars and the magnetic field for the accuracy of orientation in night migrating birds. Oikos 30, 195–206. (144)
WILTSCHKO, R. and WILTSCHKO, W. (1978b) Evidence for the use of magnetic outward-journey information in homing pigeons. Naturwiss. 65, 112–113. (176)
WILTSCHKO, R. and WILTSCHKO, W. (1980) Naturwiss. 67, 512. (81)
*WILTSCHKO, R. and WILTSCHKO, W. (1981) The development of sun compass orientation in young homing pigeons. Behav. Ecol. Sociobiol. 9, 135–141. (176)
WILTSCHKO, R., WILTSCHKO, W. and KEETON, W. T. (1978) Effect of outward journey in an altered magnetic field on the orientation of young homing pigeons. In: Schmidt-Koenig, K. and Keeton, W. T. (eds) Animal Migration, Navigation and Homing. Springer, Heidelberg. pp. 152–161. (176)
WILTSCHKO, R., NOHR, D. and WILTSCHKO, W. (1981) Pigeons with a deficient sun compass use the magnetic compass. Science, 214, 343–345. (136, 142)
WILTSCHKO, W. (1968) Über den Einfluss statischer Magnetfelder auf die Zugorientierung der Rotkehlchen (Erithacus rubecula). Z. Tierpsychol. 25, 537–558. (97, 104)
WILTSCHKO, W. (1972) The influence of magnetic total intensity and inclination on directions preferred by migrating European robins (Erithacus rubecula). In: Galler, S. R., Schmidt-Koenig, K., Jacobs, G. J. and Belleville, R. E. (eds) Animal Orientation and Navigation. NASA SP-262 US Govt. Printing Office, Washington D.C. pp. 569–578. (98, 104, 164)
WILTSCHKO, W. (1974) Der Magnetkompass der Gartengrasmücke (Sylvia borin). J. Orn. 115, 1–7. (97)
*WILTSCHKO, W. (1978) Further analysis of the magnetic compass of migratory birds. In: Schmidt-Koenig, K. and Keeton, W. T. (eds) Animal Migration, Navigation and Homing. Springer, Heidelberg. pp. 302–310. (104), 164)
*WILTSCHKO, W. (1982) The migratory orientation of garden warblers, Sylvia borin. In: Papi, F. and Wallraff, H. G. (eds) Avian Navigation. Springer, Heidelberg. pp. 50–58. (105, 138, 139, 144)
WILTSCHKO, W. and GWINNER, E. (1974) Evidence for an innate magnetic compass in Garden Warblers. Naturwiss. 61, 406. (131)
WILTSCHKO, W. and MERKEL, F. W. (1966) Orientierung zugunruhiger Rotkehlchen im statischen Magnetfeld. Verh. deutsch. Zool. Jena 1965, 362–367. (97)
WILTSCHKO, W. and MERKEL, F. W. (1971) Zugorientierung von Dorngrasmücken (Sylvia communis) im Erdmagnetfeld. Vogelwarte 26, 245–249. (97)

WILTSCHKO, W. and WILTSCHKO, R. (1972) Magnetic compass of European robins. *Science* **176**, 62–64. (96)

WILTSCHKO, W. and WILTSCHKO, R. (1975a) The interaction of stars and magnetic field in the orientation system of night migrating birds. Part I. Autumn experiments with European warblers (Gen. *Sylvia*). *Z. Tierpsychol.* **37**, 337–355. (97, 142, 144)

WILTSCHKO, W. and WILTSCHKO, R. (1975b) The interaction of stars and magnetic field in the orientation system of night migrating birds. Part II. Spring experiments with European robins (*Erithacus rubecula*). *Z. Tierpsychol.* **39**, 265–282. (143)

WILTSCHKO, W. and WILTSCHKO, R. (1976a) Interrelation of magnetic compass and star orientation in night-migrating birds. *J. Comp. Physiol.* **109**, 91–99. (143)

WILTSCHKO, W. and WILTSCHKO, R. (1976b) Die Bedeutung des Magnetkompasses fur die Orientierung der Vogel. *J. Orn.* **117**, 362–387. (163)

WILTSCHKO, W. and WILTSCHKO, R. (1978) A theoretical model for homing in birds. *Oikos*, **30**, 177–187. (15, 42, 52, 131, 206, 226)

WILTSCHKO, W. and WILTSCHKO, R. (1981) Disorientation of inexperienced young pigeons after transportation in total darkness. *Nature, Lond.* **291**, 433–434. (119, 179)

*WILTSCHKO, W. and WILTSCHKO, R. (1982) The role of outward-journey information in the orientation of homing pigeons. *In*: Papi, F. and Wallraff, H. G. (eds) *Avian Navigation*. Springer, Heidelberg. pp. 239–252. (182, 211)

WILTSCHKO, W., WILTSCHKO, R. and KEETON, W. T. (1976) Effects of a "permanent" clock-shift on the orientation of young homing pigeons. *Behav. Ecol. Sociobiol.* **1**, 229–243. (80, 134)

WILTSCHKO, W., GWINNER, E. and WILTSCHKO, R. (1980) The effect of celestial cues on the ontogeny of non-visual orientation in the Garden Warbler (*Sylvia borin*). *Z. Tierpsychol.* **53**, 1–8. (131)

WINDSOR, D. M. (1975) Regional expression of directional preferences by experienced homing pigeons. *Anim. Behav.* **23**, 335–343. (191, 195)

YEAGLEY, H. L. (1947) A preliminary study of a physical basis of bird navigation. *J. Appl. Phys.* **18**, 1035–1063. (117, 152)

YEAGLEY, H. L. (1951) A preliminary study of a physical basis of bird navigation II. *J. Appl. phys.* **22**, 746–760. (152, 160)

YEOWART, N. S. and EVANS, M. J. (1974) Thresholds of audibility for very low frequency pure tones. *J. Acoust. Soc.* **55**, 814–818. (70)

YODLOWSKI, M. L., KREITHEN, M. L. and KEETON, W. T. (1977) Detection of atmospheric infrasound by homing pigeons. *Nature, Lond.* **265**, 725–726. (70, 71)

YORKE, E. D. (1981) Time-varying magnetic field effects on hypothetical magnetite magnetoreceptors. *EOS* **62**, 850. (120)

ZOEGER, J. and FULLER, M. (1980) Magnetic material in the head of a dolphin. *EOS* **61**, 225. (120)

Subject index

Age, and accuracy of orientation 182–4
Albatross 74
Anaesthesia
 general 175
 local 64, 66–9
Anas crecca (Teal) 89
Anas platyrhynchos (Mallard) 26–7, 88–9
Angles, judgement of 39, 42
Ants 85
Apodemus sylvaticus (Woodmouse) 23, 122–3
Arctosa spp. (Wolf spiders) 79
Aristotle 1
Aviary experiments 63–6, 142, 186, 193–5
Axis of rotation (of night sky) 46–7, 90–91, 138–9

Bacteria 120
Bar magnets 106, 112, 134–6, 141, 160, 163
Barometric pressure
 detection of 149
 weather forecasting, and 149
Barrington, Daines 2
Bats 55–6, 75
Bed orientation 109–111, 115–6
Bees 7, 80, 82, 85, 120
Biological clock 35–6
Bird ringing (banding) 3
Blindfolded birds 215
Breeding chorus, as landmark 75
Bus experiments 160, 172–3

Callithrix sp. (Marmoset) 123
Cap-and-collar coils 140–1
cardiac conditioning 71, 84, 89, 105
Carduelis carduelis (goldfinch) 203
Cats 4, 55
Cattle 4
Celestial equator 46–7
Celestial rotation 46–7
Chair experiments 105–112, 115
Chitons (Polyplacophora) 120
Chronometer
 of birds 50, 168–9
 of mariners 49
Ciconia ciconia (White stork) 128–30
Circannual clock 138, 202–4

Circular statistics 22
 second-order analysis 96–7
Clock-and-compass model (of migration) 204–7, 209, 219
Clock, biological 35–6
 circadian 82
 circannual 138, 202–4
 lunar 92
Clock-shifting experiments 36, 57, 62, 77–8, 81, 136, 141–2, 166, 168–9, 174, 192, 198–9
Cochlea 71
Columba livia (Homing pigeon, Racing pigeon, Rock dove) 5–7, 11–15, 21–2, 25–6, 40–41, 44, 56–8, 61–74, 78–86, 105, 112–4, 119–121, 123–4, 126, 134–42, 152–64, 169–201, 228, 230
Columba palumbus (Wood pigeon) 60, 138
Columella 71
Compass
 innate 127–39
 in 'non-migrants' 133–4
 magnetic, 95–127, 134–145, 148–9, 163–4
  disruption of, 103–116, 160–4
  experimental difficulties 103–5
  polarity vs inclination 97–102
 moon 92–4, 125
  mimicry by artificial lights 92–4
  time compensation 92
 polarization patterns 82–5, 125, 147
  detection by ultra violet light 85
 star 87–91, 125, 138–9, 142–9, 210
  development of 88–90, 137–9
  time compensation 88–90
 sun 35–6, 77–82, 124–6, 134–6, 140–2, 145–9, 168–9, 200
  accuracy of 81–2
  development of 80–81, 134–7
  time compensation 78–81, 134–6
 wind 149–50
Compass directions 22, 26, 29, 35, 37, 39, 57, 70, 156, 172
Conjugate point 51–2, 153
Cook, Captain 35
Coriolis force (see: Latitude)
Corvus brachyrhynchos (Prairie crow) 128
Corvus corone cornix (Hooded crow) 130–1, 225–6
Cuculus canorus (Cuckoo) 4, 127–8, 131, 134

Subject index

Cuvier, Baron 2

Darwin, Charles 4
Day length 138, 204
Declination (of magnetic field) 52–3, 165, 170
Deer 55
Deflector lofts 64–6
Deliberate error, as navigational strategy 49
Detour experiments 174–6
Diomedea exulans (Wandering albatross) 16
Direction maintenance 144–5
Disorientation, zone of 195–6
Displacement–release experiments 10
  as enforced exploration 19–20
  (see also: Homing experiments)
Distance of release 21, 195–7
Distant landmarks 40–43, 56–60
Dogs 4, 55
Dolichonyx oryzivorus (Bobolink) 80
Dolphins 120
Doppler shift 72
Downwind migration 217
Drift (see: Wind-drift)

Earth's magnetic field (see: Geomagnetic field)
Echolocation 55, 75
Electrostatic fields 107–9
  and clothing 108–9
Environmental gradients 52
Equinoxial tests (of sun-arc hypothesis) 153–5
Erithacus rubecula (European robin) 2, 31, 96–9, 101–4, 119, 143–4, 147, 164, 222
E-vector 83–5
Experimental design 66–9
  (see also: Redundancy, principle of)
Exploration 15, 17, 33, 36–7, 41, 127, 172, 206, 209–15,
  by adults 226
Exploration model (of migration) 52, 204–27

Familiar area 14–19, 34, 37, 42–3, 183–4
Familiar area map 42–5, 49, 54, 182, 206, 211
Familiarity, effect on accuracy 192
Ficedula albicollis (Collared flycatcher) 202
Ficedula hypoleuca (Pied flycatcher) 103, 131, 137–8, 202, 212
Fidelity (to sites) 15
Finches (Fringillidae) 227
Fish 80, 83, 118

Fog 58, 61
Food, training to 32, 78–80
Forster, John Reinhold 2
Fregata spp. (Frigate birds) 6
French Revolution 3
Frogs 55, 75
Frosted-glass contact lenses 61–3, 158, 160

Geese 174–5
Geomagnetic field 39
  reversal of, 99–102
  (see also: Declination; Inclination; Intensity; Polarity; Pole)
Goal-area navigation model (of migration) 52, 204–207, 219
Goal orientation 10, 14, 25–6, 33
Glaucomys volans (Flying squirrel) 82
Gradients 52, 171–2
Green turtle 120
Grid maps 45–52, 151–70, 195
  (see also: Latitude, Longitude)
Grus grus (Crane) 60
Guinea pig 124
Gyroscope, internal 38–9

Harderian gland 121
Hard-weather migration 220–21
Hawks 62
'Heavy' water 169
Helmholtz coils 28–9, 96, 176, 178
Hibernation 2
Hierarchy of cues 125–7
Hippocampus, and mental maps 34
Hirundo rustica (Barn swallow) 2–3, 6–7, 15, 21
Home range 15, 20, 43
Homer 1
Homeward component (see: Circular statistics)
Homing, by pigeons at night 56
Homing experiments 19–21, 37
Homing pigeons (see: *Columba livia*)
Homing speed 13–14, 160
Homing success 20
Homo sapiens (Man) 4–8, 14–15, 22, 34–5, 43, 54–6, 62, 68, 70, 72, 75, 83–4, 90, 105, 108–123, 127, 144, 160, 172–3, 195
  Aborigines 35
  city dwellers 35, 127
  mariners 44, 50
  Polynesians 35, 46, 75
Horizon
  distance of 43
  internal 155

Hummingbirds 85
Hylocichla fuscescens (Veery) 224

Inclination (of geomagnetic field) 48, 97–101
Inertial navigation (see: Navigation)
Infrasounds 44, 70–73, 149, 210
Insects 85, 107, 118
Instinct 4
Intensity (of geomagnetic field) 48, 50–51, 97, 104, 113, 157–64
  diel variation in 51
  magnetic storms, and 113–5, 158–64
Invertebrates 6, 14, 17, 82

Jenner, Edward 2

Kinaesthesis 62
K-index (of magnetic storm activity), 114, 158–60

Lagena 71
Landmarks 34, 36, 38–45, 75–6, 127, 142, 190–2, 195
  acoustic 44, 56, 69–73, 192, 210
  gravity 73
  magnetic 73–4
  oceanic 75
  olfactory 42, 44, 56, 63–9, 74–5, 192–5
  visual 38, 52–4, 56–62, 75
    at night 59, 60, 75
  wind 75
Landmark scanning
  migration, during 60
  release, on 57–8
Laniarius erythrogaster (Black-headed gonolek) 82
Lanius collurio (Red-backed shrike) 145
Larus argentatus (Herring gull) 58, 215
Larus delawarensis (Ring-billed gull) 114, 137–8, 160, 164
Larus fuscus (Lesser black-backed gull) 207–9, 218
Larus spp 222
Latitude, determination of 45–48
  by Coriolis force 46–8, 152–3, 170
  by geomagnetic field 48, 51, 158–64
  by stars 46, 157–8
  by sun 45, 153–7, 170
Leading lines 214
Least navigation hypothesis 125–6, 140
Lighthouses 93–4

Light traps 92
Lizards 81, 85
Loft, mobile 153
Longitude, determination of 49–50
  by geomagnetic field 50, 165–6
  by gravity gradients 165
  by stars 49–50, 166
  by sun 49–50, 166

Magnetic anomalies 160–4, 198
Magnetic maps 158–66, 229–30
Magnetic sense (see: Compass; Latitude; Longitude; Magnetoreception)
Magnetic storms 113–5, 158–64
Magnetite 120–124
  manipulation of particles? 109–113, 116
Magnetoreception 4, 176–9
  conditioning, and 105
  disruption of 103–116
  improvement of 113
  induction, by 107, 117–8
  Man, by 105–112, 119–122
  movement, the necessity for 105
  optical resonance, by 118–9
  sensitivity of 158–62
Magnus, Olaus 2
Magnetoreceptor, the 68, 116–124
Mammals 4, 23, 34, 134
Map-and-compass 172
Mental image (of habitat) 220, 222
Mean vector (in circular statistics) 24
Mental maps 34–36, 55, 171
Mice 7, 22, 55, 68, 120
Microtus pennsylvanicus (Meadow vole) 21
Midnight sun 79
Migration circuit 16
Migration threshold 220
Migratory 'divide' 129–30
Migratory restlessness 30–32, 87, 138, 203, 206, 220
  in spring 220
Mirror experiments 77, 155
Monarch butterfly 120
Moon (see: Compass)
Moon phase 93–4, 165
Moonwatching 58
Mosaic map 34–6, 76, 171, 199, 211, 215, 227–8
Moths 92, 94, 144
Motivation to home 20
Myotis lucifugus (Little brown bat) 21

Nasal tubes 67–8
Navigation, (see also: Grid maps; Landmarks)

definition of 9
geodetic 32, 41–2
inertial 38, 172
location-based 37, 41–5
route-based 37–42, 49, 62, 156–7, 173–4, 184, 197–8, 211, 213–5
by magnetism 176–9, 182
by smell 42–44, 180–2
Navigation, while homing 197–9
Nerve bisection and olfaction 66–8
Newton, Alfred 4
Newts 85
Noctua pronuba (Large yellow underwing moth) 92
Nonsense orientation 26–7, 185

Oceanic landmarks (see: Landmarks)
Oil droplets, retinal 84
Olfaction (see also: Landmarks; Navigation, route-based)
interspecific differences 74–5
Olive oil 63
Operant conditioning 82, 156–7, 169
Orchestoides sp. 92
Orientation 10, 12–13, 25–6
vs. goal orientation 10–12, 25–6
Orientation cages 22–3, 28–32, 57, 96, 131, 145
Overcast 60, 134–7, 146–7, 218

Palisade experiments 193
Passer domesticus (House sparrow) 7
Passerculus sandwichensis (Savannah sparrow) 97, 131, 137–8, 145–6, 149
Passerina cyanea (Indigo bunting) 87, 90–91, 97, 138, 166
Passerines 32, 103, 202, 216, 222
Pecten 117
Petrels 75
Phase-setters 35–6, 203–4
Phoenicurus phoenicurus (Redstart) 2
Photoperiod (see: Day length)
Phylloscopus collybita (Chiffchaff) 204
Phylloscopus trochilus (Willow warbler) 204
Phylloscopus spp. (Warblers) 202
Pigeon post 5
Pigeon racing 6, 12–14, 56
Pilotage 17, 44
definition of 9
Pineal body (and magnetoreception) 124
Planetarium experiments 88–91
Pluvialis apricaria (Golden plover) 3
Pluvialis dominica (American golden plover) 226–7

Polarization, band of maximum 83
Polarized light, 65–6 (see also: Compass)
detection of, 117
Pole, geographic vs. geomagnetic 49, 51, 152
Pole star 46–7
false 90
Polarity (of magnetic lines of force) 97–101 (see also: Geomagnetic field)
Post-fledging migration 210–13
Preferred compass direction (PCD) 26, 30, 88, 127–33, 185–6, 190–92, 195, 204, 213–4, 224–5
Programmed restlessness 202–4, 209–210
Programmed orientation 127–32, 203
Pseudacris triseriata (Chorus frog) 69–70
Pseudodrift 216–7
Pseudosounds 71–2

Radar 58–9, 150, 215, 218
Radio-tracking 13, 58, 224
Random search 20–21
Rays 118
Recalibration (of sun and star compasses) 142–4, 192
Redirected morning flights 219
Redundancy, principle of 61–3, 126, 164
Release-site bias 58, 187–93, 230
and landscapes 190–3, 230
Reversal migration 215, 218–9
Riparia riparia (Bank swallow/Sand martin) 22, 57, 188, 212, 227
Rock dove 6–7
vs. homing pigeon 200
Rodents 23, 209
Route-based navigation (see: Navigation)
Route reversal 17, 40
Running wheel 209–10

St. Hilaire, Geoffroy de 2
Salamanders 32, 55, 83
Salmon 7, 20
Seasonal migration 201–228
Seabirds 74–5
Seiurus noveboracensis (Water thrush) 222
Shadows 82
Sharks 118
Shearwaters 74–5
Sheep 4
Sinuses, and magnetoreception 123
'Sixth sense' 4
Snails 21
Solar wind 113
Speed, of racing pigeons 14

Subject index

Star maps 47
Star navigation (see: Latitude; Longitude; Zenith stars)
Star orientation (see: Compass)
  development of 88–90
Star patterns, learning of 89–90
Steeplechasing 40, 199
Storm petrels 75
Sturnus vulgaris (Starling) 12, 29, 32, 77–81, 130, 134, 137, 205, 225
Sun (see: Compass; Latitude; Longitude)
Sun-arc hypothesis 49–50, 166
Sunset 85, 145–9, 218
Sunspot activity (and magnetic storms) 113, 160
Sylvia atricapilla (Blackcap) 97, 99, 102, 145, 147–8, 158, 166
Sylvia borin (Garden warbler) 97, 99, 102, 131–4, 137–9, 142, 144, 157–8, 202–3
Sylvia cantillans (Subalpine warbler) 97
Sylvia communis (Whitethroat) 97, 145, 157
Sylvia curruca (Lesser whitethroat) 157–8, 166
Sylvia warblers 202

Talitrus sp. (Sandhopper) 92
Talorchestia (Sandhopper) 92
Target areas 38, 184, 212
Thunderstorms 70
Time, and longitude 167
Time, sense of 35–6, 82, 167–9
Time compensation (see: Compass)
Toads 55, 75, 85
Training (of pigeons) 12–13
Transequatorial displacement 79–80, 157–8
Transequatorial migration 99–100, 144
Transmutation 2
Tuna fish 7, 68, 120, 123
Tupaia 35
Turdus iliacus (Redwing) 227
Turntable experiments 175

Turpentine 63

ultra violet detection 85–6, 117

Vanellus vanellus (Lapwing) 221
Vanishing points 21, 57–8, 64
Vector navigation 179
Vertebrates 6, 14, 17, 83, 85, 123
  possibility of common magnetoreceptor 123
Visibility, effects of 43, 58
V-test 25
'VW' effect 176

Walkabout experiments 160, 172–3
Warblers 87–8
  (see also: *Sylvia*; *Phylloscopus*)
Waves (at sea) 60
Weather forecasting, and bird migration 70, 149
White, Gilbert 2
Wildfowl 88, 217
Wind direction, and navigation by smell 42, 63–6
Wind drift 60, 66, 214–5, 219
  compensation for 60, 214
Winter range
  recognition of 219–222
  homing to 222–3

Zeitgeber (see: Phase-setters)
Zenith stars 46
Zonotrichia albicollis (White-throated sparrow) 85, 145, 219
Zonotrichia atricapilla (Golden-crowned sparrow) 222–3
Zonotrichia leucophrys (White-crowned sparrow) 222–3
Zugunruhe (see: Migratory restlessness)